New Directions in Civil Engineering

SERIES EDITOR: W. F. CHEN *Purdue University*

Published and Forthcoming Titles

Response Spectrum Method in Seismic Analysis and Design of Structures
Ajaya Kumar Gupta *North Carolina State University*

Stability Design of Steel Frames
W. F. Chen *Purdue University* and E. M. Lui *Syracuse University*

Concrete Buildings: Analysis for Safe Construction
W. F. Chen *Purdue University* and K. H. Mossallam *Ministry of Interior, Saudi Arabia*

Stability and Ductility of Steel Structures under Cyclic Loading
Yuhshi Fukumoto *Osaka University* and George C. Lee *State University of New York at Buffalo*

Unified Theory of Reinforced Concrete
Thomas T. C. Hsu *University of Houston*

Flexural-Torsional Buckling of Structures
N. S. Trahair *University of Sydney*

Advanced Analysis of Steel Frames: Theory, Software, and Applications
W. F. Chen *Purdue University* and S. Toma *Hokkaigakuen University*

Water Treatment Processes: Simple Options
S. Vigneswaran *University of Technology, Sydney* and C. Visvanathan *Asian Institute of Technology*

Simulation-Based Reliability Assessment for Structural Engineers
Pavel Marek *San José State University,* Milan Guštar *Prague* and Thalia Anagnos *San José State University*

Analysis and Software of Cylindrical Members
W. F. Chen *Purdue University* and S. Toma *Hokkaigakuen University*

Fracture Mechanics of Concrete
Zdeněk P. Bažant *Northwestern University* and Jaime Planas *Technical University, Madrid*

Artificial Intelligence and Expert Systems for Engineers
C.S. Krishnamoorthy *Indian Institute of Technology* and S. Rajeev *Indian Institute of Technology*

Introduction to Environmental Geotechnology
Hsai-Yang Fang *Lehigh University*

Buckling of Thin-Walled Structures
J. Rhodes *University of Strathclyde*

Fracture Processes of Concrete
Jan G.M. van Mier *Delft University of Technology*

Limit Analysis and Concrete Plasticity (2nd Edition)
M.P. Nielsen *Technical University of Denmark*

Aseismic Testing of Building Structures
Bo-Long Zhu *Tongji University, Shanghai*

Winter Concreting
B.A. Krylov *Design and Technology Institute (NIIZhB), Moscow*

Contaminated Soils and Sediments
Raymond N. Yong *McGill University*

The Finite Strip Method
Y.K. Cheung and L.G. Tham *University of Hong Kong*

Analysis and Design of Steel Concrete Composite Structures
Edoardo Cosenzo *University of Naples, Italy* and Riccardo Zandonini *University of Trento, Italy*

Theory of Adaptive Structures
Senol Utku *Duke University*

LRFD Steel Design Using Advanced Analysis
W. F. Chen *Purdue University* and Seung-Eock Kim *Yonsei University, Korea*

LRFD
Steel Design *using* Advanced Analysis

W. F. Chen
Purdue University
Seung-Eock Kim
Yonsei University, Korea

CRC Press
Boca Raton New York

Acquiring Editor: *Tim Pletscher*
Senior Project Editor: *Susan Fox*
Cover Design: *Denise Craig*
Marketing Manager: *Susie Carlisle*
Direct Marketing Manager: *Becky McEldowney*

Library of Congress Cataloging-in-Publication Data

Chen, Wai-Fah, 1936-
LRFD steel design using advanced analysis / Wai-Fah Chen, Seung
-Eock Kim.
p. cm. — (New directions in civil engineering)
Includes bibliographical references and index.
ISBN 0-8493-7432-4 (alk. paper)
1. Building, Iron and steel. 2. Structural frames—Design.
3. Load factor design. 4. Steel, Structural. I. Kim, Seung-Eock.
II. Title. III. Series.
TA684.C4368 1997
624.1′773—dc21 96-36957
CIP

International Standard Book Number 0-8493-7432-4
Library of Congress Card Number 96-36957
Printed in the United States of America 1 2 3 4 5 6 7 8 9 0
Printed on acid-free paper

Contents

Preface

This book presents a simple, concise, and direct method of designing steel frames, using a practical, advanced technique that produces member sizes very close to those given by the LRFD method. The main advantage of advanced analysis is that the laborious and sometimes confused member capacity checks needed to satisfy the AISC-LRFD specifications are avoided. Advanced analysis captures the limit state strength and stability of a structural system and its component members directly. While elastic analysis is still the standard in engineering practice, a new generation of codes is expected to adopt advanced techniques and thus significantly reduce cost and time. In recent years, the rapid development of computer hardware and software, coupled with growing knowledge of structural behavior has made it possible to introduce advanced analysis to the design office.

This book is written for both academics in universities and engineers in the industry. The reader need have no specific knowledge of this subject, but should be familiar with methods of elastic analysis and conventional LRFD design. Advanced analysis has been presented here in a practical and simple manner for both analysis and design with emphasis placed on implementing advanced analysis in engineering practice. This book describes not only this general method in as simple a manner as possible, but also provides introduction on how this new method can be used in frame design.

The scope of the book is indicated by its contents. Chapter 1 presents the development of various methods of advanced analysis and a review of the AISC-LRFD method. The concept and formulation of advanced analysis are discussed in Chapter 2. Chapter 3 addresses verification studies for a wide range of frames. The step-by-step procedure required to perform this analysis is described in Chapter 4. Chapter 5 introduces an educational version of this advanced technique coded in FORTRAN. A disk is included on the back cover to exercise this new tool. In Chapter 6, design examples using advanced analysis are presented in detail, and also compared to the LRFD method. The design examples cover frames including braced, unbraced, and semi-rigid frames. The input data for each design example are provided so engineers can easily reproduce their own input.

The authors thank Purdue University Teaching Assistant Elizabeth M. Webster and Graduate Student I-hong Chen for their help in proofreading this manuscript and their valuable comments. The authors also extend special thanks to Mrs. Molly Stetler for her outstanding job in typing the manuscript.

Wai-Fah Chen
Seung-Eock Kim

chapter one

Trend Toward Advanced Analysis

1.1 Introduction

The steel design methods used in the U.S. are Allowable Stress Design (ASD), Plastic Design (PD), and Load and Resistance Factor Design (LRFD). In ASD, the stress computation is based on a first-order elastic analysis, and the geometric nonlinear effects are implicitly accounted for in the member design equations. In PD, a first-order plastic-hinge analysis is used in the structural analysis. Plastic design allows inelastic force redistribution throughout the structural system. Since geometric nonlinearity and gradual yielding effects are not accounted for in the analysis of plastic design, they are approximated in member design equations. In LRFD, a first-order elastic analysis with amplification factors or a direct second-order elastic analysis is used to account for geometric nonlinearity, and the ultimate strength of beam-column members is implicitly reflected in the design interaction equations. All three design methods require separate member capacity checks including the calculation of the K-factor. In the following, the characteristics of the LRFD method are briefly described.

The strength and stability of a structural system and its members are related, but the interaction is treated separately in the current AISC-LRFD Specification [4]. In current practice, the interaction between the structural system and its members is represented by the effective length factor. This aspect is described in the following excerpt from SSRC Technical Memorandum No.5 [56]:

> Although the maximum strength of frames and the maximum strength of component members are interdependent (but not necessarily coexistent), it is recognized that in many

> structures it is not practical to take this interdependence into account rigorously. At the same time, it is known that difficulties are encountered in complex frameworks when attempting to compensate automatically in column design for the instability of the entire frame (for example, by adjustment of column effective length). Therefore, SSRC recommends that, in design practice, the two aspects, stability of separate members and elements of the structure and stability of the structure as a whole, be considered separately.

This design approach is marked in Figure 1.1 as the indirect analysis and design method.

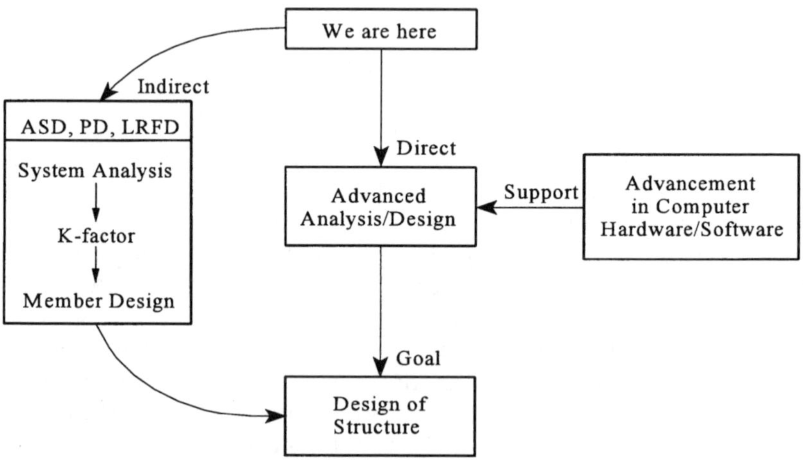

Figure 1.1 Analysis and design methods.

In the current AISC-LRFD Specification [4], first-order elastic analysis or second-order elastic analysis is used to analyze a structural system. In using first-order elastic analysis, the first-order moment is amplified by B_1 and B_2 factors to account for second-order effects. In the Specification, the members are isolated from a structural system, and they are then designed by the member strength curves and interaction equations as given by the Specifications, which implicitly account for the effects of second-order, inelasticity, residual stresses, and geometric imperfections [17]. The column curve and beam curve were developed by a curve-fit to both theoretical solutions and experimental data, while the beam-column interaction equations were determined by a curve-fit to the so-called "exact" plastic-zone solutions generated by Kanchanalai [30].

In order to account for the influence of a structural system on the strength of individual members, the effective length factor is used as illustrated in Figure 1.2.

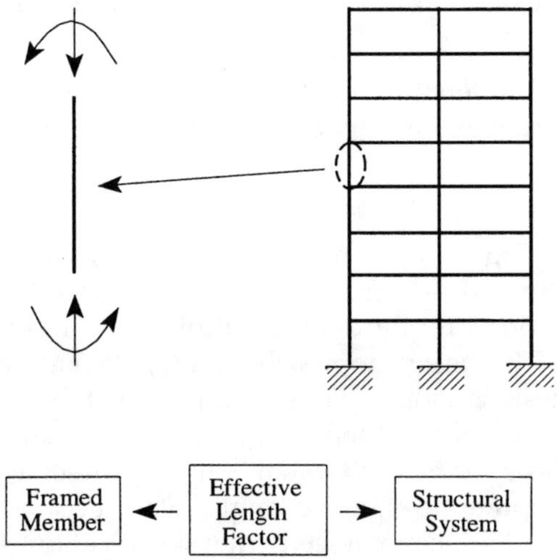

Figure 1.2 Interaction between a structural system and its component members.

The effective length method generally provides a good design of framed structures. However, several difficulties are associated with the use of the effective length method as follows:

1. The effective length method cannot accurately account for the interaction between the structural system and its members. This is because the interaction in a large structural system is too complex to be represented by the simple effective length factor K. As a result, this method cannot accurately predict the actual required strengths of its framed members.

2. The effective length method cannot capture the inelastic redistributions of internal forces in a structural system, because the first-order elastic analysis with B_1 and B_2 factors accounts only for second-order effects but not the inelastic redistribution of internal forces. The effective length method provides a conservative estimation of the ultimate load-carrying capacity of a large structural system.

3. The effective length method cannot predict the failure modes of a structural system subject to a given load. This is because the LRFD interaction equation does not provide any information about failure modes of a structural system at the factored loads.
4. The effective length method is not user-friendly for a computer-based design.
5. The effective length method requires a time-consuming process of separate member capacity checks involving the calculation of K-factors.

With the development of computer technology, two aspects (the stability of separate members and the stability of the structure as a whole) can be treated rigorously for the determination of the maximum strength of the structures. This design approach is marked in Figure 1.1 as the direct analysis and design method. The development of the direct approach to design is called advanced analysis or, more specifically, "Second-Order Inelastic Analysis for Frame Design". In this direct approach, there is no need to compute the effective length factor because separate member capacity checks encompassed by the specification equations are not required. With the current available computing technology, it is feasible to employ advanced analysis techniques for direct frame design. This method has been considered impractical for design office use in the past. The purpose of this book is to present a practical, direct method of steel frame design, using advanced analysis, that will produce almost identical member sizes as those of the LRFD method.

In the following, we shall present an encapsulation of various analysis approaches for frame design ranging from a simplified indirect member-by-member design approach to a sophisticated direct structural system design approach. The aim is to provide a smooth transition from the current indirect LRFD analysis and design practice to the direct advanced analysis method suitable for adoption in engineering practice.

1.2 Design Formats

At present, design practice is based on one of three design philosophies: (1) ASD, (2) PD, or (3) LRFD. Recently, an emerging technology called advanced analysis has been developed. This approach has the potential to unlock the power of computers for a direct assessment of system and member strength and stability without the need to make individual member capacity checks. The advanced analysis approach and the various design approaches upon which current practice is based are discussed separately in the following sections.

1.2.1 Allowable Stress Design (ASD)

In ASD, it is ensured that the stresses in a structure under working or service loads do not exceed designated allowable values. The allowable values are obtained by dividing the yield stress or ultimate stress of the material by a factor of safety. The general format for an allowable stress design is:

$$\frac{R_n}{F.S.} \geq \sum_{i=1}^{m} Q_{ni} \tag{1.1}$$

where

R_n = nominal resistance of the structural member expressed in units of stress
Q_n = nominal working or service stresses computed under working loads
$F.S.$ = factor of safety (e.g., 1.5 for beams, 1.67 for tension members)
i = type of load (i.e., dead load, live load, wind load, etc.)
m = number of load type

The left-hand side of Equation (1.1) represents the allowable stress of the structural member or component under a given loading condition (e.g., tension, compression, bending, or shear). The right-hand side of the equation represents the combined stress produced by various load combinations (e.g., dead, live, or wind load). Formulas for the allowable stresses for various types of structural members under various types of loadings are specified in the AISC specification [6]. A satisfactory design is reached when the stresses in the member, computed using a first-order elastic analysis under working load conditions, do not exceed their allowable values. The effects of second-order moments and inelasticity are considered indirectly. One should realize that in allowable stress design, the factor of safety is applied only to the resistance term, and safety is evaluated at the service load. Thus, ASD is characterized by the use of unfactored "working" loads in conjunction with a single factor of safety applied to the resistance. Because of the greater variability and unpredictability of the live load and other loads in comparison with the dead load, a uniform reliability is not possible with ASD.

1.2.2 Plastic Design (PD)

In PD (with load factors), it is ensured that factored load combinations do not exceed the maximum plastic strength of the structure or component.

It takes the form:

$$R_n \geq \lambda \sum_{i=1}^{m} Q_{ni} \tag{1.2}$$

where

R_n = nominal plastic strength of the member
Q_n = nominal load effect (e.g., axial force, shear force, bending moment)
λ = load factor [e.g., 1.70 for $(D + L)$, 1.30 for $(D + L + W)$]
i = type of load (D = dead load, L =live load, W = wind load)
m = number of load types

Note that in this method, safety is incorporated only in the load term and is evaluated at the ultimate (plastic strength) limit state. Applying a factor of safety to the load term is more appropriate than ASD because uncertainty associated with loads is higher than that associated with resistances. A uniform reliability cannot be fully achieved with PD because only a single factor of safety (here called load factor) is applied to all loads.

PD is superior to ASD in the sense that it considers moment redistribution in beams and frames in a more direct way. However, the analysis relies on the ability of steel to sustain considerable deformation in excess of yielding. Therefore, the regions of the structure in which significant plasticity (plastic hinges) develops must have adequate deformation capacity to carry the load. To achieve this, members with plastic hinges must satisfy the limitations on the flange and web proportions for plastic design sections. In addition, sufficient lateral and torsional restraint must be provided to prevent the member from buckling laterally. Although material nonlinearity has been considered in first-order plastic hinge analysis, the effects of the spread of yielding and member stability are not accounted for. Thus, like ASD, any significant nonlinear effects have to be accounted for approximately in member strength equations. In other words, member capacity checks are required when plastic analysis is used so as to prevent instability type of failure.

1.2.3 Load and Resistance Factor Design (LRFD)

In LRFD, it is ensured that the factored load effects do not exceed the factored nominal resistance of the structural member or component. There are two safety factors: one applied to the loads, the other to the resistance. This is a more rational approach because both the loads and the resistances have different uncertainties. Thus, the LRFD takes the form:

$$\phi R_n \geq \sum_{i=1}^{m} \gamma_i Q_{ni} \tag{1.3}$$

where

R_n = nominal resistance of the structural member
Q_n = nominal load effect (e.g., axial force, shear force, bending moment)
ϕ = resistance factor ($\leq$ 1.0) (e.g., 0.9 for beams, 0.85 for columns)
γ_i = load factor (usually > 1.0) corresponding to Q_{ni} (e.g., $1.4D$ and $1.2D + 1.6L + 0.5S$)
i = type of load (e.g., D = dead load, L = live load, S = snow load)
m = number of load type

Note that LRFD uses separate factors for each load and can therefore reflect the degree of uncertainty of different loads and combinations of loads. As a result, more uniform reliability can be achieved.

In the 1986 and 1994 LRFD Specifications, the resistance factors were developed mainly through a calibration with ASD [51], whereas the load factors are based on a statistical analysis [10], [25]. The factors and load combinations used by LRFD are summarized in Tables 1.1 and 1.2, respectively. A satisfactory design is one in which the probability of

Table 1.1 Resistance Factors

Member type and limit state	ϕ
Tension member, limit state: yielding	0.90
Tension member, limit state: fracture	0.75
Pin-connected member, limit state: tension	0.75
Pin-connected member, limit state: shear	0.75
Pin-connected member, limit state: bearing	0.75
Column; compression	0.85
Beams; flexure	0.90
High strength bolts, limit state: tension	0.75
High strength bolts, limit state: shear	0.75

exceeding a limit state of the structural member (for example, yielding, fracture, or buckling) is minimal. Based on a first-order second-moment probabilistic analysis [9], the safety of the structural member

Table 1.2 Load Factors and Load Combinations

$1.4D$
$1.2D + 1.6L + 0.5(L_r \text{ or } S \text{ or } R)$
$1.2D + 1.6(L_r \text{ or } S \text{ or } R) + (0.5L \text{ or } 0.8W)$
$1.2D + 1.3W + 0.5L + 0.5(L_r \text{ or } S \text{ or } R)$
$1.2D \pm 1.0E + 0.5L + 0.2S$
$0.9D \pm (1.3W \text{ or } 1.0E)$

Note: D = dead load, L = live load, L_r = roof load, W = wind load, S = snow load, E = earthquake load, and R = nominal load due to initial rainwater on ice exclusive of the ponding contribution.

is measured by a reliability or safety index [4] defined as:

$$\beta = \frac{\ln\left(\bar{R}_n/\bar{Q}_n\right)}{\sqrt{V_R^2 + V_Q^2}} \tag{1.4}$$

where

$\bar{R}$ = mean resistance
$\bar{Q}$ = mean load effect
V_R = coefficient of variation of resistance ($= \sigma_R/\bar{R}$)
V_Q = coefficient of variation of load effect ($= \sigma_Q/\bar{Q}$)

in which σ equals the standard deviation.

The physical interpretation of the reliability index β is shown in Figure 1.3. The shaded area in the figure represents the probability of $\ln(R/Q)$ being less than zero, i.e., the probability that the resistance will

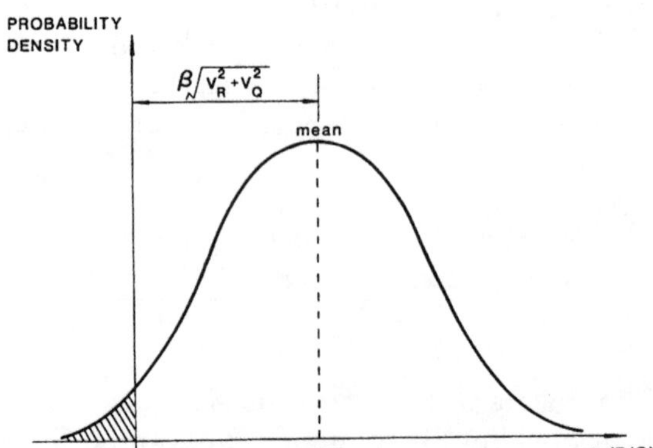

Figure 1.3 Reliability index.

be smaller than the load effect indicates that a limit state has been exceeded. The larger the value of β, the smaller the shaded area, so that it becomes more improbable that a limit state will be exceeded. Thus, the magnitude of β reflects the safety of the member. In the development of the current LRFD Specification [4] [5], the target values of β were selected as 3.0 for members and 4.5 for connectors under dead plus live and/or snow loading. A higher value of β for connectors ensures that the connections are designed to be stronger than their adjoining members.

In this design method, a designer has the option to either carry out a second-order elastic analysis and directly compute the load effect (moments and forces) in the members or carry out the first-order elastic analysis and estimate the second-order effects by using the moment amplification factors B_1 and B_2 provided in the specifications [4]. The design interaction equations implicitly account for the effects of the spread of yielding and member stability. Unfortunately, this treatment does not satisfy compatibility between the actual inelastic member (as loaded by the design forces and accounted for in the interaction equation) and the elastic system as assumed in the frame analysis [41].

1.2.4 *Advanced Analysis/Design*

Since the power of personal computers and engineering workstations is rapidly increasing, it is now feasible to employ advanced analysis techniques for direct design use. Advanced Analysis indicates any method that can sufficiently capture the limit state strength and stability of a structural system and its individual members so that separate member capacity checks encompassed by the specification equations are not required. This is because advanced analysis accounts for material and geometric nonlinearities directly. It is expected that the use of advanced analysis will simplify the design process considerably. Conventional design methods such as ASD, PD, and LRFD require separate member capacity checks which are often tedious and confusing to those in engineering practice. They account for nonlinearities in their design equations rather than in their analyses.

The main distinction between advanced analysis and conventional methods is that advanced analysis can predict the structural system strength, whereas others can predict only the member strengths. Advanced analysis uses a format associated with the structural system rather than individual members. Since advanced analysis provides the most benefit in frame design, specification provisions of this type are required for practical use. The provisions for advanced analysis in the Eurocode [22] and Australian limit-state specifications [57] postulate that if all the significant planar behavioral effects are modeled properly in the analysis, checking of conventional beam-column equations is not required.

Advanced analysis uses the same resistance factors and load factors as those of the LRFD method. In this way, a uniform reliability can be achieved by reflecting the degree of uncertainty of different loads and combinations of loads.

1.3 AISC-LRFD Design Method

In this section, the conventional LRFD method is discussed in detail because this method has been used as the basis for the present development and calibration of advanced analysis. In the AISC-LRFD Specification, a first or second-order elastic analysis is used to analyze a structural system, and limit state design equations are then used for separate member capacity checks. To account for the influence of a structural system on the strength of individual members, the effective length factor "K" is employed. In fact, the updated LRFD Specification [4] adopted a new K-factor formula in order to improve the leaning column effect. Since the use of first-order analysis is the norm of the current engineering practice, an unbraced frame is generally decomposed into a non-sway and sway component as shown in Figure 1.4, and two separate sets of K-factors and member forces are involved in the design computation. They must be treated correctly in further calculations of several factors including B_1, B_2, C_m, and C_b, among others. The first part of this section reviews frame design procedures in the LRFD to highlight the difficulties associated with the member capacity checks. The concept of the column curve, interaction equations, the effective length factor, and moment amplification factors are then discussed. Finally, case studies of the columns in two unbraced frames, using this procedure, are presented to clarify possible confusion associated with the LRFD implementation.

1.3.1 Overview of AISC-LRFD Design Equations

Interaction Equation

Chapter H of the AISC-LRFD Specification [4] provides the following interaction equations for the design strength of beam-columns:

$$\frac{P_u}{\phi_c P_n} + \frac{8}{9}\left(\frac{M_{ux}}{\phi_b M_{nx}} + \frac{M_{uy}}{\phi_b M_{ny}}\right) \leq 1.0 \quad \text{for} \quad \frac{P_u}{\phi_c P_n} \geq 0.2 \tag{1.5a}$$

$$\frac{P_u}{2\phi_c P_n} + \left(\frac{M_{ux}}{\phi_b M_{nx}} + \frac{M_{uy}}{\phi_b M_{ny}}\right) \leq 1.0 \quad \text{for} \quad \frac{P_u}{\phi_c P_n} < 0.2 \tag{1.5b}$$

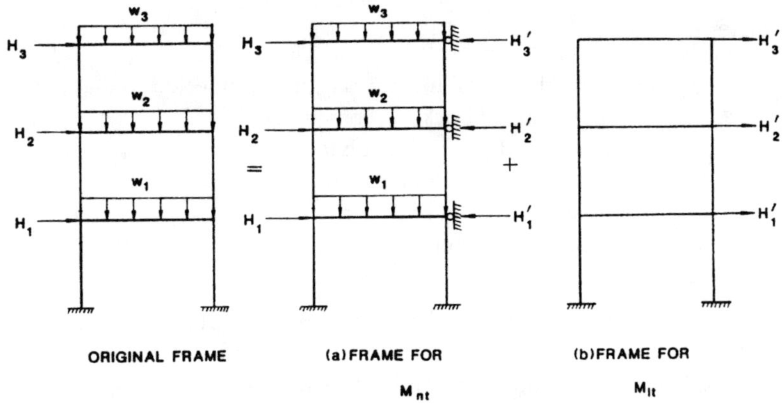

Figure 1.4 Decomposition of unbraced frame.

where

P_u = required compressive strength
P_n = nominal compressive strength
M_u = required flexural strength
M_n = nominal flexural strength
x = subscript relating symbol to strong-axis bending
y = subscript relating symbol to weak-axis bending
ϕ_c = resistance factor for compression $= 0.85$
ϕ_b = resistance factor for flexure $= 0.9$

These interaction equations have been developed based on a curve-fit to the "exact" plastic-zone solution of 82 beam-columns generated by Kanchanalai [30].

Effective Length Factor K

To obtain the effective length factor K of a column using the alignment chart method, the relative stiffness at top and bottom of a column must be calculated according to the following definition:

$$G = \frac{\sum\left(\frac{\tau I_c}{L_c}\right)}{\sum\left(\frac{I_g}{L'_g}\right)} \tag{1.6}$$

where

$\sum$ = summation of all members rigidly connected to that joint in the plane of buckling considered
τ = inelastic stiffness reduction factor
I_c = moment of inertia of column
I_g = moment of inertia of girder
L_c = unsupported length of column
L'_g = modified unsupported length of girder

The term τ accounts for inelastic stiffness reduction of a column. If a column is loaded into the inelastic range while the adjacent beams remain elastic, the beams are more effective in restraining the column against buckling. As a result, the effective length of the column is reduced, and the column will have a higher load-carrying capacity with the inelasticity adjustment factor τ:

$$\tau = -3.205 \left(\frac{P_u}{P_y}\right) \ln \left(\frac{P_u/P_y}{0.85}\right) \quad \text{for} \quad \frac{P_u}{P_y} > \frac{1}{3} \tag{1.7a}$$

$$\tau = 1.0 \quad \text{for} \quad \frac{P_u}{P_y} \leq \frac{1}{3} \tag{1.7b}$$

where

P_u = required compressive strength, summation of axial forces in sway and non-sway cases
P_y = squash load

The alignment chart method assumes that all joints are rigidly connected. If the girders have different boundary conditions at far ends, the girder length must be modified as follows:

For braced frames:

$$L'_g = L_g/2 \quad \text{for the far end of the girder is fixed} \tag{1.8a}$$

$$L'_g = L_g/1.5 \quad \text{for the far end of the girder is hinged} \tag{1.8b}$$

For unbraced frames:

$$L'_g = 1.5L_g \quad \text{for the far end of the girder is fixed} \tag{1.8c}$$

$$L'_g = 2.0L_g \quad \text{for the far end of the girder is hinged} \tag{1.8d}$$

The Commentary of the AISC Specification [4] suggests the use of a modified L'_g for a more accurate determination of the G values for unbraced frames:

$$L'_g = L_g \left(2 - \frac{M_F}{M_N}\right) \tag{1.9}$$

where

M_F = moment at the far end of the girder due to lateral load
M_N = moment at the near end of the girder due to lateral load
M_F/M_N = positive in double curvature bending

The Commentary of the AISC Specification [4] recommends the use of a modified K'_i to account for the leaning effect of columns. The modified effective length factor of the i-th column of a story is:

$$K'_i = \sqrt{\frac{\sum P_u}{P_{ui}} \frac{I_i}{\sum (I_i/K_i^2)}} \geq \sqrt{\frac{5}{8}} K_i \tag{1.10}$$

where

K'_i = effective length factor with story stability effect for the i-th non-leaning (rigid) column
K_i = effective length factor for the i-th rigid-column based on alignment chart for sway case
I_i = moment of inertia in plane of bending for the i-th rigid column
P_{ui} = required axial compressive strength for the i-th rigid column, sum of axial forces in sway and non-sway cases
$\sum P_u$ = required axial compressive strength of all columns in a story
$\sum (I_i/K_i^2)$ = sum of (I_i/K_i^2) of rigid columns in a story

Column Strength P_n

In order to determine P_n, Chapter E of the LRFD Specification [4] has adopted the following column strength curves based on experimental and theoretical studies.

$$P_n = A_g F_y \left(0.658^{\lambda_c^2}\right) \quad \text{for } \lambda_c \leq 1.5 \tag{1.11a}$$

$$P_n = A_g F_y \left(\frac{0.877}{\lambda_c^2}\right) \quad \text{for } \lambda_c > 1.5 \tag{1.11b}$$

where

A_g = gross area of member
F_y = specified yield stress
λ_c = slenderness parameter

The column curves account implicitly for second-order effects, residual stresses, and an initial out-of-straightness of 1/1500. λ_c is expressed as:

$$\lambda_c = \frac{KL}{\pi r}\sqrt{\frac{F_y}{E}} \tag{1.12}$$

where

K = effective length factor
L = length of member
r = radius of gyration
E = modulus of elasticity

KL/r should be taken as the larger of the effective slenderness ratios for strong- and weak-axis buckling.

Flexural Strength M_n

Chapter F of the AISC Specification recommends that the nominal flexural strength M_n be taken as the lowest obtained according to the limit state of: yielding, lateral-torsional buckling, flange local buckling, and web local buckling. If a compact section is assumed, lateral torsional buckling (as well as yielding) should be considered. For I-shaped members subject to bending about the strong-axis, M_n is determined by:

$$M_n = M_p \quad \text{for} \quad L_b \leq L_p \tag{1.13a}$$

$$M_n = C_b\left[M_p - \left(M_p - M_r\right)\tfrac{(L_b - L_p)}{(L_r - L_p)}\right] \leq M_p \quad \text{for} \quad L_p < L_b \leq L_r \tag{1.13b}$$

$$M_n = C_b\frac{\pi}{L_b}\sqrt{EI_yGJ + \left(\frac{\pi E}{L_b}\right)^2 I_yC_w} \leq M_p \quad \text{for} \quad L_b > L_r \tag{1.13c}$$

where

M_p = plastic moment = F_yZ, where Z is plastic section modulus
M_r = F_LS_x, where F_L is smaller of $\left(F_{yf} - F_r\right)$ or F_{yw}
F_{yf} = yield stress of flange
F_{yw} = yield stress of web

F_r = compressive residual stress in flange; 10 ksi for rolled shape, 16.5 ksi for welded shape
L_b = unbraced length of the member in the out-of-plane bending
L_p = limiting unbraced length for full plastic bending capacity
L_r = limiting unbraced length of inelastic lateral torsional buckling
I_y = moment of inertia about weak-axis
G = shear modulus of elasticity of steel (11,200 ksi)
J = torsional constant
C_w = warping constant

The limiting unbraced plastic and elastic lengths (L_p and L_r) shall be determined, respectively, as follows:

$$L_p = \frac{300 r_y}{\sqrt{F_{yf}}} \tag{1.14a}$$

$$L_r = \frac{r_y X_1}{F_L}\sqrt{1 + \sqrt{1 + X_2 F_L^2}} \tag{1.14b}$$

where

$X_1 = \frac{\pi}{S_x}\sqrt{\frac{EGJA}{2}}$
$X_2 = 4\frac{C_w}{I_y}\left(\frac{S_x}{GJ}\right)^2$
S_x = section modulus about major axis

It is noted that the lateral-torsional buckling limit state is applicable to members subject to strong-axis bending not weak-axis bending, square or circular shapes. The terms M_p, M_r, L_p, and L_r may be found with the aid of Beam Design Table in the AISC-LRFD Specification without using Equations (1.13a through 1.13c, 1.14a and 1.14b) described above. C_b is a modification for non-uniform moment diagrams. The physical meaning of C_b is that it represents the amount of an increase in load-carrying capacity when compared with the critical uniform loading case. An empirical formula for C_b is expressed as [4]:

$$C_b = \frac{12.5 M_{max}}{2.5 M_{max} + 3M_A + 4M_B + 3M_C} \tag{1.15}$$

where

$M_{\max}$ = absolute value of maximum moment in the unbraced segment, sum of moments in sway and non-sway cases
M_A = absolute value of moment at quarter point of the unbraced segment, sum of moments in sway and non-sway cases
M_B = absolute value of moment at centerline of the unbraced segment, sum of moments in sway and non-sway cases
M_C = absolute value of moment at three-quarter point of the unbraced segment, sum of moments in sway and non-sway cases

C_b is allowed to be taken conservatively as 1.0 for all cases.

Second-Order Moment M_u

Chapter C of the LRFD Specification recommends that a member's maximum moment M_u be determined by second-order elastic analysis or from the following approximate second-order analysis procedure:

$$M_u = B_1 M_{nt} + B_2 M_{lt} \tag{1.16}$$

where

M_{nt} = required flexural strength in member assuming there is no lateral translation (NT) of the frame
M_{lt} = required flexural strength in member as a result of lateral translation (LT) of the frame only
B_1 = P-δ moment amplification factor
B_2 = P-Δ moment amplification factor

The term B_1 accounts for the amplification of the first-order NT moment associated with member curvature effects. B_1 is defined by:

$$B_1 = \frac{C_m}{1 - \frac{P_u}{P_{e1}}} \tag{1.17}$$

where

C_m = equivalent moment factor
P_u = required axial compressive strength for the member under consideration, summation of axial forces in both sway and non-sway cases
P_{e1} = $\pi^2 EI/(KL)^2$, where K is an effective length factor in the plane of bending based on the assumption that side-sway is prevented

The term C_m adjusts B_1 to account for cases other than uniform bending. For the purpose of design, one needs to know the location of the maximum moment if it occurs away from member ends. To eliminate this calculation, the concept of "equivalent moment" is introduced in design practice. The equivalent moment factor, C_m, is a function of the ratios of the end moments and the axial force. A simplified expression for C_m that eliminates its dependency on the axial force has been proposed by Austin [12] for compression members not subject to transverse loading between the ends. It takes the simple form:

$$C_m = 0.6 - 0.4\frac{M_1}{M_2} \tag{1.18}$$

where M_1/M_2 is the ratio of the smaller to larger end moment in non-sway cases. M_1/M_2 is positive when the member is bent in reverse curvature and negative when bent in single curvature. For compression members subject to transverse loads, the value of C_m is determined by a "rational analysis", or by the use of 0.85 for members whose ends are restrained and 1.00 for members whose ends are unrestrained.

B_2 accounts for the amplification of the member end moments associated with lateral translation of the story as expressed by:

$$B_2 = \frac{1}{1 - \sum P_u \left[\frac{\Delta_{oh}}{\sum HL}\right]} \tag{1.19a}$$

or

$$B_2 = \frac{1}{1 - \frac{\sum P_u}{\sum P_{e2}}} \tag{1.19b}$$

where

$\sum P_u$ = required axial strength of all columns in a story
Δ_{oh} = lateral inter-story deflection
$\sum H$ = summation of all story horizontal forces producing Δ_{oh}
L = story height
$\sum P_{e2}$ = $\sum \left[\pi^2 EI/(KL)^2\right]$, where K is an effective length factor not including the leaning column effect in the plane of bending assuming that side-sway is free

1.3.2 Column Curves

Structural Stability Research Council Curves

Both the ASD curve and PD curve originate from the CRC curve, which was developed on the basis of the bifurcation concept that assumes columns are perfectly straight. Although the effect of residual stresses is explicitly accounted for, the effect of geometrical imperfections is only accounted for implicitly by applying a variable factor of safety to the basic strength curve.

Realizing that perfectly straight columns are rarely encountered in real life, researchers [14] [54] investigated theoretically and experimentally the strength and stability of initially crooked imperfect columns. It is evident from the results of these studies that the strengths of different types of steel columns, which are the result of different manufacturing and fabrication processes, different sizes and steel grades, and different axes of bending, may vary considerably, and multiple-design curves are desirable.

On the basis of a computer model developed for a geometrically imperfect column with an initial out-of-straightness at midheight equal to 0.001L, and with actual measured values of residual stresses, a set of three multiple-column strength curves (from a total of 112 columns) was developed [14]. Each of these curves is representative of the strength of a related category of columns. In the categories covered by these column curves we find hot-rolled and cold-straightened members, wide-flange and box shapes, as well as round bars and members composed of welded plates.

The Structural Stability Research Council (SSRC), in its third edition of the Guide [29], presents these three column curves along with the CRC curve. The expressions for the three SSRC curves are as follows:

Curve 1

$$\frac{P}{P_y} = \begin{cases} 1 \text{ (yield level)} & (0 \le \lambda_c \le 0.15) \\ 0.990 + 0.122\lambda_c - 0.367\lambda_c^2 & (0.15 \le \lambda_c \le 1.2) \\ 0.051 + 0.801\lambda_c^{-2} & (1.2 \le \lambda_c \le 1.8) \\ 0.008 + 0.942\lambda_c^{-2} & (1.8 \le \lambda_c \le 2.8) \\ \lambda_c^{-2} \text{ (Euler buckling)} & (\lambda_c \ge 2.8) \end{cases} \tag{1.20a}$$

Curve 2

$$\frac{P}{P_y} = \begin{cases} 1 \text{ (yield level)} & (0 \le \lambda_c \le 0.15) \\ 1.035 - 0.202\lambda_c - 0.222\lambda_c^2 & (0.15 \le \lambda_c \le 1.0) \\ -0.111 + 0.636\lambda_c^{-1} + 0.087\lambda_c^{-2} & (1.0 \le \lambda_c \le 2.0) \\ 0.009 + 0.877\lambda_c^{-2} & (2.0 \le \lambda_c \le 3.6) \\ \lambda_c^{-2} \text{ (Euler buckling)} & (\lambda_c \ge 3.6) \end{cases} \tag{1.20b}$$

Curve 3

$$\frac{P}{P_y} = \begin{cases} 1 \text{ (yield level)} & (0 \leq \lambda_c \leq 0.15) \\ 1.093 - 0.622\lambda_c & (0.15 \leq \lambda_c \leq 0.8) \\ -0.128 + 0.707\lambda_c^{-1} - 0.102\lambda_c^{-2} & (0.8 \leq \lambda_c \leq 2.2) \\ 0.008 + 0.792\lambda_c^{-2} & (2.2 \leq \lambda_c \leq 5.0) \\ \lambda_c^{-2} \text{ (Euler buckling)} & (\lambda_c \geq 5.0) \end{cases} \quad (1.20c)$$

These equations were obtained by curve-fitting a parabola or hyperbola to the designated characteristic column curves that are the arithmetic mean curves of the three divided categories. These column curves are used in conjunction with the Column Selection Table shown in Table 1.3.

The curves in each category were developed on the basis of a stability analysis of an imperfect column. In a stability analysis, the complete load-deflection behavior of a column is traced from the onset of load to its ultimate state. For this reason, a stability analysis is known as a load-deflection analysis. The peak point of this load-deflection curve is the maximum load the column can carry. As mentioned previously, a stability analysis is quite different from a bifurcation analysis of a perfect column. In a bifurcation analysis, the load corresponding to the state of bifurcation of equilibrium is calculated using an eigenvalue analysis. However, for columns that are initially crooked, lateral deflection begins as soon as the load is applied and so there is no distinct point of bifurcation. Usually, recourse must be made to numerical methods in stability analyses.

For comparison, the three SSRC curves are plotted with the CRC, ASD, and PD curves in Figure 1.5. It can be seen that these curves "belly down" in the intermediate slenderness range ($0.75 < \lambda_c < 1.25$) because of the combined detrimental effect of residual stress and initial crookedness on the column strength as predicted by the computer model. Tests on real columns have shown that the detrimental effects of residual stresses and initial crookedness are not always synergistic and so the SSRC curves with "belly down" in the intermediate slenderness range will be too conservative for most columns in building frames.

AISC LRFD Curve

The AISC LRFD Specification [4] [5] uses the following curve to represent column strengths:

$$\frac{P}{P_y} = \begin{cases} \exp\left[-0.419\lambda_c^2\right] & \lambda_c \leq 1.5 \\ 0.877\lambda_c^{-2} & \lambda_c > 1.5 \end{cases} \quad (1.21)$$

Note that only one curve is given for the whole range of column strengths. In the development of this curve, the following assumptions were made:

Table 1.3 SSRC Multiple Column Curve Selection Table

Fabrication details			Bending axis	A7 A36	A 242	A572 50	A572 65	A 441	A 514	Hybrid A514 FL A441 WEB	Hybrid A514 FL A36 WEB	Hybrid A441 FL A36 WEB
Rolled wide-flange		Light	Major	2	1	(1)	(1)	(1)	1	—	—	—
			Minor	2	2	(2)	(1)	(2)	1	—	—	—
		Heavy	Major	3	(2)	(2)	(1)	(2)	1	—	—	—
			Minor	3	(2)	(2)	(2)	(2)	1	—	—	—
Welded built-up H	Light	Flame-cut	Major	2	(2)	2	(1)	(2)	1	1	1	2
			Minor	2	(2)	2	(2)	(2)	2	1	1	2
		Universal mill	Major	3	(2)	(2)	(2)	(2)	1	(1)	(1)	2
			Minor	3	(3)	(3)	(2)	(3)	2	(2)	(2)	2
	Heavy	Flame-cut	Major	2	(2)	2	(2)	2	(1)	(1)	(1)	(2)
			Minor	2	(2)	2	(2)	2	(2)	(2)	(2)	(2)
		Universal mill	Major	3	(2)	(2)	(2)	2	(1)	(2)	(2)	(2)
			Minor	3	(2)	(2)	(2)	2	(2)	(2)	(2)	(3)
Circular	Massive solid		—	(2)	(1)	(1)	(1)	(1)	(1)	—	—	—
	Tubes	Extruded rolled	—	(1)	(1)	(1)	(1)	(1)	(1)	—	—	—
		Welded	—	(2)	(1)	(1)	(1)	(1)	(1)	—	—	—
Box	Extruded rolled		Major	(1)	(1)	(1)	(1)	(1)	(1)	—	—	—
			Minor	(1)	(1)	(1)	(1)	(1)	(1)	—	—	—
	Welded	Flame-cut	Major	2	(2)	(2)	(1)	(2)	1	—	—	—
			Minor	2	(2)	(2)	(1)	(2)	1	—	—	—
		Universal mill	Major	2	(2)	(2)	(1)	(2)	(1)	—	—	—
			Minor	2	(2)	(2)	(1)	(2)	(1)	—	—	—
Stress-relieved shapes (all types)			Major	1	1	1	1	1	1	1	1	1
			Minor	1	1	1	1	1	1	1	1	1

Cold-straightened (gagged or roller-straightened) are designed according to the column curve immediately above the curve for the corresponding unstraightened shape. This is not valid for shapes already assigned to curve 1.

Numbers in parentheses may be subject to later change

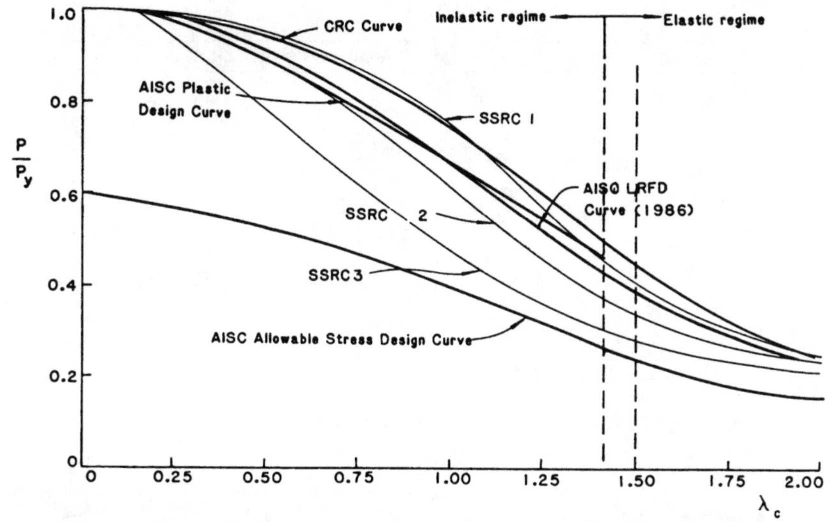

Figure 1.5 Comparison of various column curves.

1. The column has small end restraints corresponding to an end-restraint parameter $G = 10$ or an effective length factor $K = 0.96$.
2. The column has an initial crookedness sinusoidal in shape and has an amplitude of $(1/1500)L$ at midheight.
3. The axial force is applied at the centroid of the column end cross-sections.

This LRFD curve is plotted in Figure 1.5 together with all other curves described above. Note that the LRFD column curve as represented by Equation (1.21) is comparable to the SSRC Curve 2, especially in the range $0 \leq \lambda \leq 1.0$.

The expressions for various AISC column curves described above are summarized in Table 1.4. The LRFD Specification recommends using only one column curve for rolled and welded sections of W, WT and HP shapes, pipe, and structural tubing. The LRFD curve is plotted in Figure 1.6.

1.3.3 *Beam-Column Interaction Equations*

In the current AISC-LRFD Specification [4], beam-columns are designed using beam-column interaction equations with the effective length factor. The LRFD interaction equations are developed based on plastic-zone solutions with the following guidelines [63]:

Table 1.4 Summary of Column Curves

Column curves	Column equations	
CRC curve	$\frac{P}{P_y} = 1 - \frac{\lambda_c^2}{4}$	$\lambda_c \le \sqrt{2}$
	$\frac{P}{P_y} = \frac{1}{\lambda_c^2}$	$\lambda_c > \sqrt{2}$
AISC-ASD curve	$\frac{P}{P_y} = \frac{1-\frac{\lambda_c^2}{4}}{\frac{5}{3}+\frac{3}{8}\left(\frac{\lambda_c}{\sqrt{2}}\right)-\frac{1}{8}\left(\frac{\lambda_c}{\sqrt{2}}\right)^3}$	$\lambda_c \le \sqrt{2}$
	$\frac{P}{P_y} = \frac{12}{23}\frac{1}{\lambda_c^2}$	$\lambda_c > \sqrt{2}$
AISC-PD curve	$\frac{P}{P_y} = \frac{1.7\left(1-\frac{\lambda_c^2}{4}\right)}{\frac{5}{3}+\frac{3}{8}\left(\frac{\lambda_c}{\sqrt{2}}\right)-\frac{1}{8}\left(\frac{\lambda_c}{\sqrt{2}}\right)^3} \le 1.0$	$\lambda_c \le \sqrt{2}$
AISC-LRFD curve	$\frac{P}{P_y} = \exp\left(-0.419\lambda_c^2\right)$	$\lambda_c \le 1.5$
	$\frac{P}{P_y} = \frac{0.877}{\lambda_c^2}$	$\lambda_c > 1.5$

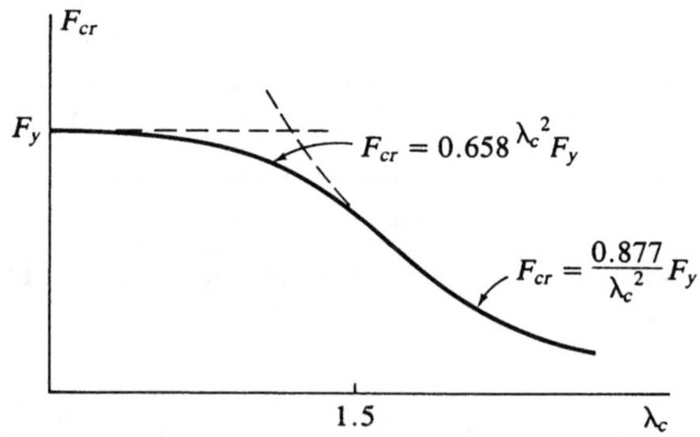

Figure 1.6 AISC-LRFD column curve.

1. The equations should be applicable to a wide range of problems: strong and weak-axis bending, sway and non-sway frames, laterally loaded columns, columns with various slenderness ratios and degrees of end restraint, and leaning column systems. They should account for both inelastic behavior and second-order effects.

2. The equations should separate the second-order member forces and the strength so that second-order elastic analysis can be permitted.

3. The equations should be based on elastic second-order analysis because second-order inelastic analysis is not readily accessible for design office use.

4. The equations should not be more than 5% unconservative when compared to exact inelastic second-order theoretical solutions.

5. The equations should not need to consider strength and stability separately because, in general, all columns of finite length fail by some combination of inelastic bending and stability effects.

6. The same ultimate strength should be predicted for mathematically identical members in both the elastic and inelastic regions, such as the four problems shown in Figure 1.7.

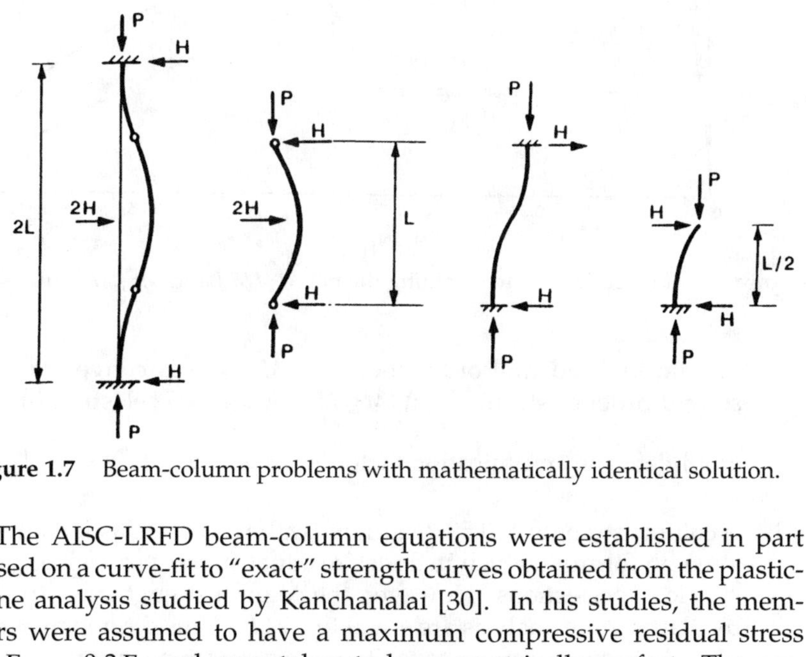

Figure 1.7 Beam-column problems with mathematically identical solution.

The AISC-LRFD beam-column equations were established in part based on a curve-fit to "exact" strength curves obtained from the plastic-zone analysis studied by Kanchanalai [30]. In his studies, the members were assumed to have a maximum compressive residual stress of $F_{rc} = 0.3F_y$ and were taken to be geometrically perfect. The procedure used in developing the LRFD beam-column interaction equations [38] [62] is outlined below.

1. The exact strength curves obtained from second-order plastic-zone analyses are plotted in terms of the normalized axial load (P/P_y) vs. the normalized first-order moment (M_1/M_p). This data applies to a perfectly straight member bent about the weak-axis with $L/r = 100$ as shown in Figure 1.8.

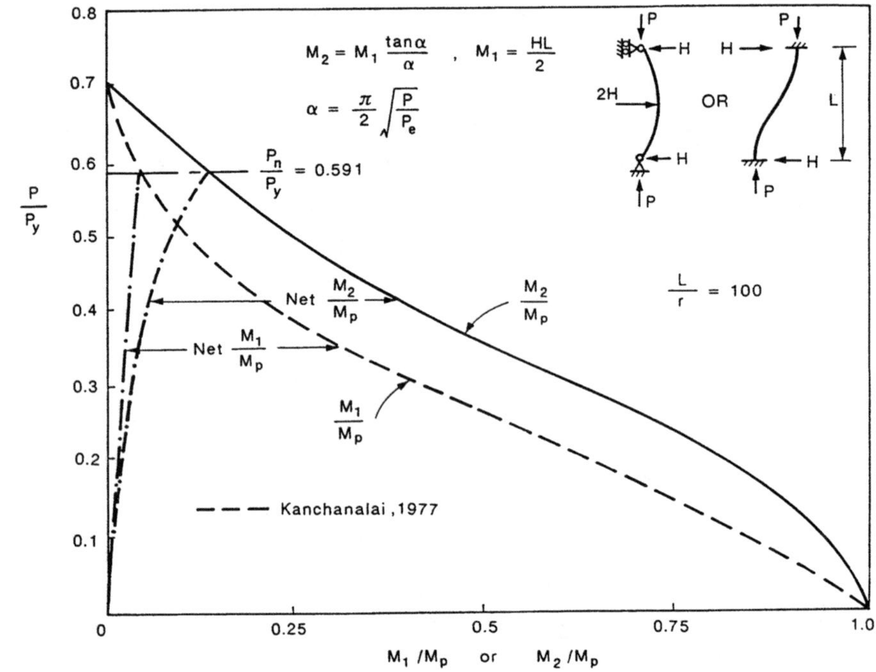

Figure 1.8 Procedure for determining the net M_1/M_p and M_2/M_p curves.

2. The normalized first-order moment (M_1/M_p) is converted to a second-order elastic moment (M_2/M_p) by the exact elastic moment amplifier. The amplification factor is $\tan\alpha/\alpha$, where $\alpha = \frac{\pi}{2}\sqrt{\frac{P}{P_e}}$.

3. A column strength of $0.59P_y$ is given by the LRFD inelastic column-strength formula which includes residual stresses as well as geometric imperfections. However, the column strengths computed by plastic-zone analysis are equal to $0.71P_y$. This discrepancy implies that some moment capacity is used up for geometric imperfections.

4. Assuming that the "used-up" capacity is a linear function of the axial load, the moment can be subtracted from a first-order moment (M_1/M_p). This value, when multiplied by the elastic moment amplification factor, yields the net usable second-order moment capacity shown as the net M_1/M_p values in Figure 1.8.

5. The net results are plotted as shown in Figure 1.9. The LRFD equations are obtained by a curve-fit to the net second-order moment. Similar net M_2/M_p curves have been generated based on

Kanchanalai's data for a wide variety of cases including strong- and weak-axis bending, sway and non-sway, and leaning column systems. The result is the current interaction equations with 8/9 and 1/2 factors.

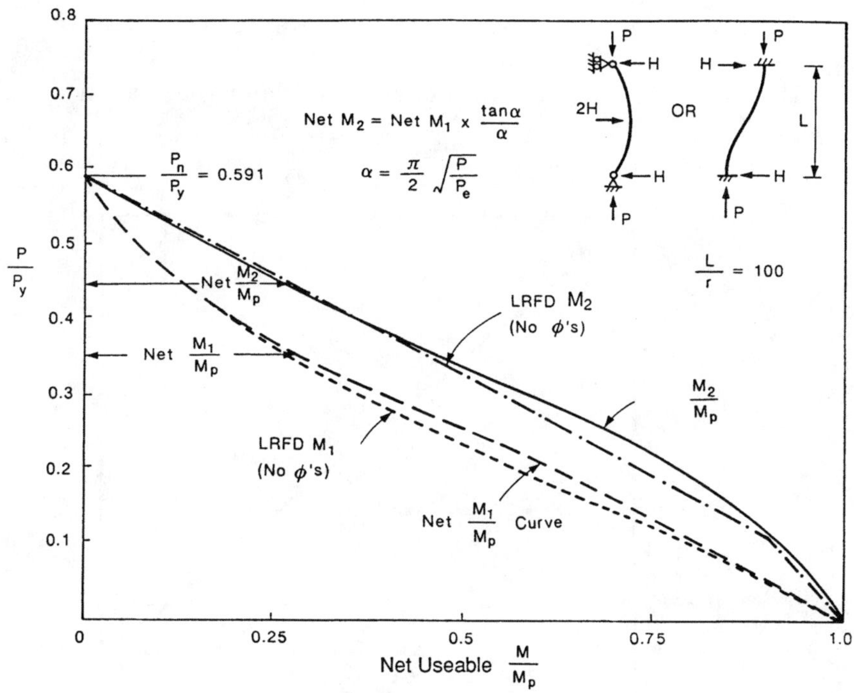

Figure 1.9 Curve-fit of the AISC-LRFD beam-column equations.

6. The final LRFD expressions are obtained by incorporating the resistance factors ϕ_b and ϕ_c.

As described above, the AISC-LRFD Specification [4] [5] developed the following interaction equations based on a curve-fit to the "exact" plastic-zone solution of 82 beam-columns generated by Kanchanalai [30].

$$\frac{P_u}{\phi_c P_n} + \frac{8}{9}\left(\frac{M_{ux}}{\phi_b M_{nx}} + \frac{M_{uy}}{\phi_b M_{ny}}\right) \leq 1.0 \quad \text{for} \quad \frac{P_u}{\phi_c P_n} \geq 0.2 \qquad (1.22a)$$

$$\frac{P_u}{2\phi_c P_n} + \frac{M_{ux}}{\phi_b M_{nx}} + \frac{M_{uy}}{\phi_b M_{ny}} \leq 1.0 \quad \text{for} \quad \frac{P_u}{\phi_c P_n} < 0.2 \qquad (1.22b)$$

where

P_u	= required axial strength
P_n	= nominal axial compressive strength
M_{ux}, M_{uy}	= required flexural strength
M_{nx}, M_{ny}	= nominal flexural strength
ϕ_c, ϕ_b	= resistance factors for compression and flexure

The curves described by the interaction equations are shown in Figure 1.10.

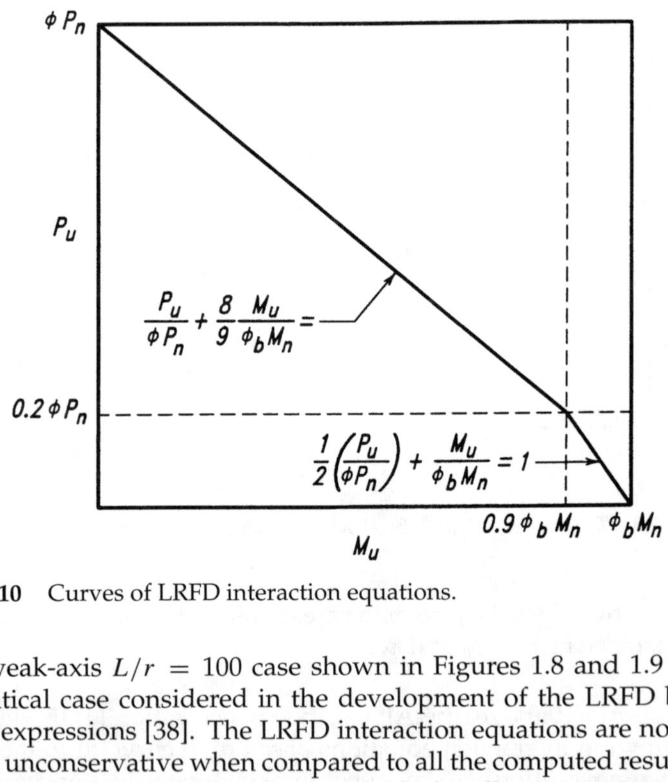

Figure 1.10 Curves of LRFD interaction equations.

The weak-axis $L/r = 100$ case shown in Figures 1.8 and 1.9 is the most critical case considered in the development of the LRFD beam-column expressions [38]. The LRFD interaction equations are no more than 5% unconservative when compared to all the computed results. In general, the LRFD equations give a good fit for all strong-axis cases of the type shown in Figure 1.9 for L/r values ranging from 0 to 100, but are very conservative for the weak-axis cases when L/r ranges from 0 to about 40. They become moderately conservative for both axes when L/r is greater than 120 [38]. The AISC-LRFD Specification recommends the use of the interaction equations for doubly and singly symmetric members including W, WT, and HP shapes, pipe, and structural tubing even though the equations were developed based on W shapes.

1.3.4 Effective Length Factor

The effective length of an end-restrained column is defined as the length of an equivalent pinned-ended column that will give the same critical load as the end-restrained column. Physically, the effective length can be visualized as the distance between the two inflection points (real or imaginary) of the buckled shape of the end-restrained member. The effective length of columns with different end restraints is shown in Figure 1.11. The theoretical values of K for various boundary conditions are summarized in Table 1.5.

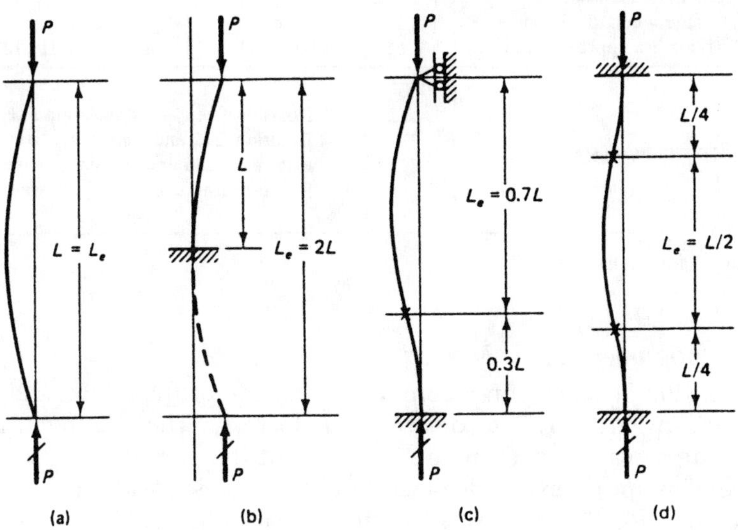

Figure 1.11 Effective lengths of columns with different restraints.

The effective length factor accounts for the interaction between the structural system and its component members. The effective length (K) concept is considered an essential part of many analysis procedures and can handle many cases occurring in all structures. The concept is valid for ideal structures, but its implementation involves several assumptions. The development and implementation of effective length factors have undergone several stages. Recently, a number of methods have been proposed by LeMessurier [39] and Lui [47], among others. A brief summary of these methods together with an illustrative example follows.

Table 1.5 Theoretical and Recommended K Values for Idealized Columns

	(a)	(b)	(c)	(d)	(e)	(f)
Buckled shape of column is shown by dashed line	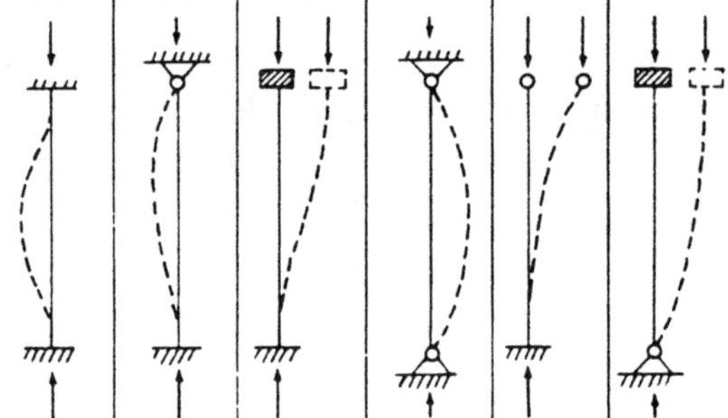					
Theoretical K value	0.5	0.7	1.0	1.0	2.0	2.0
Recommended design value when ideal conditions are approximated	0.65	0.80	1.2	1.0	2.10	2.0
End condition code	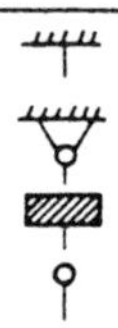	**Rotation fixed and translation fixed** **Rotation free and translation fixed** **Rotation fixed and translation free** **Rotation free and translation free**				

Adopted from AISC, 1994

Alignment Chart Method

The LRFD Specification recommends the use of the alignment charts shown in Figure 1.12 to determine K-factors. The alignment charts provide a way to determine K easily, but they are developed under several assumptions of idealized conditions which seldom exist in real structures [4] [27]. These assumptions are as follows:

1. Behavior is purely elastic.
2. All members have constant cross-section.
3. All joints are rigid.
4. For braced frames, rotations at opposite ends of beams are equal in magnitude, producing single-curvature bending.
5. For unbraced frames, rotations at opposite ends of the restraining beams are equal in magnitude, producing reverse-curvature bending.
6. The stiffness parameters $L\sqrt{(P/EI)}$ of all columns are equal.

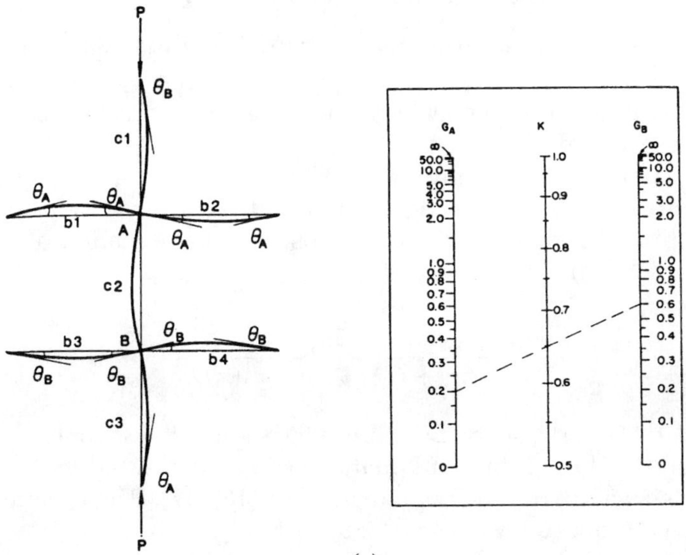

(a)

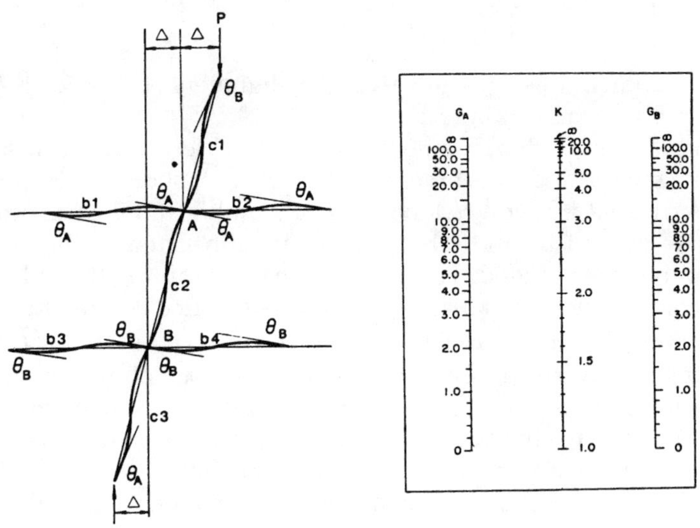

(b)

Figure 1.12 Alignment charts. (a) Braced frame and (b) unbraced frame.

7. Joint restraint is distributed to the column above and below the joint in proportion to I/L of the two columns.
8. All columns buckle simultaneously.
9. No significant axial compression exists in the girders.

The unexpected errors may be produced when the real conditions violate these assumptions.

The charts are based on the buckling of the subassemblage shown in Figure 1.12. The buckling solution of the idealized subassemblage for unbraced frames results in a transcendental equation of the form (Figure 1.12(b)):

$$\frac{G_A G_B (\pi/K)^2 - 36}{6(G_A + G_B)} - \frac{\pi/K}{\tan(\pi/K)} = 0 \tag{1.23}$$

in which G_A and G_B are the column-to-beam stiffness ratios at the column ends. The solution of Equation (1.23) is expressed in the form of the side-sway permitted alignment chart [16] [17]. The transcendental equation for braced frames is (Figure 1.12(a)):

$$\frac{G_A G_B}{4}\left(\frac{\pi}{K}\right)^2 + \left(\frac{G_A + G_B}{2}\right)\left[1 - \frac{\pi}{K\tan(\pi/K)}\right] + \frac{2K\tan(\pi/2K)}{\pi} = 1 \tag{1.24}$$

The solution of this equation gives the alignment chart effective length factors for a braced frame.

The alignment chart method is the most widely used because it provides a simple and direct means to obtain K. It should be appreciated, however, that it involves a number of simplifications or assumptions that are unrealistic and the results may be inaccurate when these assumptions are violated. For example, the assumption that all columns in a story buckle simultaneously cannot be satisfied because in any practical frame it is unlikely that the same stiffness parameter $L\sqrt{P/EI}$ will be achieved for all columns in a particular story. Columns may vary in geometry or dimensions and, if they are the same, axial loading may be distributed unevenly between columns. More active research on flexible connections is in progress and, as a result, increased use of Partially Restraint (PR) construction with semi-rigid connections could violate the assumption that all joints are rigid. Moreover, K is not influenced by members in adjacent bays.

LeMessurier's Method

A more accurate method to compute K was given by LeMessurier [39], who proposed an approach in which the lateral restraint between columns can be accounted for. This approach considers the fact that all columns in a story buckle simultaneously, that a strong column or a column with low axial force will brace a weak column or a column carrying high axial load, or that some columns lean on others in the same story. The effective length factor for column "i" of a story in accordance with LeMessurier can be obtained by using the expression

$$K_i^2 = \frac{\pi^2 E I_i}{P_{ui} L_i^2} \left[\frac{\sum P_u + \sum C_L P_u}{\sum P_L} \right] \tag{1.25}$$

where

EI_i = flexural rigidity of column "i"
L_i = actual height of column "i"
P_{ui} = required axial compressive strength for the i-th rigid column
$\sum P_u$ = required axial compressive strength of all columns in a story
I_i = moment of inertia in plane of bending for "i"-th rigid column
P_L = $\frac{\beta EI}{L^2}$
β = $\frac{6(G_A+G_B)+36}{2(G_A+G_B)+G_A G_B+3}$
C_L = $\left(\beta \frac{K^2}{\pi^2} - 1\right)$
K = K-factor obtained from the sidesway permitted alignment chart

Equation (1.25) accounts directly for leaning columns that have been sized for gravity loads only. A conservative and simple design approximation using a modified elastic effective length factor K given by:

$$K_i'^2 = \frac{I_i}{P_{ui}} \frac{\sum P_u}{\sum \left(I_i / K_i^2\right)}, \qquad P_{ui} > 0 \tag{1.26}$$

is suggested in the revised AISC LRFD Manual [4].

This method accounts accurately for the fact that all the rigid columns in a story resist sidesway, and that stronger columns (smaller $L\sqrt{P/EI}$ ratio) brace the weaker columns (larger $L\sqrt{P/EI}$) until sidesway buckling occurs. The method accounts directly for the presence of gravity or leaning columns in the story. LeMessurier has outlined an iterative procedure for calculating an inelastic K.

There are several deficiencies in the alignment chart method and these have been addressed by Lui [47] and others. Even methods such as LeMessurier's, which were proposed to rectify inadequacies, require the use of alignment charts that may not be convenient for computer implementation.

LeMessurier's New Method

In a recent paper by LeMessurier [37], a new method to determine the K-factor was developed for unbraced frames as follows:

$$K_A = K_\alpha \times K_N \tag{1.27}$$

where

$$\begin{aligned}
K_N &= 1 + (1-\alpha)^4 \\
K_\alpha &= \sqrt{1 + \tfrac{5}{6}(N-1)} \\
\alpha &= M_S/M_L \\
N &= \tfrac{P_{n1}+P_{n2}+\sum P_{n0}}{P_{n1}+P_{n2}} \\
M_S &= \text{smaller moment from first-order analysis result} \\
M_L &= \text{larger moment from first-order analysis result} \\
P_{n1}, P_{n2} &= \text{loads on non-leaning column} \\
\sum P_{n0} &= \text{loads on leaning column}
\end{aligned}$$

Equation (1.27) provides estimates for K for columns that account for both member and frame instabilities with unequal distribution of gravity loads and leaning columns. The K_α, K_N accounts for the member instability effect and the frame instability effect, respectively. The advantage of this formula over conventional LRFD methods is that no special charts or iterations are required to calculate K, and this facilitates computer implementation. This formula is valid for $S_L = \dfrac{N \sum P_y}{\sum P_L} \leq \dfrac{4}{9}$ where P_y is the squash load and P_L is the lateral stiffness defined as $\frac{HL}{\Delta_{OH}}$, in which H is the horizontal force, L is the column height, and Δ_{OH} is the story drift.

Lui's Method

A simple and elegant method that accounts for both member instability and frame instability in the calculation of effective length factors was proposed recently by Lui [47]. Member instability, referred to as the $P-\delta$ effect, is considered in terms of stability functions [17] which are simplified to a great extent by using a Taylor series expansion. Frame instability, referred to as the $P-\Delta$ effect, is accounted for by the use of a story stiffness concept. The two effects are explicitly combined into one formula, for which K_i factor for a member "i" in a frame can be determined as

$$K_i^2 = \frac{\pi^2 E I_i}{P_i L_i^2} \sum \frac{P}{L} \left(\frac{1}{5 \sum \eta} + \frac{\Delta_{oh}}{\sum H} \right) \tag{1.28}$$

where

P_i	= compressive axial force in member i
$\sum \frac{P}{L}$	= sum of the axial force to length ratio of all members in a story
$\sum H$	= sum of the story lateral forces at and above the story under consideration
Δ_{oh}	= inter-story deflection, i.e., relative displacement between adjacent stories
η	= $\frac{\left(3+4.8m+4.2m^2\right)EI}{L^3}$
m	= M_A/M_B
M_A/M_B	= member end moments with $M_A < M_B$
$\sum \eta$	= sum of η of all members in the story being considered

The formula given by Lui's method is simple and needs only a first-order frame analysis. No special charts or iterative procedures are required and this renders it more suitable for computer-based design. It also takes into account explicitly the member instability ($P - \delta$) effects and frame instability ($P - \Delta$) effects. It has been shown that K can be predicted by this formula with sufficient accuracy for columns in unbraced frames with unequal distribution of lateral stiffness and gravity loads, and for frames with leaning columns [47].

The first term within the bracket on the right side of Equation (1.28), $(1/5 \sum \eta)$, represents member instability effect, and the second term $(\Delta_{oh}/ \sum H)$ accounts for the frame instability effect. The first step in the procedure is to carry out a first-order elastic analysis of the frame in question to determine the horizontal deflection at each story level. The only loads applied to the frame are the lateral loads calculated as a fraction (say 0.1%) of the applied story gravity load for each story. In practice, any value can be chosen because the quantities ($\Delta_{oh}/ \sum H$ and M_A/M_B) required in Equation (1.28) are not affected by the value of the lateral load used. In a first-order elastic analysis, all quantities vary linearly with the applied lateral load leaving the ratio of the quantities unaffected.

The second step in the procedure is to calculate the value of $\Delta_{oh}/ \sum H$ for each story. In the third step, the values of m, η, P/L for each column in a story are evaluated. Equation (1.28) is then used to obtain the value of K for each rigid column in that story. The final step is repeated for each story. All calculations can be carried out easily in a tabular form for each story.

An Illustrative Example

The following example of a frame with a leaning column illustrates the computation of K using the four methods described previously. The frame is subject to a gravity load P on each column as shown in Figure 1.13(a). K for the right column AB is evaluated in what follows.

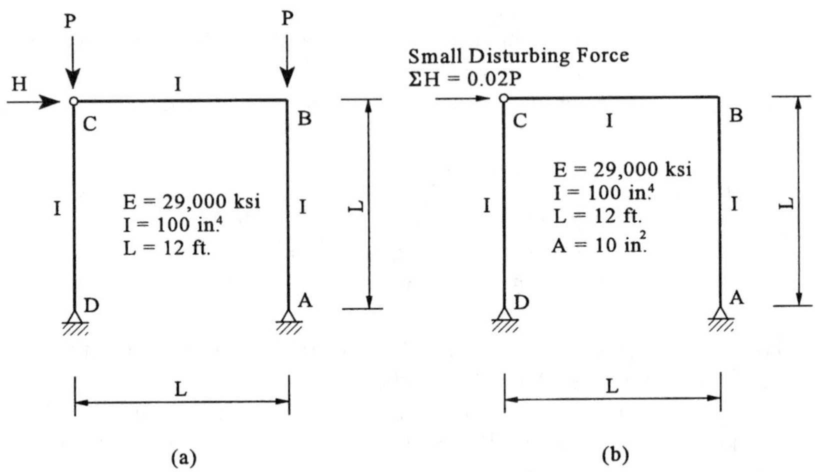

Figure 1.13 A leaning column frame for illustration.

Alignment Chart Method

$$G_A = \infty \qquad G_B = \frac{\sum(EI/L)_{\text{col}}}{\sum(EI/L)_{\text{beam}}} = 1.0$$

Since the far end of the beam is hinged, G_B is adjusted using Equation (1.8d)

$$G_B = (2)(1.0) = 2.0$$

we have

$$K = 2.6$$

K should be adjusted to account for the leaning column effect using Equation (1.10)

$$K' = \sqrt{\frac{2P}{P}(2.6)^2} = 3.68$$

$$K' = \sqrt{\frac{5}{8}}(2.6) = 2.06 < 3.68$$

We use $K' = 3.68$.

LeMessurier Method

For this frame, since only Column AB provides stability to the system, LeMessurier's Equation (1.25) reduces to [39]

$$K_{AB}^2 = \frac{\pi^2}{P_{AB}}\left[\frac{\sum P + (C_L P)_{AB}}{\beta_{AB}}\right] \tag{1.29}$$

From the expressions for $(C_L P)_{AB}$ and β_{AB} as given in Equation (1.25) with $G_A = \infty$, $G_B = 2.0$, and $K = 2.6$, one obtains these values, respectively, as

$$\beta_{AB} = \frac{6(G_A + G_B) + 36}{2(G_A + G_B) + G_A G_B + 3} = \frac{6}{2+2} = 1.5$$

$$(C_L P)_{AB} = \left(\beta_{AB}\frac{K^2}{\pi^2} - 1\right)P = \left[1.5\left(\frac{2.6}{\pi}\right)^2 - 1\right]P = 0.0274P$$

$$P_{AB} = P, \qquad \sum P = 2P$$

Substituting these values in Equation (1.29), we get

$$K_{AB}^2 = \frac{\pi^2}{P}\left[\frac{2P + 0.0274P}{1.5}\right] = 13.34$$

For the right Column AB, we have

$$K_{AB} = 3.65$$

LeMessurier's New Method

$$\alpha = M_s/M_L = 0/M_L = 0$$

$$N = 2P/P = 2$$

$$K_N = 1 + (1-\alpha)^4 = 1 + (1)^4 = 2$$

$$K_\alpha = \sqrt{1 + \frac{5}{6}(N-1)} = \sqrt{1 + \frac{5}{6}(2-1)} = 1.35$$

From Equation (1.27), we have

$$K_A = K_\alpha \times K_N = 1.35 \times 2 = 2.7$$

We have

$$S_L = \frac{N \sum P_y}{\sum P_L} = \frac{(2)(10)(36)}{(1)\left(12^2\right)/0.681} = 3.405 > 4/9$$

where $\Delta_{OH} = 0.681$ is obtained from first-order analysis under the horizontal load 1 kip. The new method is not valid for this example.

Lui's Method

In the first step, a small lateral load equal to 1% of the gravity loads ($2P = 100$ kips), i.e., 1.0 kip is applied as shown in Figure 1.13(b) to initiate sway. In the second step, the sway deflection, Δ_{oh}, from a first-order analysis is found to be 0.681 in. In the third step, the remaining calculations to determine the K-factor for the right column AB can be carried out as summarized in Table 1.6. It should be noted that η for

Table 1.6 Calculation of Parameters for Lui's K-Factor Column

Column	$I(in.^4)$	$L(in.)$	$P(kips)$	P/L	m	η
AB	100	144	50	0.347	0	$\frac{3EI}{L^3}$
CD	100	144	50	0.347	0	$\frac{2.4EI}{L^3}$
Σ			100	0.694		$\frac{5.4EI}{L^3}$

leaning column CD is $2.4EI/L^3$ corresponding to $m = 0$. Using Equation (1.28) and the above table, we have

$$K^2 = \frac{\pi^2 E I_i}{P_i L_i^2} \sum \frac{P}{L}\left(\frac{1}{5\sum \eta} + \frac{\Delta_{oh}}{\sum H}\right)$$

$$= \frac{\pi^2(29000)(100)}{50(144)^2}\left[0.694\left(\frac{144^3}{(5)(5.4)(29,000)(100)} + \frac{0.681}{1}\right)\right]$$

$$= 13.78$$

From which we obtain $K_{AB} = 3.71$.

Concluding Remarks

A comparison of K-factors determined by the methods discussed is summarized in Table 1.7. All methods except the LeMessurier's New Method predict similar values of K-factors. LeMessurier's New Method underpredicts the K-factor value by 26% compared with that of the LeMessurier's original method. Alignment charts are popular because they provide a simple and direct means of getting K-factors. It does, however, produce inaccurate results when the assumptions are violated. Although the other two methods provide reliable K-factors for columns in frames with unequal distribution of column stiffnesses, gravity loads, and leaning columns, they do not account for the inelastic stiffness reduction of a column (τ) as defined by Equations (1.7a and 1.7b) and may underestimate the column strength when P_u/P_y is greater than 1/3.

Table 1.7 Comparison of K-Factor Determined by Various Methods

Method	Alignment chart (AISC,1994)	LeMessurier (1977)	New LeMessurier (1995)	Lui (1992)
K-factor	3.68	3.65	2.70	3.71

The current frame design methods require K calculations which are tedious and not user-friendly for a computer-based design. Advanced Analysis eliminates the K calculation by directly evaluating the strength of the whole system. Practical computer-based advanced analysis methods are expected to gradually replace the conventional methods of frame design in the future.

1.3.5 Moment Amplification Factor

The maximum second-order elastic moment in a beam-column member, M_u, can be determined by a direct second-order elastic analysis or a linear first-order elastic analysis with modified moments by appropriate moment amplification factors. As discussed previously, the LRFD Specification requires that a first-order analysis be separated into NT and LT cases. For the NT analysis, the gravity loads are applied to the structural system and the analysis is performed with artificial side-sway restraints at every level. The LT analysis is performed based on the lateral loads applied to the system plus the reverse of the reactions from the artificial restraints used in the NT analysis.

The design moment (M_u) is obtained using elastic B_1 and B_2 amplification factors and the beam-column moments obtained from the NT and LT analyses. The first term of Equation (1.16) represents the component of M_u associated with the $P - \delta$ effect. For a framed beam-column subject only to end forces with a B_1 term evaluated in Equation (1.17) equal to 1.0, the maximum second-order elastic moment occurs at the end of the member. If B_1 is greater than 1.0, the maximum $P - \delta$ moment occurs within the span of the member.

The second term of Equation (1.16), $B_2 M_{lt}$, represents the second-order elastic moment associated with side-sway of the frame. This moment always occurs at the end of a member. The simple addition of $B_1 M_{nt}$ and $B_2 M_{lt}$ in Equation (1.16) is a conservative way of estimating the maximum second-order elastic moment in the member because the amplified moment $B_1 M_{nt}$ occurs within the span of the member when $B_1 \geq 1.0$, and the moment $B_2 M_{lt}$ is always at the member end.

B_1 Factor

For a beam-column subject to an axial force P and end moments M_1 and M_2, as shown in Figure 1.14, a closed-form expression for the maximum second-order elastic moment in the member may be derived based on

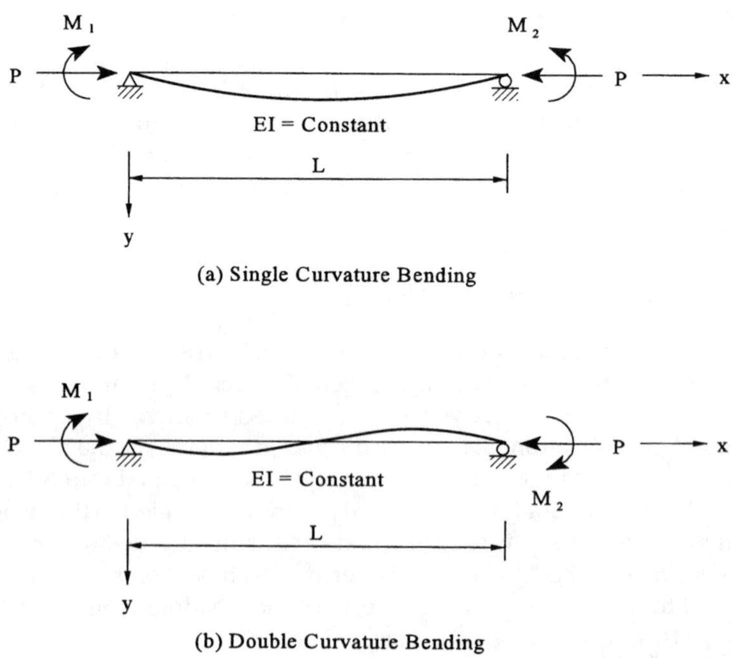

Figure 1.14 Maximum moment in a beam-column subject to end moments and axial force.

the governing differential equation for a beam-column [17] and it can be condensed into the form:

$$M_u = A_F \mid M_2 \mid \tag{1.30}$$

where M_2 is the larger of the two end moments, and A_F is an amplification factor defined as:

for $0 \leq X_c \leq L$

$$A_F = \sqrt{\frac{(M_1/M_2)^2 + 2\,(M_1/M_2)\cos kL + 1}{\sin^2 kL}} \tag{1.31a}$$

for $X_c < 0$ or $X_c > L$

$$A_F = 1.0 \tag{1.31b}$$

The term X_c is the location of the maximum moment along the length of the member and can be determined by:

$$\tan\,(kX_c) = -\frac{(M_1/M_2)\cos kL + 1}{(M_1/M_2)\sin kL} \tag{1.32}$$

where

$$k = \sqrt{\frac{P_u}{EI}}$$

M_1 and M_2 are the smaller and larger end moments, respectively. M_1/M_2 is positive when the member is bent in double curvature and negative when in single curvature.

Equation (1.30) is the general expression for the maximum moment of a beam-column subject to prescribed end forces. The maximum moment may occur at the member's end and thus is equal to the larger of the two end moments; or, it may occur within the member and its magnitude and location can be computed by Equations (1.30) and (1.32). It should be noted that in calculating the location of the maximum moment, the kX_c term in Equation (1.32) should lie within the range of 0 to π when $\tan(kX_c) < 0$, and kX_c should be greater than π when $\tan(kX_c) > 0$.

The "exact" amplification factor given in Equations (1.31a) and (1.31b) may be expressed as:

$$A_F = C_m \sec\left(\frac{kL}{2}\right) \tag{1.33}$$

where C_m is the equivalent moment factor given by

for $0 \leq X_c \leq L$

$$C_m = \sqrt{\frac{(M_1/M_2)^2 + 2\,(M_1/M_2)\cos kL + 1}{2(1 - \cos kL)}} \tag{1.34}$$

for $X_c < 0$ or $X_c > L$

$$C_m = \frac{1}{\sec\left(\frac{kL}{2}\right)} = \cos\left(\frac{kL}{2}\right) \tag{1.35}$$

Two approximations may be introduced to simplify Equations (1.33) and (1.34). First, Equation (1.34) is replaced by the simple expression [12]:

$$C_m = 0.6 - 0.4\,(M_1/M_2) \tag{1.36}$$

Massonnet [49] proposed a better approximate expression for C_m factor:

$$C_m = \sqrt{0.3\,(M_1/M_2)^2 - 0.4\,(M_1/M_2) + 0.3} \tag{1.37}$$

Figure 1.15 shows a comparison of theoretical and approximate C_m factors. Second, $\sec\left(\dfrac{kL}{2}\right)$ in Equation (1.33) is replaced by

$$\sec\left(\frac{kL}{2}\right) \approx \frac{1}{1 - \frac{P_u}{P_e}} \tag{1.38}$$

These theoretical and approximate values are shown in Figure 1.16. The resulting amplification factor is

$$A_F = \frac{C_m}{1 - \frac{P_u}{P_e}} \geq 1.0 \tag{1.39}$$

which is similar to the B_1 employed by the AISC-LRFD.

When considering the above equations, it is important to note that the "exact" amplification factor given by Equations (1.31a) and (1.31b) has been derived from the governing differential equation for an isolated beam-column segment, as shown in Figure 1.14, with force boundary conditions prescribed at its ends. Thus, for a simply supported beam-column with applied moments at its ends, M_1 and M_2 are simply the applied end moments. However, for a beam-column that is a part of a

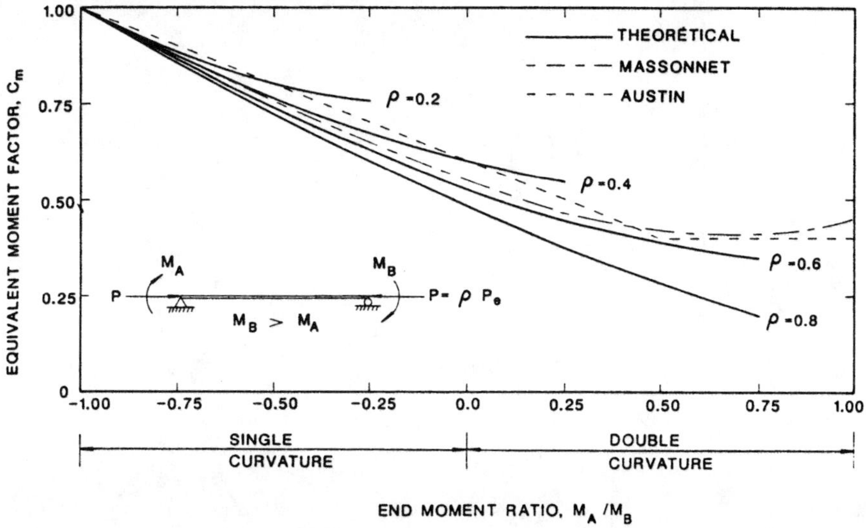

Figure 1.15 Theoretical and approximate C_m factors.

general framing system, the M_1 and M_2 values must be the second-order elastic moments calculated at the member ends. In contrast, when using the LRFD B_1 factor, the amplified end moment is not the second-order moment but the first. To account for this aspect, the LRFD Specification replaces the P_e term in Equation (1.39) with the P_{e1} value determined using K. This accounts for the effect of end restraint by reducing the magnitude of M_u in the beam-column member. Incorporating P_{e1}, the B_1 factor may be written as:

$$B_1 = \frac{C_m}{1 - \frac{P_u}{P_{e1}}} \tag{1.40}$$

B_2 Factor

The $P-\Delta$ moment amplification factor B_2 in Equation (1.19a) was developed based on a story stiffness concept [19] [52]. By assuming that each story behaves independently of all other stories, and that the additional moments in the columns caused by the $P - \Delta$ effect are equivalent to that caused by a lateral force of $\sum P\Delta/L$ (Figure 1.17), the sway stiffness of the story can be defined as:

$$S_F = \frac{\text{horizontal force}}{\text{lateral displacement}}$$

$$= \frac{\sum H}{\Delta_0} = \frac{\sum H + \sum P\Delta/L}{\Delta} \tag{1.41}$$

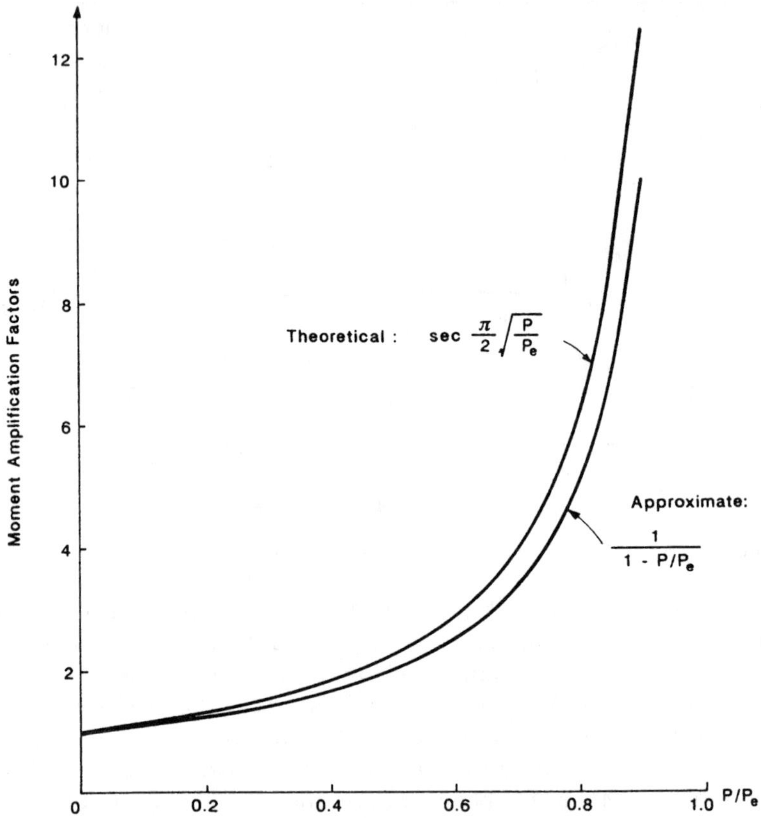

Figure 1.16 Theoretical and approximate moment amplification factors.

where

Δ_0 = first-order deflection
Δ = second-order deflection

Solving Equation (1.41) for Δ gives:

$$\Delta = \left(\frac{1}{1 - \frac{\sum P\Delta_0}{\sum HL}} \right) \Delta_0 \tag{1.42}$$

If rigid connections and elastic behavior are assumed, the moment in members under sway conditions is proportional to the lateral deflection.

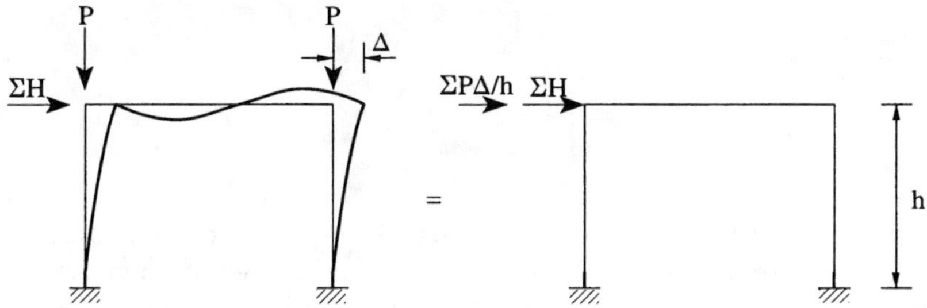

Figure 1.17 $P - \Delta$ effect of unbraced frame.

Therefore, we can write the amplified sway moment M_{alt} as

$$M_{alt} = \left(\frac{1}{1 - \frac{\sum P\Delta_0}{\sum HL}} \right) M_{lt} \tag{1.43}$$

The alternate expression for B_2 in Equation (1.19b) was developed based on a multiple-column buckling concept. When a simply supported beam-column is acted upon by a concentrated lateral load Q at midspan, the theoretical maximum deflection can be written approximately as [17]:

$$y_{\max} \approx y_0 \left[\frac{1}{1 - \left(\frac{P}{P_e}\right)} \right] \tag{1.44}$$

where

y_0 = a first-order deflection at midspan

By assuming that when instability is to occur in a story, all columns in that story will become unstable simultaneously. As a result, a direct extension of Equation (1.44) is justified by replacing the term P/P_{e2} with $\sum P / \sum P_{e2}$ where the summation is carried over all columns in a story. Thus,

$$\Delta = \left(\frac{1}{1 - \frac{\sum P}{\sum P_{e2}}} \right) \Delta_0 \tag{1.45}$$

Using the same argument that if rigid joints and elastic behavior are assumed the sway moments are directly proportional to the lateral deflections of the story, we can now write the amplified sway moment as

$$M_{alt} = \left(\frac{1}{1 - \frac{\sum P}{\sum P_{e2}}} \right) M_{lt} = B_2 M_{lt} \tag{1.46}$$

It should be mentioned that Equation (1.16) is a conservative estimate of the maximum moment. This is because the amplified moment resulting from the $P - \delta$ effect (i.e., the term $B_1 M_{nt}$) and the amplified moment resulting from the $P - \Delta$ effect (i.e., the term $B_2 M_{lt}$) do not necessarily occur at the same location. For elastic behavior, the $P - \Delta$ effect usually magnifies the end moments. Nevertheless, because of the assumptions involved in developing the $P - \delta$ and $P - \Delta$ amplification factors, as well as the difficulties involved in locating the exact location of each of the magnified moments in the member, Equation (1.16) gives a justifiable estimation of the design moment.

1.3.6 *Illustrative Example 1: Two-Bay Unbraced Frame*

Frame Configuration

Figure 1.18 shows a fixed-supported two-bay portal frame subject to gravity and lateral loads. The frame is braced against out-of-plane motion at each story height and in the middle of Beam EF. For simplicity, the same member size is used for both beams and columns. A36 steel is used for all members. The load condition is shown in Figure 1.18.

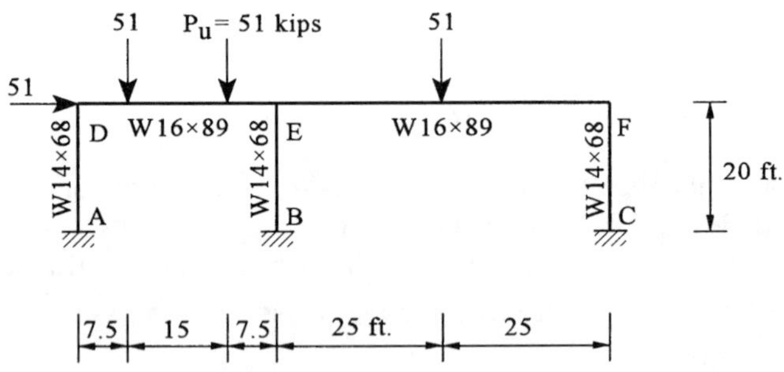

Figure 1.18 Frame configuration and load condition of two-bay frame.

Analysis

In the LRFD method, a first-order elastic analysis is performed on the structure with assumed member sizes shown in Figure 1.18. The frame is decomposed into non-sway and sway components, and the corresponding moments are shown in Figure 1.19. In the following, the adequacy of Column CF and Beam DE are checked.

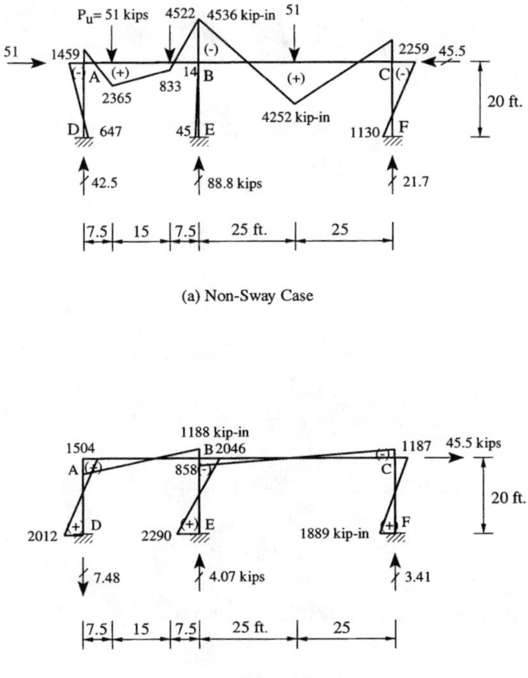

(a) Non-Sway Case

(b) Sway Case

Figure 1.19 First-order analysis for two-bay frame.

Check Adequacy of Column CF (W14×68)

- Determine P_n

  To obtain accurate K-factor, we shall use the inelastic stiffness reduction factor τ [Equations (1.7a) and (1.7b)].

$$P_u = 21.7 + 3.41 = 25.11 \text{ kips}$$

$$\frac{P_u}{P_y} = \frac{25.11}{(36)(20)} = 0.035 < \frac{1}{3}$$

Therefore,

$$\tau = 1$$

This implies that Column CF remains elastic. Further correction factor L'_g for far-end beam condition is given by Equation (1.9).

$$L'_g = 50\left(2 - \frac{858}{1,187}\right) = 64 \text{ ft}$$

Using the member section given in Figure 1.18, we have:

$$G_F = \frac{(1)(723)/(20)}{(1,300)/(64)} = 1.78$$

$$G_C = 0$$

To account for the fact that the fixed-end is not ideal, use:

$$(G_C)_{\text{adjust}} = 1.0$$

From the side-sway inhibited alignment chart

$$K_x = 1.43$$

The slenderness parameter can be calculated from Equation (1.12):

$$\lambda_{cx} = \frac{(1.43)(20)(12)}{(\pi)(6.01)}\sqrt{\frac{36}{29,000}} = 0.64 < 1.5$$

For weak-axis bending, $K_y = 1$ and

$$\lambda_{cy} = \frac{(1.0)(20)(12)}{(\pi)(2.64)}\sqrt{\frac{36}{29,000}} = 1.09 < 1.5$$

Comparing λ_{cx} and λ_{cy}, the weak-axis bending controls. Using Equation (1.11a), P_n is:

$$P_n = (20.0)(36)(0.658)^{(1.09)^2} = 438 \text{ kips}$$

$$\phi_c P_n = (0.85)(438) = 372 \text{ kips}$$

- Determine M_n
  The braced length (L_b) of 20 ft falls between $L_p = 10.3$ ft and $L_r = 37.3$ ft. Both L_p and L_r can be found in the Beam Design Tables of AISC Specification [4] or calculated by Equations (1.14a

and 1.14b). To determine M_n, C_b for Member CF is calculated by Equation (1.15).

$$M_{\max} = 2,259 + 1,187 = 3,446 \text{ kip-in}$$

$$M_A = 283 + 1,120 = 1,403 \text{ kip-in}$$

$$M_B = 565 - 351 = 214 \text{ kip-in}$$

$$M_C = 1,412 + 418 = 1,830 \text{ kip-in}$$

$$C_b = \frac{(12.5)(3,446)}{(2.5)(3,446) + (3)(1,403) + (4)(214) + (3)(1,830)} = 2.25$$

From either Equations (1.13a through 1.13c) or the beam design table, we have

$$\phi_b M_p = 311 \text{ kip-ft}$$

$$\phi_b M_r = 201 \text{ kip-ft}$$

M_n is determined as:

$$\phi_b M_n = 2.25\left[311 - (311 - 201)\frac{20 - 10.3}{37.3 - 10.3}\right]$$

$$= 611 \text{ kip-ft} > \phi_b M_P = 311 \text{ kip-ft}$$

$$\text{Use } \phi_b M_n = 311 \text{ kip-ft}$$

- Determine P_u

$$P_u = 21.7 + 3.41 = 25.11 \text{ kips}$$

- Determine M_u
To calculate B_1, we use Equations (1.17 and 1.18).

$$C_m = 0.6 - 0.4\frac{1,130}{2,259} = 0.4$$

From the side-sway inhibited alignment chart, we have:

$$G_F = \frac{723/20}{1,300/50} = 1.39$$

$$(G_C)_{\text{adjust}} = 1.0$$

$$K_x = 0.79$$

and so,

$$P_{e1} = \frac{(\pi^2)\,(29,000)(723)}{[(0.79)(20)(12)]^2} = 5,757 \text{ kips}$$

$$B_1 = \frac{0.4}{1 - \frac{25.11}{5,757}} = 0.4 < 1.0$$

Since B_1 factor is less than 1.0, use $B_1 = 1$. Note that B_1 factor may be taken as 1.0 because it is rarely greater than 1.0 in a sway frame. To calculate B_2, we use Equation (1.19b).

$$\sum P_u = (51)(3) = 153 \text{ kips}$$

$$K_{CF} = 1.43$$

$$K_{AD} \approx 1.30; \left[G_D = \frac{723/20}{1,300/30} = 0.834, \quad G_A = 0, \quad (G_A)_{\text{adjust}} = 1.0 \right]$$

$$K_{BE} \approx 1.25; \left[G_E = \frac{723/20}{1,300/30 + 1,300/50} = 0.521 \right]$$

$$P_{e2} = \frac{(\pi^2)\,(29,000)(723)}{[(20)(12)]^2} \left(\frac{1}{1.43^2} + \frac{1}{1.30^2} + \frac{1}{1.25^2} \right) = 6,182 \text{ kips}$$

$$B_2 = \frac{1}{1 - \frac{153}{6,182}} = 1.025$$

Thus, we have:

$$M_u = (1)(-2,259) + (1.025)(-1,187) = 3,476 \text{ kip-in.} \qquad \text{at top}$$

$$M_u = (1)(1130) + (1.025)(1,889) = 3,066 \text{ kip-in.} \qquad \text{at bottom}$$

- Check the interaction equation

$$\frac{P_u}{\phi_c P_n} = \frac{25.11}{372} = 0.0675 < 0.2$$

$$\frac{P_u}{2\phi_c P_n} + \frac{M_{ux}}{\phi_b M_{nx}} = \frac{0.0675}{2} + \frac{3,476}{(311)(12)} = 0.965 < 1.0$$

Therefore, W14x68 is adequate.

Check Adequacy of Beam DE(W16×89)

Since the axial force in a beam is negligible, we need only check the flexural strength of the beam.

- Determine M_n
  From the beam design table,

$$L_p = 10.4 \text{ ft}$$

$$L_r = 38.6 \text{ ft}$$

$$L_b = 30 \text{ ft}$$

$$L_p < L_b < L_r$$

From Figure 1.19,

$$M_{\max} = 4,522 + 1,188 = 5,710 \text{ kip-in}$$

$$M_A = 2,365 + 831 = 3,196 \text{ kip-in}$$

$$M_B = 1,599 + 158 = 1,757 \text{ kip-in}$$

$$M_C = 833 - 515 = 318 \text{ kip-in}$$

From Equation (1.15),

$$C_b = \frac{(12.5)(5.710)}{(2.5)(5.710) + (3)(3.196) + (4)(1.757) + (3)(318)} = 2.24$$

From the beam design table,

$$\phi_b M_p = 473 \text{ kip-ft}$$

$$\phi_b M_r = 302 \text{ kip-ft}$$

Use Equation (1.13b),

$$\phi_b M_n = (2.24)\left[473 - (473 - 302)\frac{30 - 10.4}{38.6 - 10.4}\right]$$

$$= 793 \text{ kip-ft} > \phi_b M_p = 473 \text{ kip-ft}$$

$$\text{Use } \phi_b M_n = 473 \text{ kip-ft}$$

- Determine M_u
  Note that the sway moment of the beam must be amplified by B_2 to account for second-order effects.

$$M_u = (1)(4,522) + (1.025)(1,188) = 5,740 \text{ kip-in}$$

- Check the interaction equation

$$\frac{P_u}{2\phi_c P_n} + \frac{M_u}{\phi_b M_n} = 0 + \frac{5,740}{(473)(12)} = 1.01$$

Therefore, W16x89 is adequate.

1.3.7 *Illustrative Example 2: Leaning Column Frame*

Frame Configuration

Figure 1.20 shows a frame with a leaning column. The exterior columns lean to the interior column which provides lateral rigidity of the frame. The frame is braced against out-of-plane bending at story height of every column. The beam is fully braced against lateral torsional buckling. A36 steel is used for all members.

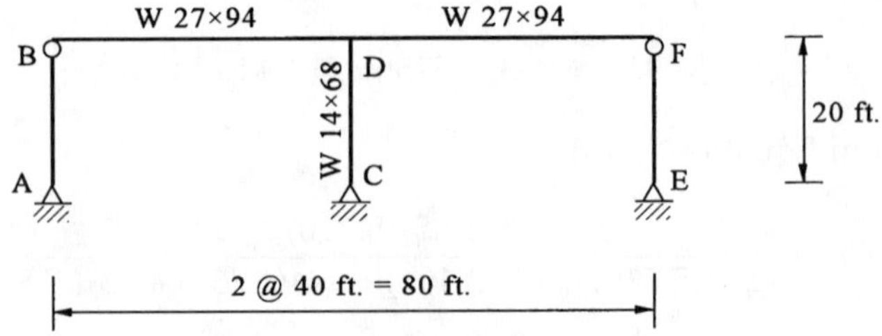

Figure 1.20 Frame configuration of leaning column frame.

Load Condition

Assume the loadings are:

$$D = 0.9 \text{ kip/ft}$$

$$L = 1.6 \text{ kip/ft}$$

$$W = 0.8 \text{ kip/ft}$$

For these loads, the pertinent load combinations are:

(1) $1.4D$
(2) $1.2D + 1.6L$
(3) $1.2D + 0.5L$
(4) $1.2D + 1.3W + 0.5L$
(5) $0.9D - 1.3W$

The load combinations (2) and (4) are the most severe for gravity and combined gravity and lateral loadings, respectively. As a result, only (2) and (4) need to be checked.

Load Combination (2): $1.2D + 1.6L$

Figure 1.21 shows the result of a first-order analysis. Since the first-order moment in Column CD is zero, the second and third terms of the interaction equations [Equation (1.22a) and (1.22b)] vanish. We need only check the first term for Column CD.

Check Adequacy of Column CD(W14×68)

- Determine P_n

The inelastic stiffness reduction factor τ is:

$$\frac{P_u}{P_y} = \frac{182}{(36)(20)} = 0.253 < \frac{1}{3}$$

$$\tau = 1$$

Using the end condition given in Figure 1.20, we have:

$$G_c = \infty \text{ (pinned-end)}$$

Although the theoretical G value for pinned-end is infinity, for design purposes it is customary to use $G = 10$ to account for the fact that an ideal pinned-ended condition does not exist. Therefore,

$$(G_C)_{\text{adjust}} = 10$$

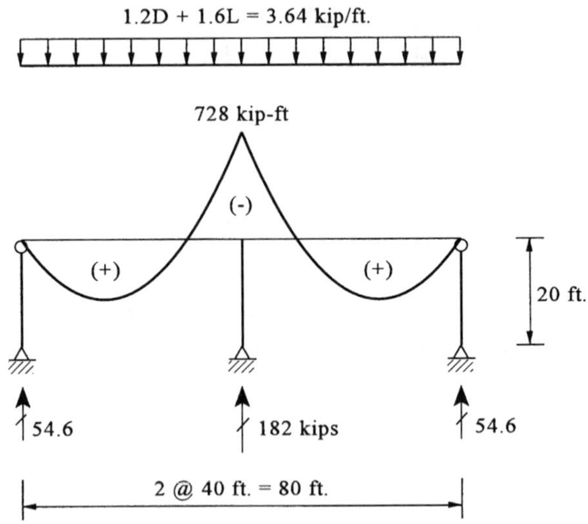

Figure 1.21 First-order analysis of leaning column frame for load combination $1.2D + 1.6L$.

To account for the far end hinge of the beam, L'_g may be taken by Equation (1.8d) as:

$$L'_g = (2)(40) = 80 \text{ ft}$$

We have:

$$G_D = \frac{723/20}{(2)(3,270)/(80)} = 0.44$$

Note that this frame should be considered as an unbraced frame regardless of the gravity loading condition and the symmetry of the frame because this frame could produce lateral displacement due to out-of-plumbness of columns. Therefore, the K-factor should be taken from the alignment chart in sway case as:

$$K_x = 1.73$$

Since the exterior pinned-ended columns lean on Column CD, K_x must be adjusted using Equation (1.10).

$$K'_x = \sqrt{\frac{291.2}{182}\frac{723}{723/1.73^2}} = 2.19 > \sqrt{\frac{5}{8}}(1.73) = 1.37$$

The slenderness parameter can be calculated from:

$$\lambda_{cx} = \frac{(2.19)(20)(12)}{(\pi)(6.01)}\sqrt{\frac{36}{29,000}} = 0.981 < 1.5$$

For weak-axis bending, we have

$$\lambda_{cy} = \frac{(1)(20)(12)}{(\pi)(2.46)}\sqrt{\frac{36}{29,000}} = 1.09 < 1.5$$

Comparing λ_{cx} and λ_{cy}, it is concluded that weak-axis bending controls. Using Equation (1.11a), P_n is:

$$P_n = (20.0)(36)(0.658)^{(1.09)^2} = 438 \text{ kips}$$

$$\phi_c P_n = 372 \text{ kips}$$

- Determine P_u
  From Figure 1.21,
  $$P_u = 182 \text{ kips}$$
- Check the interaction equation
  $$\frac{P_u}{\phi_c P_n} = \frac{182}{372} = 0.489 < 1.0$$

Therefore, W14x68 is adequate.

Check Adequacy of Beam DF(W27×94)

- Determine M_n
  Since the beam is laterally braced, the full plastic moment M_p is developed:
  $$\phi_b M_p = (0.9)(278)(36) = 9,007 \text{ kip-in} = 751 \text{ kip-ft}$$
- Determine M_u
  From Figure 1.21,
  $$M_u = 728 \text{ kip-ft}$$
- Check the interaction equation
  $$\frac{M_u}{\phi_b M_n} = \frac{728}{751} = 0.97 < 1.0$$

Therefore, W27x94 is adequate.

Load Combination (4): $1.2D + 1.3W + 0.5L$

Figure 1.22 shows load condition for the leaning column frame. The wind load is assumed to distribute to the windward and leeward sides of the frame in a 7:3 ratio. Figure 1.23 shows the result of a first-order analysis of the non-sway and sway components of the frame. Herein, the column size (not the beam size) is checked, since the beam size is controlled by load combination (2).

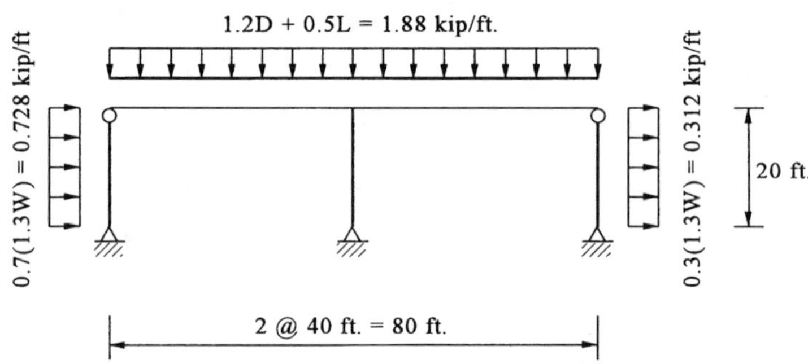

Figure 1.22 Load condition of leaning column frame for load combination $1.2D + 1.3W + 0.5L$.

Check Adequacy of Column CD(W14×68)

- Determine P_n

  Here, as in load combination (2), we have:

$$K_x = 1.73$$

Since the exterior pinned-ended columns lean on Column CD, K_x must be adjusted using Equation (1.10)

$$K'_x = \sqrt{\frac{150.4}{94} \cdot \frac{723}{723/(1.73)^2}} = 2.19$$

and so,

$$\lambda_{cx} = \frac{(2.19)(20)(12)}{(\pi)(6.01)} \sqrt{\frac{36}{29,000}} = 0.981$$

From load combination (2), we have:

$$\lambda_{cy} = 1.09$$

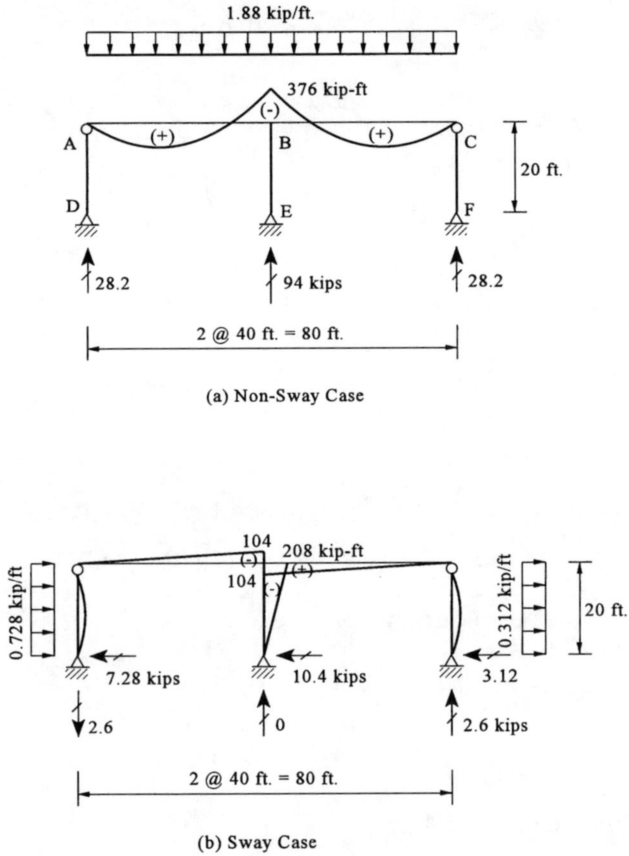

Figure 1.23 First-order analysis of leaning column frame for load combination $1.2D + 1.3W + 0.5L$.

Since $\lambda_{cy} > \lambda_{cx}$, P_n is the same as that of the load combination (2).

$$P_n = 438 \text{ kips}$$

$$\phi_c P_n = 372 \text{ kips}$$

- Determine M_n
  From the beam design table of AISC Specification, we have

$$L_p = 10.3 \text{ ft}$$

$$L_r = 37.3 \text{ ft}$$

$$L_b = 20 \text{ ft}$$

$$L_p < L_b < L_r$$

C_b is determined as:

$$M_{\max} = 208$$

$$M_A = 52$$

$$M_B = 104$$

$$M_C = 156$$

$$C_b = \frac{(12.5)(208)}{(2.5)(208) + (3)(52) + (4)(104) + (3)(156)} = 1.67$$

From the beam design table, we have

$$\phi_b M_p = 311 \text{ kip-ft}$$

$$\phi_b M_r = 201 \text{ kip-ft}$$

Use Equation (1.13b),

$$\phi_b M_n = 1.67 \left[311 - (311 - 201) \frac{20 - 10.3}{37.3 - 10.3} \right] =$$

$$= 453 \text{ kip-ft} > \phi_b M_p = 311 \text{ kip-ft}$$

Use $\phi_b M_n = 311$ kip-ft

- Determine P_u
  From Figure 1.23, we have

$$P_u = 94 + 0 = 94 \text{ kips}$$

- Determine M_u
  To calculate B_1, we use Equations (1.17 and 1.18).

$$C_m = 0.6 - 0.4 \frac{0}{208} = 0.6$$

$$(G_c)_{\text{adj}} = 10$$

$$L'_g = \frac{L_g}{1.5} = 26.7$$

$$G_D = \frac{723/20}{(2)(3,270)/(26.7)} = 0.148$$

$$K_x = 0.73 \text{ from the alignment chart for a braced frame}$$

$$P_{e1} = \frac{(\pi^2)(29,000)(723)}{[(0.73)(20)(12)]^2} = 6,742 \text{ kips}$$

$$B_1 = \frac{0.6}{1 - \frac{94}{6,742}} = 0.608 < 1.0$$

Since B_1 factor is less than 1.0, use $B_1 = 1$. To calculate B_2,

$$\sum P_u = 150.4 \text{ kips}$$

$$K_x = 1.73$$

Note that K-factor used in the calculation of $\sum P_{e2}$ should not include leaning column effect.

$$P_{e2} = \frac{(\pi^2)(29,000)(723)}{[(1.73)(20)(12)]^2} = 1,200 \text{ kips}$$

$$B_2 = \frac{1}{1 - \frac{150.4}{1,200}} = 1.14$$

Thus, we have:

$$M_u = (1.0)(0) + (1.14)(208) = 238 \text{ kip-ft}$$

- Check the interaction equation

$$\frac{P_u}{\phi_c P_n} = \frac{94}{372} = 0.253 > 0.2$$

$$\frac{P_u}{\phi_c P_n} + \frac{8}{9}\frac{M_{ux}}{\phi_b M_{nx}} = 0.253 + \frac{8}{9} \cdot \frac{238}{311} = 0.933 < 1.0$$

Therefore, W14x68 is adequate.

1.3.8 Semi-Rigid Frames

Conventional methods of steel frame analysis use idealized joint models such as the rigid-joint model or the pinned-joint model. The behavior of joints will naturally fall between these two extremes, and more attention has been directed in recent years to develop more accurate models. The extensive research into flexible connections provided a database from which changes were made to design provisions.

Section A2.2 of the LRFD Specification identifies Type PR (partially restrained) construction as one of the two basic types of construction. PR joints provide some restraint to the members they connect. When the flexibility of the connections is considered in analysis and design, the LRFD Specification permits such evaluation by rational analytical or empirical means. The second type of construction (Type FR or fully restrained) is considered in the specification by using a simplified second-order elastic analysis with B_1 and B_2 amplification factors. In contrast to this, the design specification provides for Type PR only general principles for analysis and design. It is up to the engineer to implement these principles in a quantitative manner.

Herein, a simplified procedure proposed by Barakat and Chen [13] for the design analysis of frames with semi-rigid connections is presented. The procedure is based on the elastic analysis permitted by Section A5 of the AISC LRFD Specification; that is, using the B_1 and B_2 amplification factors in conjunction with the effective length concept. Recall that B_1 and B_2 require the structural system to behave linearly. The proposed procedure addresses this by linearizing (especially the moment-rotation $[M - \theta_r]$ relationship) of the connection. It uses the fact that a linear moment rotation relationship in the connection in the form of secant stiffness, then the "B_1 and B_2 method" may be applied to analyze Type PR joints as long as flexibility is considered in the first-order analysis. The design procedure is outlined in the following steps.

Step 1: Determine Connection Stiffness Values R_{ko} and R_{kb}

To linearize the analysis of flexible frames and turn the "B_1 and B_2 method" into a form that accounts for connection flexibility, the following linear $M - \theta_r$ relationships are proposed using two secant stiffnesses. The first is a reduced initial stiffness, or modified initial stiffness R_{ko}. The second is a secant stiffness R_{kb} determined using the beam-line method. These stiffnesses resemble average connection behavior under gravity and horizontal loads much like the B_1 and B_2 method does. The first stiffness is used in a first-order analysis of the nonsway frame (Figure 1.4(a)), in which the principal outcome is M_{nt}. The second stiffness is used in a first-order analysis of the sway frame (Figure 1.4(b)) and leads to a moment of M_{lt}.

Modified Initial Connection Stiffness R_{ko} Existing connection data [35] shows that the initial values of connection tangent stiffnesses are high compared to those at later loading stages. Researchers have used the initial connection stiffness R_{ki} extensively in analysis of flexible frames due mainly to its ease of use as it can be determined graphically or analytically. In spite of the convenience to use R_{ki}, it is fair to say that R_{ki} is too high a value for use throughout the analysis. Among others, Goto and Chen [28] showed the inadequacy of using a unique initial stiffness value throughout the analysis of flexible frames. A stiffness moderately softer than the initial R_{ki} was the proposed solution.

The modified initial stiffness R_{ko} is the secant modulus corresponding to an initial rotation θ_0 (Figure 1.24) which corresponds to the intersection of the initial tangent modulus R_{ki} and the ultimate moment M_u. The

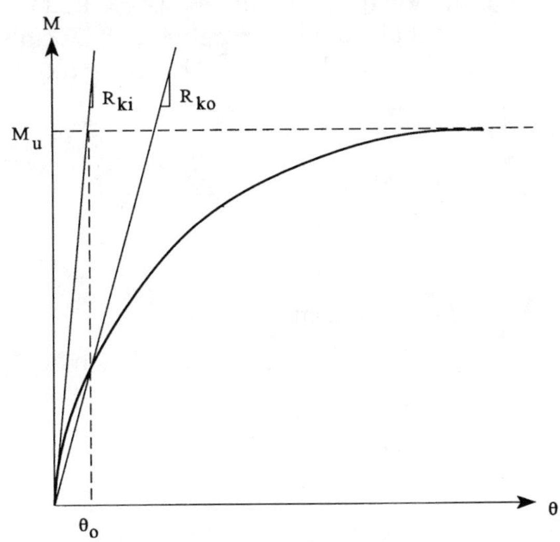

Figure 1.24 Initial and modified initial stiffness for M_{nt} calculations.

proposed stiffness is evaluated using the $M-\theta_r$ curve models proposed by Lui and Chen [48] and Kishi et al. [36] and presented in Section 2.5. The later model provides the connection properties (i.e., the $M - \theta_r$ relation, R_{ki} and M_u) required to determine the secant modulus R_{ko}. The adoption of this model is encouraged by the fact that the connection parameters are determined directly from its material and geometric properties. The procedure is systematic and can be easily coded.

Using the initial connection stiffness R_{ki} and the ultimate moment capacity M_u of the connection, the moment-rotation $M - \theta_r$ relationship can be adequately represented by [36]:

$$M = \frac{R_{ki}\theta_r}{[1 + (\theta_r/\theta_0)^n]^{1/n}} \tag{1.47}$$

Second Stiffness R_{kb} In the advanced stages of loading, the connection sustains large rotations and consequently loses stiffness especially in sway (unbraced) frames. If any idealized connection stiffness is to be considered in the analysis of a frame under sway loads (determination of M_{lt}), such stiffness should be less than that used for determining M_{nt} moments caused by non-sway loads. The other major factors that may influence the degree to which the connection stiffness has decreased before lateral loads commence are the intensity of gravity loads and the stiffness of the beams. With these considerations in mind, the secant stiffness determined by the beam-line is chosen as the average connection stiffness that approximates the connection behavior in the analysis for M_{lt}. The determination of this stiffness is presented in the following.

For elastic behavior, the relationship between the beam end moment M_E and end rotation ϕ_b can be determined by using the slope-deflection equation and is expressed by:

$$M_E = M_F - \frac{2EI}{L}\phi_b \tag{1.48}$$

where

M_F = the fixed-end moment

For a uniformly loaded beam, Equation (1.48) can be written as

$$M_E = \frac{wL^2}{12}\left(1 - \frac{\phi_b}{\phi_{bo}}\right) \tag{1.49}$$

where

ϕ_{bo} = $\frac{wL^3}{24EI}$, the end rotation for a uniformly loaded, simply supported beam

For any value of w, Equation (1.49) represents a straight line (a beam-line) on an end moment vs. end rotation diagram (Figure 1.25).

Figure 1.25 is a schematic showing the moment-rotation relationships for the three components of a beam-joint-column assembly [50].

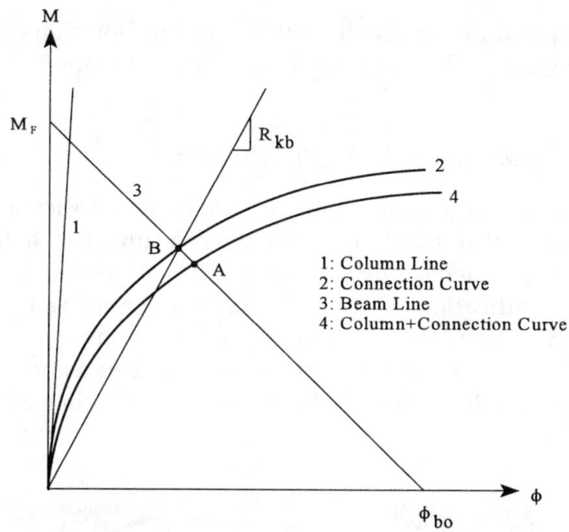

Figure 1.25 Moment-rotation curves for joint components and the determination of R_{kb} for M_{lt}.

Curve 1 represents the moment-rotation relationship at the column end, curve 2 is the connection $M - \theta_r$ relation, and curve 3 is the beam line. Compatibility of rotational deformation at a joint combining these elements will be satisfied at the intersection of the beam line (curve 3) with curve 4 which combines the connection and column rotations (curve 1 + curve 2). This is shown as point A in the figure. Noting that the end rotation of the column is usually very small compared to that of the connection, curve 4 can be approximated by curve 2 for the determination of point A which is, consequently, replaced by point B. The simplified linear model of the connection $M - \theta_r$ relation is chosen as the secant modulus that passes through the intersection point of the beam line and the moment-rotation curve of connection (point B). This modulus is denoted here as R_{kb} and will be implemented in the first-order analysis of the sway frame of Figure 1.4(b) for the determination of M_{lt} moments.

Having prepared the two linear connection models (R_{ko} and R_{kb}) intended for use in the analysis procedure, we now proceed to deal with the question of the effective length of columns under flexible restraint conditions.

Step 2: Determine First-Order Moments, M_{nt} and M_{lt}

A first-order elastic analysis using the linear connection stiffness R_{ko} is performed for the frame with loads and boundary conditions shown in Figure 1.4(a). This step determines the moment M_{nt} for nonsway frames. Another first-order elastic analysis using the linear connection stiffness

R_{kb} is performed on the frame with loads and boundary conditions as shown in Figure 1.4(b). This step determines the moment M_{lt} for sway frames.

Step 3: Determine the Modified K-factor

The effects of elastic joint flexibility can be taken into account by using a reduced beam stiffness $(EI')_b$. Since the modulus for steel E is considered, the change affects only second moment of inertia of the beam element. The modification is accomplished by examining a general case of a beam member restrained by different connections at its ends. Further, special cases of a member with the same end connections will be considered. Figure 1.26(c) shows a beam-column member with elastic end

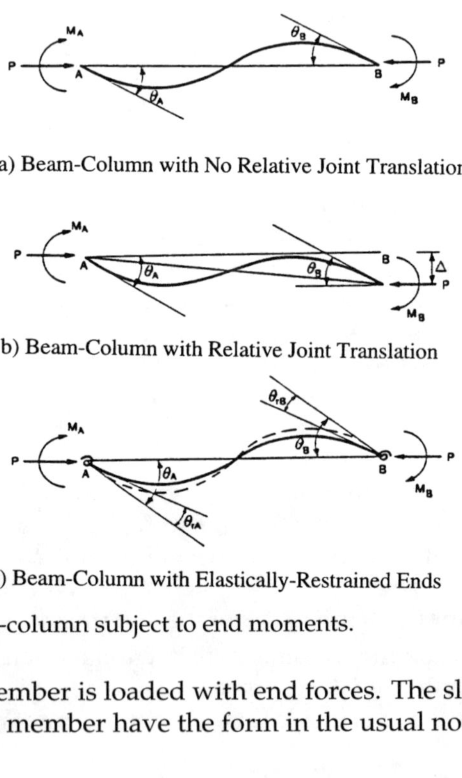

a) Beam-Column with No Relative Joint Translation

b) Beam-Column with Relative Joint Translation

c) Beam-Column with Elastically-Restrained Ends

Figure 1.26 Beam-column subject to end moments.

restraints. The member is loaded with end forces. The slope-deflection equations for this member have the form in the usual notations

$$M_A = \frac{EI}{L}\left[s_{ii}\left(\theta_A - \theta_{rA}\right) + s_{ij}\left(\theta_B - \theta_{rB}\right)\right] \tag{1.50a}$$

$$M_B = \frac{EI}{L}\left[s_{ij}\left(\theta_A - \theta_{rA}\right) + s_{ii}\left(\theta_B - \theta_{rB}\right)\right] \tag{1.50b}$$

Substituting $\theta_{rA} = \frac{M_A}{R_{kA}}$ and $\theta_{rB} = \frac{M_B}{R_{kB}}$, we get:

$$M_A = \frac{EI}{L}\left[s_1'\theta_A + s_2'\theta_B\right] \tag{1.51a}$$

$$M_B = \frac{EI}{L}\left[s_2'\theta_A + s_3'\theta_B\right] \tag{1.51b}$$

where s_1', s_2', and s_3' are the modified stability functions due to the presence of flexible connections, expressed as:

$$s_1' = \frac{R_{kA}'\left(R_{kB}'s_{ii} + s_{ii}^2 - s_{ij}^2\right)}{R_{kA}'R_{kB}' + s_{ii}\left(R_{kA}' + R_{kB}'\right) + s_{ii}^2 - s_{ij}^2} \tag{1.52a}$$

$$s_2' = \frac{R_{kA}'R_{kB}'s_{ij}}{R_{kA}'R_{kB}' + s_{ii}\left(R_{kA}' + R_{kB}'\right) + s_{ii}^2 - s_{ij}^2} \tag{1.52b}$$

$$s_3' = \frac{R_{kB}'\left(R_{kA}'s_{ii} + s_{ii}^2 - s_{ij}^2\right)}{R_{kA}'R_{kB}' + s_{ii}\left(R_{kA}' + R_{kB}'\right) + s_{ii}^2 - s_{ij}^2} \tag{1.52c}$$

where

$$R_{kA}' = \frac{L}{EI}R_{kA} \tag{1.53a}$$

$$R_{kB}' = \frac{L}{EI}R_{kB} \tag{1.53b}$$

Equations (1.51a and 1.51b) are the modified slope-deflection equations accounting for linear flexible connections at the member ends. Similar expressions have been presented by Lothers [46], where an incremental formulation was used for a nonlinear analysis.

Two idealized beam members will now be considered. These cases express the two deflection modes associated with the bifurcation of braced frames (single curvature bending) and unbraced frames (double curvature bending).

Modified K-Factor for Braced Frames For a beam member with identical end connections ($R_{kA} = R_{kB} = R_k$) and single curvature bending, the end rotations are equal in magnitude and opposite in direction ($\theta_B = -\theta_A$). Considering that $s_{ii} = 4$ and $s_{ij} = 2$ for beam members,

Equations (1.52a through 1.52c) can be reduced to

$$s_1' = \frac{4R_k'^2 + 12R_k'}{R_k'^2 + 8R_k' + 12} \tag{1.54a}$$

$$s_2' = \frac{2R_k'^2}{R_k'^2 + 8R_k' + 12} \tag{1.54b}$$

$$s_3' = s_1' \tag{1.54c}$$

The end moments (Equations (1.51a and 1.51b) are determined by

$$M_A = \frac{2E}{L} I \left[\frac{1}{1 + \frac{2EI}{R_k L}} \right] \theta_A \tag{1.55a}$$

$$M_B = -M_A \tag{1.55b}$$

Comparing Equation (1.55a) with the equations derived for a rigidly connected beam member in which the moment at each end is expressed by

$$M = \frac{2EI}{L} \theta \tag{1.56}$$

we can write Equation (1.55a)

$$M_A = \frac{2EI_{nt}'}{L} \theta_A \tag{1.57}$$

where

$$I_{nt}' = \left[\frac{1}{1 + \frac{2EI}{R_{ko} L}} \right] I \tag{1.58}$$

in which R_{ko} indicates rotation stiffness in braced case and is equal to R_k. I_{nt}' is the modified moment of inertia for a beam element with linearly flexible connections at both ends and for which loading conditions produce single curvature bending. Consequently, the modified relative stiffness factor can be written directly in the form

$$G' = \frac{\sum \left(\frac{EI}{L} \right)_c}{\sum \left(\frac{EI_{nt}'}{L} \right)_b} \tag{1.59}$$

It can be seen that the modified relative stiffness factor G' is larger than G which reflects the reduced restraint at one column end. This ultimately leads to higher values of the effective length factor K. The modified K-factor may be obtained by entering an alignment chart for a braced frame with the modified values G'.

Modified K-Factor for Unbraced Frame The modified stability functions s_1', s_2', and s_3' are still expressed as they were in Equations (1.54a through 1.54c). The only difference in this case is that $\theta_A = \theta_B = \theta$. Substituting these values into Equations (1.51a and 1.51b) gives

$$M_A = \frac{6E}{L} I \left[\frac{1}{1 + \frac{6EI}{R_k L}} \right] \theta_A \tag{1.60a}$$

$$M_B = M_A \tag{1.60b}$$

Comparing this equation with that derived for a rigidly connected beam member in which the moment at each end is expressed by

$$M = \frac{6EI}{L} \theta \tag{1.61}$$

Equation (1.60a) can be written as

$$M_A = \frac{6EI_{lt}'}{L} \theta_A \tag{1.62}$$

where

$$I_{lt}' = \left[\frac{1}{1 + \frac{6EI}{R_{kb} L}} \right] I \tag{1.63}$$

in which R_{kb} indicates rotation stiffness in unbraced case and is equal to R_k. I_{lt}' is the modified moment of inertia for a beam element with elastic end restraints and bent in double curvature. The modified relative stiffness factor can then be directly written in the following form

$$G' = \frac{I \left(\frac{EI}{L} \right)_c}{I \left(\frac{EI_{lt}'}{L} \right)_b} \tag{1.64}$$

The modified K-factor can then be obtained by entering an alignment chart for unbraced frames with the values G'.

Step 4: Member Capacity Checks
The values of P_{ek} for each compression member can now be determined using the modified K-factors. The amplification factors B_1 and B_2 may be computed by Equations (1.17, 1.19a and 1.19b). The member capacity checks may be conducted using the interaction Equations [(1.5a and 1.5b)].

1.4 Methods of Advanced Analyses

Over the past 20 years, extensive research has been made to develop and validate several advanced analysis methods. Advanced analysis refers to any method that captures the strength and stability of a structural system and its individual members in such a way that separate member capacity checks are not required. Five different types of advanced analysis methods are discussed in the following:

1. Plastic-zone method
2. Quasi-plastic hinge method
3. Elastic-plastic hinge method
4. Notional-load plastic hinge method
5. Refined-plastic hinge method

These different methods are based on the degree of refinement in representing the plastic yielding effects. The plastic-zone method uses the greatest refinement while the elastic-plastic hinge method allows a drastic simplification. The quasi-plastic hinge method is somewhere in between these two methods. The notional-load plastic hinge method and the refined-plastic hinge method are an improvement on the elastic-plastic hinge method for practical application. The load-deformation characteristics of the plastic analysis methods are illustrated in Figure 1.27, while the spread of plasticity is illustrated schematically in Figure 1.28.

1.4.1 Plastic-Zone Method

In the plastic-zone method, frame members are discretized into finite elements, and the cross-section of each finite element is subdivided into many fibers as shown in Figure 1.29. The deflection at each division point along a member is obtained by numerical integration. The incremental load-deflection response at each loading step, which updates the geometry, captures the second-order effects. The residual stress in each fiber is assumed constant since the fibers are small enough. The

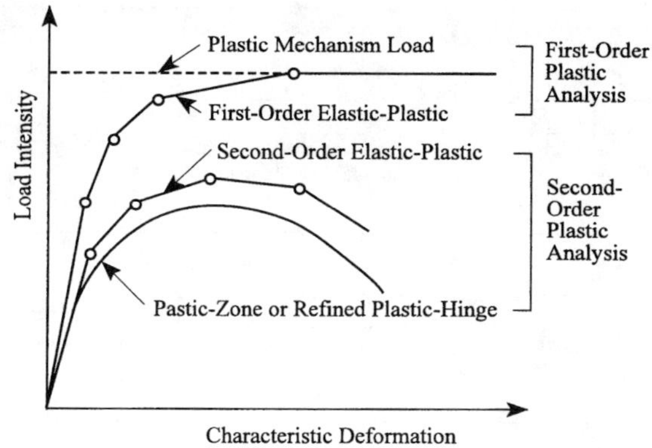

Figure 1.27 Load-deformation characteristics of plastic analysis methods.

stress state at each fiber can be explicitly traced so the gradual spread of yielding can be captured. The plastic-zone analysis eliminates the need for separate member capacity checks since it explicitly accounts for second-order effects, spread of plasticity, and residual stress. As a result, the plastic-zone solution is known as an "exact solution". The AISC-LRFD beam-column equations were established in part based on a curve-fit to the "exact" strength curves obtained from the plastic-zone analysis by Kanchanalai [30].

There are two types of plastic-zone analyses. The first involves the use of three-dimensional finite shell elements in which the elastic constitutive matrix in the usual incremental stress-strain relations is replaced by an elastic-plastic constitutive matrix when yielding is detected. Based on a deformation theory of plasticity, the effects of combined normal and shear stresses may be accounted for. This analysis requires modeling of structures using a large number of finite three-dimensional shell elements and numerical integration for the evaluation of the elastic-plastic stiffness matrix. The three-dimensional spread-of-plasticity analysis, when combined with second-order theory which deals with frame stability, is computational intensive. Therefore, it is best suited for analyzing small-scale structures and providing the detailed solutions for member local instability and yielding behavior if they are required. Since a detailed analysis of local effects in realistic building frames is not common practice in engineering design, this approach is considered too expensive for practical use.

The second approach for second-order plastic-zone analysis is based on the use of beam-column theory, in which the member is discretized into line segments and the cross-section of each segment is subdivided into finite elements. Inelasticity is modeled considering normal stress only. When the computed stress at the centroid of any fiber reaches the uniaxial normal strength of the material, the fiber is considered to have

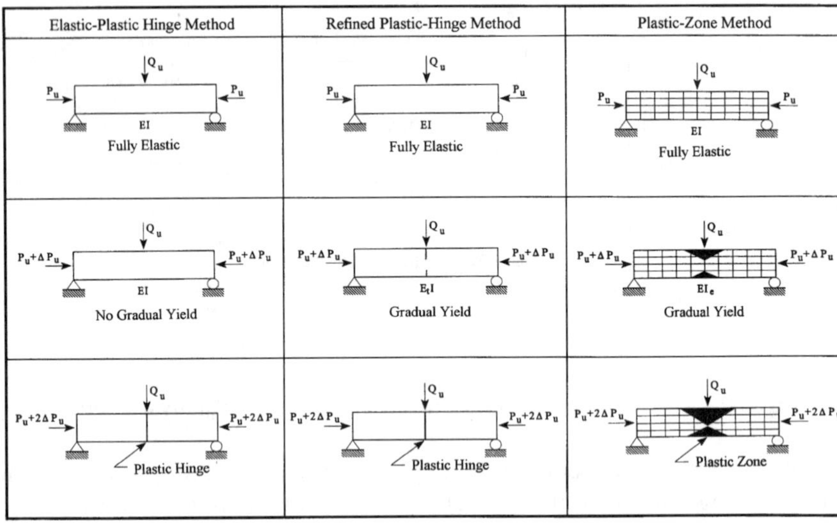

Figure 1.28 Concept of spread of plasticity for various advanced analysis methods.

yielded. Also, compatibility is treated by assuming that full continuity is retained throughout the volume of the structure in the same manner as elastic range calculations. Although quite sharp curvature may exist in the vicinity of inelastic portions of the structure, "plastic hinges" can never develop. In plastic-zone analysis, the calculation of forces and deformations in the structure after yielding requires an iterative trial-and-error process because of the nonlinearity of the load-deformation response and the change in cross-section effective stiffness in inelastic regions associated with the increase in the applied loads and the change in structural geometry. Although most plastic-zone analysis methods have been developed for planar analyses [8] [20] [26] [58] [61], three-dimensional plastic-zone techniques are also available [15] [59].

A plastic-zone analysis that includes the spread of plasticity, residual stresses, initial geometric imperfections, and any other significant second-order effects, would eliminate the need for checking individual member capacities in the frame. Therefore, this type of method is classified as advanced inelastic analysis in which the checking of beam-column interaction equations is not required. In fact, the member interaction equations in modern limit-states specifications were developed, in part, by curve-fit to results from this type of analysis. In reality, some significant behaviors such as joint and connection's performances tend to defy precise numerical and analytical modeling. In such cases, a simpler method of analysis that adequately represents the significant behavior would be sufficient for engineering application.

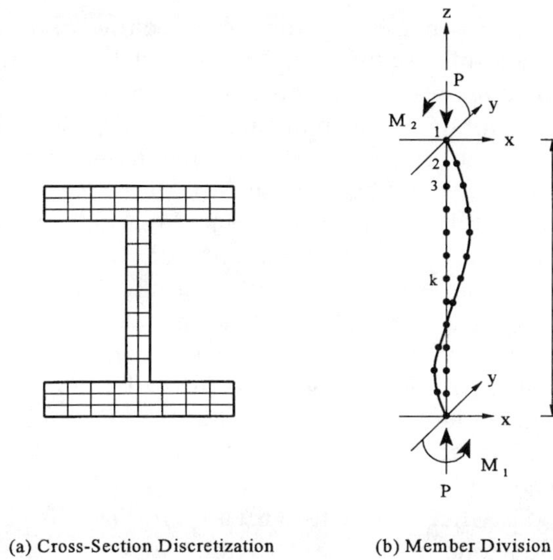

Figure 1.29 Model of plastic-zone analysis.

Whereas the plastic-zone solution is regarded as an "exact solution", the method may not be used in daily engineering design because it is too intensive in computation. Its applications are limited to [23] [62]:

1. the study of detailed structural behavior
2. verifying the accuracy of simplified methods
3. providing comparison with experimental results
4. deriving design methods or generating charts for practical use
5. applying for special design problems

The plastic-zone method will be used here as a benchmark for the development of the practical advanced analysis methods.

1.4.2 *Quasi-Plastic Hinge Method*

The quasi-plastic hinge method developed by Attala [11] is an intermediate approach between the plastic-zone and the elastic-plastic hinge methods. It requires less computation but its results are very similar to those of the plastic-zone method. For this reason, it is called a quasi-plastic hinge method.

An element, developed from equilibrium, kinematic, and constitutive relationships, accounts for gradual plastification under combined bending and axial force. Inelastic force-strain model of the cross-section is developed by fitting nonlinear equations to data of the moment-axial force-curvature response. Using the inelastic cross-section model, flexibility coefficients for the full member are obtained by successive integrations along its length. An inelastic-element stiffness matrix is obtained by use of the incremental flexibility relationships.

Initial yield and full plastification surface are used to analytically represent gradual yielding effect of the cross-section. Ketter's residual stress pattern [31] is used to determine an initial yield surface. Ketter's pattern has peak compressive residual stresses at the flange tips equal to $0.3F_y$ with a linear transition of stress from the flange tips to the web-joint and constant tensile stress through the web. A fully plastic surface is generated by calibration to a plastic-zone solution [53]. The parameters of the full plastification equation are determined by a curve-fit procedure.

This method predicts strengths with an error less than 5% compared with the plastic-zone method for a wide range of case studies. The accuracy of this method is thus compatible with the plastic-zone method and less computational effort is necessary.

However, it is difficult to extend this method to three-dimensional analysis because the formulation is based on flexibility relationships. As a result, it does not meet one of the requirements of advanced analysis of the SSRC task force report [55], which states "The model should be readily extensible to three-dimensional analysis. That is, the framework of the model should accommodate the formulation of three-dimensional elements." Moreover, this model does eliminate the necessity of the refined model through the cross-section but still requires many elements along the member.

1.4.3 Elastic-Plastic Hinge Method

A more simple and efficient approach for representing inelasticity in frames is the elastic-plastic hinge method. It assumes that the element remains elastic except at its ends where zero-length plastic hinges form. This method accounts for inelasticity but not the spread of yielding or plasticity at sections nor the residual stress effect between two plastic hinges.

The elastic-plastic hinge methods may be divided into first- and second-order plastic analyses. For first-order elastic-plastic hinge analysis, the nonlinear geometric effects are neglected and not considered in the formulation of the equilibrium equations. As a result, the method predicts the same ultimate load as conventional rigid-plastic analyses.

In second-order elastic-plastic hinge analysis, the deformed structural geometry is considered. The simple way to account for the geometric nonlinearity is to use the stability function which enables only one beam-column element per member to capture the second-order effect. This provides an efficient and economical method of frame analysis, and has a clear advantage over the plastic-zone method. This is particularly true for structures in which the axial force in component members is small and the dominated behavior is bending. In such cases, second-order elastic-plastic hinge analysis may be used to describe the inelastic behavior sufficiently, assuming that lateral-torsional and local buckling modes of failure are prevented [40].

The second-order elastic-plastic hinge analysis is only an approximate method. When used to analyze a single beam-column element subject to combined axial load and bending moment, it may overestimate the strength and stiffness of the element in the inelastic range. Although elastic-plastic hinge approaches provide essentially the same load-displacement predictions as plastic-zone methods for many frame problems, they may not be classified as advanced analysis methods in general [41] [43] [60].

However, research by Ziemian [64] [65] has shown that the elastic-plastic hinge analysis can be classified as an advanced inelastic analysis because it is accurate for matching the strength and load-displacement response of several building frames from plastic-zone analysis. Many cases considered in Ziemian's work, especially when the axial load is less than $0.5P_y$, are not sensitive benchmarks for determining the accuracy and the possible limitations of the elastic-plastic hinge method. Therefore, suitable benchmark problems should be used to provide a more in-depth study of the qualities and limitations of second-order elastic-plastic hinge method before it can be accepted as a legitimate tool in the design of steel structures.

For slender members whose dominant mode of failure is elastic instability, the method provides good results when compared with plastic-zone solutions. However, for stocky members with significant yielding, the plastic-hinge method over-predicts the actual strength and stiffness of members due to the gradual stiffness reduction as the spread of plasticity increases in an actual member [41] [42] [62]. As a result, considerable refinements must be made before it can be used for analysis of a wide range of framed structures.

1.4.4 Notional-Load Plastic-Hinge Method

One approach to advance the use of second-order elastic-plastic hinge analysis for frame design is to specify artificially large values of frame imperfections (i.e., initial out-of-plumbness). This is the approach adopted by EC3 [22] for frame design using second-order analysis. In addition to accounting for the standard erection tolerance for out-of-plumbness, these artificial large imperfections intend to account for the effect of residual stresses, frame imperfections, and distributed plasticity not considered in frame analysis. The geometric imperfections adopted by $EC3$ are a maximum out-of-plumbness of $\Psi = 1/200$ for an unbraced frame, but no maximum out-of-straightness value recommended for a braced member as shown in Figure 1.30.

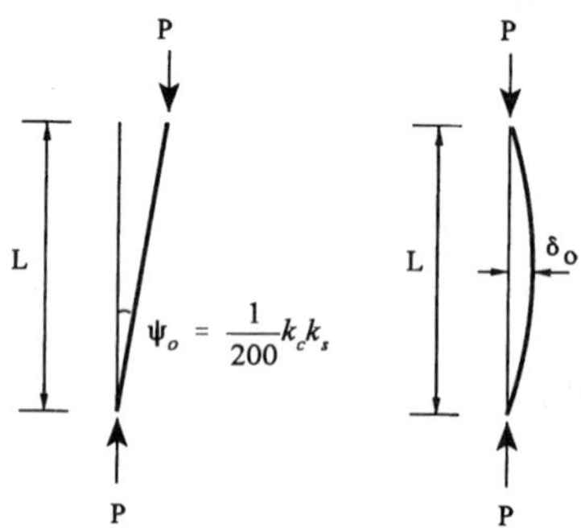

$$k_c = \sqrt{\frac{1}{2} + \frac{1}{n_c}} \leq 1.0$$

$$k_s = \sqrt{\frac{1}{5} + \frac{1}{n_s}} \leq 1.0$$

Member imperfections δ_o modeled only if $P > 0.26\pi^2 EI/L^2$

n_c = number of columns per plane
n_c = number of stories per plane

Figure 1.30 Explicit imperfection model for elastic-plastic analysis recommended by ECCS.

The notional-load plastic-hinge method is similar in concept to the "enlarged" geometric imperfection approach of the EC3. The ECCS [23] [24], the Canadian Standard [21], and the Australian Standard [57] allow use of this technique. The notional-load method uses equivalent lateral loads to approximate the effect of member imperfections and distributed plasticity. In the ECCS, the exaggerated notional loads of 0.5% times gravity loads are used to avoid over-predicting the strength of the member as does the elastic-plastic hinge method. The application of these notional loads to several example frames is illustrated in Figure 1.31. Liew's research [40] shows that this method under-predicts the strength by more than 20% in the various leaning column frames and over-predicts the strength up to 10% in the isolated beam-columns subject to the axial forces and bending moments. As a result, modification of this approach is required before it may be used in design applications.

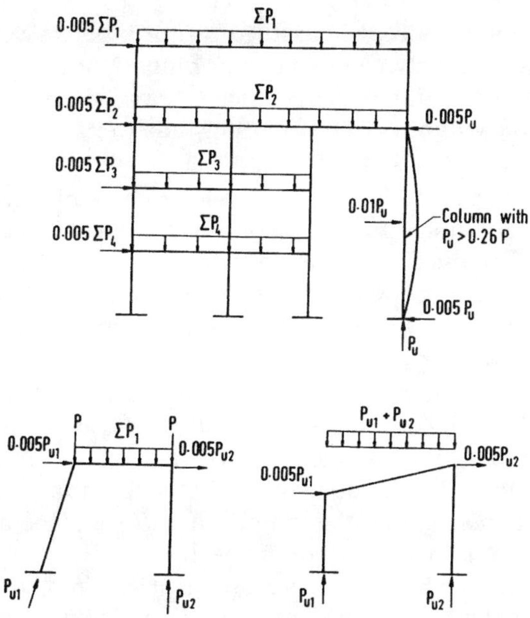

Figure 1.31 Examples on application of notional loads for second-order elastic-plastic hinge analysis.

1.4.5 *Refined-Plastic Hinge Method*

In recent work by Abdel-Ghaffar et al. [1], Al-Mashary and Chen [2], King, et al. [34], Liew and Chen [41], Liew et al. [44] [45], White et al. [61], Kim [32], and Kim and Chen [33], among others, an inelastic analysis approach based on simple refinements of the elastic-plastic hinge model has been proposed for plane frame analysis. It represents the effect of distributed plasticity through the cross-section, assuming that the plastic hinge stiffness degradation is smooth. The inelastic behavior of the member is modeled in terms of member force instead of the detailed level of stresses and strains as used in the plastic-zone analysis model. The principal merits of the refined-plastic hinge model are that it is as simple and efficient as the elastic-plastic hinge analysis approach, and it is sufficiently accurate for the assessment of strength and stability of a structural system and its component members.

The refined-plastic hinge method is based on simple modifications of the elastic-plastic hinge method. Two modifications are made to account for the gradual section stiffness degradation at the plastic hinge locations as well as gradual member stiffness degradation between the two plastic hinges. Herein, the section stiffness degradation function is adopted to reflect the gradual yielding effect in forming plastic hinges. Then, the tangent modulus concept is used to capture the residual stress effect along the member between two plastic hinges. As a result, the refined-plastic hinge method retains the efficiency and simplicity of the plastic hinge method without overestimating the strength and stiffness of a member.

An analysis approach has been developed by which connection flexibility can be accounted for in a second-order refined-plastic hinge analysis of steel frames. The connections are modeled as rotational springs physically tied to the ends of the element by enforcing equilibrium and compatibility conditions between the connection element and the beam-column element.

In the recent work by Liew [40], the LRFD tangent modulus is used to account for both the effect of residual stresses and geometric imperfections. This model does not account for geometric imperfections when P/P_y is less than 0.39 because the LRFD tangent modulus is identical to the elastic modulus in this range. As a result, the approach over-predicts the column strength by more than 5% when KL/r of the column is greater than 85 for yield stresses at 36 ksi, and when KL/r of the column is greater than 70 for yield stresses at 50 ksi. The LFRD E_t may not be an appropriate model to be used for advanced analysis [32] [33].

The CRC tangent modulus in Liew's work [40] only accounts for the effect of residual stresses. It over-predicts the strength of members by about 20% compared to the conventional LRFD solutions because the modulus does not account for the effect of geometric imperfections. However, in the CRC tangent modulus model, different members with different residual stresses can be incorporated because the effect of geometric imperfections is considered separately. As a result, CRC tangent modulus is used in refined plastic analyses.

Against this background, it can be concluded that the refined-plastic hinge method strikes a balance between the requirements for realistic representation of frame behavior and for ease of use. It is considered that in both of these respects, the method is satisfactory for general practical use. This is the method that will be adopted in this book.

1.5 Why Advanced Analysis

The advantages of advanced analysis in design use are outlined in the following:

1. Advanced analysis is another tool for structural engineers to use in steel design, and its adoption is not mandatory but will provide a flexibility of options to the designer.

2. Advanced analysis captures the limit state strength and stability of a structural system and its individual members directly, so separate member capacity checks encompassed by the specification equations are not required.
3. Compared to the LRFD and ASD, advanced analysis provides more information of structural behavior by direct inelastic second-order analysis.
4. Advanced analysis overcomes the difficulties due to incompatibility between the elastic global analysis and the limit state member design in the conventional LRFD method.
5. Advanced analysis is user-friendly for a computer-based design, but the LRFD and ASD are not because they require the calculation of K-factor on the way from their analysis to separate member capacity checks.
6. Advanced analysis captures the inelastic redistribution of internal forces throughout a structural system, and allows an economic use of material for highly indeterminate steel frames.
7. It is now feasible to employ advanced analysis techniques that have been considered impractical for design office use in the past because the power of personal computers and engineering workstations is rapidly increasing.
8. Member sizes determined by advanced analysis are close to those determined by the LRFD method because the advanced analysis method is calibrated against the LRFD column curve and beam-column interaction equations. As a result, advanced analysis provides an alternative to the LRFD.
9. Advanced analysis is time-effective because it completely eliminates tedious and often confused member capacity checks including the calculation of K-factors in the LRFD and ASD.

1.6 Summary

This chapter conveys an important message; that is, if the engineer uses an advanced analysis that captures the limit state of strength and stability, one is freed from member capacity checks, which are often tedious and confused in their procedure. Another benefit of advanced analysis is that it can accurately predict system behavior which cannot be captured by the conventional design methods (such as ASD, PD, and LRFD). In recent years, the rapid development of computer hardware and software coupled with an increasing understanding of structural behavior has made it possible to adopt the advanced analysis techniques for design office use.

Among various advanced analyses, including plastic-zone, quasi-plastic hinge, elastic-plastic hinge, notional-load plastic-hinge, and refined-plastic hinge methods, the refined-plastic hinge method is recommended because it retains the efficiency and simplicity of computation and accuracy for practical use. The method is developed by imposing simple modifications on conventional elastic-plastic hinge method. These include a simple modification to account for the gradual sectional stiffness degradation at the plastic hinge locations and to include the gradual member stiffness degradation between two plastic hinges.

In this chapter, the LRFD method is discussed for comparison and calibration purposes with the advanced analysis. In the LRFD method, first- or second-order elastic analysis is used to analyze a structural system, and separate member design is performed using limit state design equations. The effective length factor is used to account for the interaction between a structural system and its component members. However, elastic frame analysis and the limit state member strength design are not compatible. The design equations become more complicated to make up the simplified assumptions of elastic behavior in analysis. Step-by-step design procedures of two example frames are provided based on the LRFD Specification. Comments are given where incorrect implementations of various factors are often made in design procedures.

The key considerations of the conventional LRFD method and the practical advanced analysis method are compared in Table 1.8. While the LRFD method does account for key behavioral effects implicitly in its column strength and beam-column interaction equations, the advanced analysis method accounts for these effects explicitly through stability functions, stiffness degradation functions, and geometric imperfections to be discussed in detail in Chapter 2.

Advanced analysis holds many answers to real behavior of steel structures and, as such, the authors recommend the proposed design method to engineers seeking to perform frame design in efficiency and rationality, yet consistent with the present LRFD Specification. In the following chapters, we will present a practical advanced analysis method for the design of steel frame structures with LRFD. The validity of the approach will be demonstrated by comparing case studies of actual members and frames with the results of analysis/design based on exact plastic-zone solutions and LRFD designs. The wide range of case studies and comparisons should confirm the validity of this advanced method.

Table 1.8 Key Considerations of LRFD and Proposed Methods

Key considerations	LRFD	Proposed methods
Second-order effects	Column curve B_1, B_2 factor	Stability function
Geometric imperfection	Column curve	Explicit imperfection modeling method •$\Psi = 1/500$ for unbraced frame •$\delta_c = L_c/1,000$ for braced frame Equivalent notional load method •$\alpha = 0.002$ for unbraced frame •$\alpha = 0.004$ for braced frame Further reduced tangent modulus method •$E_t' = 0.85E_t$
Stiffness degradation associated with residual stresses	Column curve	CRC tangent modulus
Stiffness degradation associated with flexure	Column curve Interaction equations	Parabolic degradation function
Connection nonlinearity	No procedure	Power model/rotational spring

References

[1] Abdel-Ghaffar, M., White, D.W., and Chen, W.F. (1991) Simplified second-order inelastic analysis for steel frame design, Special Volume of Session on Approximate Methods and Verification Procedures of Structural Analysis and Design, Proceedings at Structures Congress 91, ASCE, New York, 47-62.

[2] Al-Mashary, F. and Chen, W.F. (1991) Simplified second-order inelastic analysis for steel frames, *J. Inst. Struct. Eng.*, 69(23), 395-399.

[3] AISC (1978) *Specification for the Design, Fabrication and Erection of Structural Steel for Buildings,* American Institute of Steel Construction, Chicago.

[4] AISC (1994) *Load and Resistance Factor Design Specification,* American Institute of Steel Construction, 2nd ed., Chicago.

[5] AISC (1986) *Load and Resistance Factor Design Specification for Structural Steel Buildings,* American Institute of Steel Construction, Chicago.

[6] AISC (1989) *Allowable Stress Design Specification for Structural Steel Buildings,* American Institute of Steel Construction, Chicago.

[7] AISC (1923) *Allowable Stress Design Specification,* American Institute of Steel Construction, 1st ed., Chicago.

[8] Alvarez, R.J. and Birnstiel, C. (1967) Elasto-Plastic Analysis of Plane Rigid Frames, School of Engineering and Science, Department of Civil Engineering, New York University, New York.

[9] Ang, A. H-S. and Cornell, C.A. (1974) Reliability bases of structural safety and design, *J. Struct. Div.,* ASCE, 100(ST9), 1755-1769.

[10] ANSI (1982) *Building Code Requirements for Minimum Design Loads in Buildings and Other Structures,* ANSI A58.1-1982, American National Standards Institute, New York.

[11] Attala, M.N., Deierlein, G.G., and McGuire, W. (1994) Spread of plasticity: Quasi-plastic-hinge approach, *J. Struct. Eng.,* 120(8), 2451-2473.

[12] Austin, W.J. (1961) Strength and design of metal beam-columns, *J. Struct. Div.,* ASCE, 87(ST4), 1-32.

[13] Barakat, M. and Chen, W. F. (1990) Practical analysis of semi-rigid frames, *Eng. J.,* AISC, 2nd Quarter, 54-68.

[14] Bjorhovde, R. (1972) Deterministic and probabilistic approaches to the strength of steel columns, Ph.D. dissertation, Department of Civil Engineering, Lehigh University, Bethlehem, PA.

[15] Chen, W.F. and Atsuta, T. (1977) *Theory of Beam-Columns, Vol. 2, Space Behavior and Design,* McGraw-Hill, New York, 732 pp.

[16] Chen, W.F. and Lui, E.M. (1992) *Stability Design of Steel Frames,* CRC Press, Boca Raton, FL, 380 pp.

[17] Chen, W.F. and Lui, E.M. (1986) *Structural Stability - Theory and Implementation*, Elsevier, New York, 490 pp.

[18] Chen, W.F. and Toma, S. (1994) *Advanced Analysis of Steel Frames*, CRC Press, Boca Raton, FL, 383 pp.

[19] Cheong-Siat-Moy, F. (1972) Consideration of secondary effects in frame design, *J. Struct. Div.*, 103(ST10), 2005-2019.

[20] Clarke, M.J., Bridge, R.Q., Hancock, G.J., and Trahair, N.S. (1992) Benchmarking and Verification of Second-Order Elastic and Inelastic Frame Analysis Programs, in *SSRC TG 29 Workshop and Nomograph on Plastic Hinge Based Methods for Advanced Analysis and Design of Steel Frames*, White, D.W. and Chen, W.F., Eds., SSRC, Lehigh University, Bethlehem, PA.

[21] CSA (1994) *Limit States Design of Steel Structures*, CAN/CSA-S16.1-M94, Canadian Standards Association.

[22] EC3 (1990) *Design of Steel Structures: Part I - General Rules and Rules for Buildings*, Vol. 1, Eurocode edited draft, Issue 3.

[23] ECCS (1984) Ultimate Limit State Calculations of Sway Frames with Rigid Joints, Technical Committee 8 - Structural Stability Technical Working Group 8.2 - System, Publication No. 33, 20 pp.

[24] ECCS (1991) *Essentials of Eurocode 3 Design Manual for Steel Structures in Buildings*, ECCS-Advisory Committee 5, No. 65, 60 pp.

[25] Ellingwood, B., MacGregor, J.G., Galambos, T.V., and Cornell, C.A. (1982) Probability-based load criteria load factors and load combinations, *J. Struct. Div.*, ASCE, 108(ST5), 978-997.

[26] El-Zanaty, M., Murray, D., and Bjorhovde, R. (1980) Inelastic Behavior of Multistory Steel Frames, Structural Engineering Report No. 83, University of Alberta, Alberta, Canada.

[27] Galambos, T.V., Ed. (1988) *Guide to Stability Design Criteria for Metal Structures*, Structural Stability Research Council, 4th ed., John Wiley & Sons, New York, 786 pp.

[28] Goto, Y. and Chen, W. F. (1986) Documentation of Program FL-FRM for Design Analysis of Flexibly Jointed Frames, CE-STR-86-34, School of Civil Engineering, Purdue University, West Lafayette, IN.

[29] Johnston, B.G., Ed. (1976) *Structural Stability Research Council Guide to Stability Design Criteria for Metal Structures*, 3rd ed., John Wiley & Sons, New York.

[30] Kanchanalai, T. (1977) The Design and Behavior of Beam-Columns in Unbraced Steel Frames, AISI Project No. 189, Report No. 2, Civil Engineering/Structures Research Lab., University of Texas at Austin, 300 pp.

[31] Ketter, R.L., Kaminsky, E.L., and Beedle, L.S. (1955) Plastic Deformation of Wide-Flange Beam Columns, *Transactions*, ASCE, 120,1028-1069.

[32] Kim, S.E. (1996) Practical Advanced Analysis for Steel Frame Design, Ph.D. dissertation, School of Civil Engineering, Purdue University, West Lafayette, IN, May, 271 pp.

[33] Kim, S.E. and Chen, W.F. (1996) Practical advanced analysis for steel frame design, The ASCE Structural Congress XIV Special Proceedings Volume on Analysis and Computation, Chicago, April, 19-30.

[34] King, W.S., White, D.W., and Chen, W.F. (1991) On second-order inelastic methods for steel frame design, *J. Struct. Eng.*, 118(2), 408-428.

[35] Kish, N. and Chen, W. F. (1986) Data Base of Steel Beam-to-Column Connections, CE-STR-86-26, School of Civil Engineering, Purdue University, West Lafayette, IN.

[36] Kish, N., Chen, W. F., Matsuoka, K. G., and Nomachi, S. G. (1987) Moment-rotation relation of single/double web-angle connections, Proc. of the state-of-the-art Workshop on Connections and the Behavior, Strength and Design of Steel Structures, Bjorhovde, R., Brozzetti, J., and Colson, A., Eds., Ecole Normale Superieure de Cachan, France, May 25-27, 1987.

[37] LeMessurier, W.J. (1995) Simplified K-factors for stiffness controlled design, ASCE, Structures Congress 8 at Boston, 1797-1812.

[38] LeMessurier, W.J. (1984) Personal communication.

[39] LeMessurier, W.J. (1977) A practical method of second-order analysis. II. Rigid Frames, *Eng. J.*, 14(2), 49-67.

[40] Liew, J.Y.R. (1992) Advanced Analysis for Frame Design, Ph.D. dissertation, School of Civil Engineering, Purdue University, West Lafayette, IN, May, 393 pp.

[41] Liew, J.Y.R. and Chen, W.F. (1991) Refining the plastic hinge concept for advanced analysis/design of steel frames, *J. Singapore Struct. Steel Soc.*, Steel Structure, 2(1), 13-30.

[42] Liew, J.Y.R., White, D.W., and Chen, W.F. (1991) Beam-column design in steel frameworks - insight on current methods and trends, *J. Constr. Steel Res.*, 18, 269-308.

[43] Liew, J.Y.R., White, D.W., and Chen, W.F. (1994) Notional-load plastic-hinge method for frame design, *J. Struct. Eng.*, ASCE, 120(5), 1434-1454.

[44] Liew, J.Y.R., White, D.W., and Chen, W.F. (1993a) Second-order refined plastic hinge analysis for frame design: Part I, *J. Struct. Eng.*, ASCE, 119(11), 3196-3216.

[45] Liew, J.Y.R., White, D.W., and Chen, W.F. (1993b) Second-order refine plastic hinge analysis for frame design: Part II, *J. Struct. Eng.*, ASCE, 119(11), 3217-3237.

[46] Lothers, J. E. (1951) Elastic restraint equations for semi-rigid connections, *Trans. ASCE*, 116, 480-502.

[47] Lui, E.M. (1992) A noble approach for K-factor determination, *Eng. J.*, 4th Quarter, 150-159.
[48] Lui, E. M. and Chen, W. F. (1986) Analysis and behavior of flexibly jointed frames, *Eng. Struct.*, No. 8, 107-118.
[49] Massonnet, C. (1959) Stability considerations in the design of steel columns, *J. Struct. Div.*, ASCE, 85, 75-111.
[50] Nethercot, D. A. (1985) Joint action and the design of steel frames, *Struct. Eng.*, 63A(12), 371-379.
[51] Ravindra, M.K. and Galambos, T.V. (1978) Load and resistance factor design for steel, *J. Struct. Div.*, ASCE, 104(ST9), 1337-1354.
[52] Rosenblueth, E. (1965) Slenderness effects in buildings, *J. Struct. Div.*, ASCE, 91(ST1), 229-252.
[53] Sanz-Picon, C.F. (1992) Behavior of Composite Column Cross Sections Under Biaxial Bending, M.S. thesis, Cornell University, Ithaca, New York.
[54] Sfintesco, D., Ed. (1976) *European Convention of Constructional Steelworks Manual on the Stability of Steel Structures*, Second ed., ECCS, Paris.
[55] SSRC (1993) *Plastic Hinge Based Methods for Advanced Analysis and Design of Steel Frames, An Assessment of the State-of-the-Art*, White, D.W. and Chen, W.F., Eds., SSRC, Lehigh University, Bethlehem, PA, 299 pp.
[56] SSRC (1981) General principles for the stability design of metal structures, Technical Memorandum No. 5, Civil Engineering, ASCE, February 1981, 53-54.
[57] Standards Australia (1990) AS4100-1990, *Steel Structures*, Sydney, Australia.
[58] Vogel, U. (1985) Calibrating Frames, *Stahlbau*, 10, 1-7.
[59] Wang, Y.C. (1988) Ultimate Strength Analysis of 3-D Beam Columns and Column Subassemblages with Flexible Connections, Ph.D. thesis, University of Sheffield, England.
[60] White, D.W. (1993) Plastic hinge methods for advanced analysis of steel frames, *J. Const. Steel Res.*, 24(2), 121-152.
[61] White, D.W. (1985) Material and Geometric Nonlinear Analysis of Local Planar Behavior in Steel Frames Using Iterative Computer Graphics, M.S. thesis, Cornell University, Ithaca, NY, 281 pp.
[62] White, D.W., Liew, J.Y.R., and Chen, W.F. (1991) Second-Order Inelastic Analysis for Frame Design: A Report to SSRC Task Group 29 on Recent Research and the Perceived State-of-the-Art, Structural Engineering Report, CE-STR-91-12, Purdue University, West Lafayette, IN, 116 pp.
[63] Yura, J.A. (1988) Elements for Teaching Load and Resistance Factor Design: Combined Bending and Axial Load, University of Texas at Austin, 55-71.

[64] Ziemian, R.D. (1990), Advanced Methods of Inelastic Analysis in the Limit States Design of Steel Structures, Ph.D. dissertation, School of Civil and Environmental Engineering, Cornell University, Ithaca, NY, 265 pp.

[65] Ziemian, R.D., White, D.W., Deierlein, G.G., and McGuire, W. (1990) One approach to inelastic analysis and design, *Proc. 1990 Natl. Steel Conf.*, AISC, Chicago, 19.1-19.

chapter two

Practical Advanced Analysis

2.1 Introduction

This chapter presents a practical advanced analysis method for the direct design of steel frames by eliminating separate member capacity checks by the specification. The refined-plastic hinge method was developed and refined by simply modifying the conventional elastic-plastic hinge method to achieve both simplicity and a realistic representation of actual behavior [3] [36] [55] [56] [80]. Verification of the method will be given in the next chapter to provide final confirmation of the validity of the method.

Connection flexibility can be accounted for in advanced analysis. Conventional analysis and design of steel structures are usually carried out under the assumption that beam-to-column connections are either fully rigid or ideally pinned. However, most connections in practice are "semi-rigid" and their behavior lies between these two extreme cases. In the AISC-LRFD Specification [5], two types of construction are designated: Type FR (fully restrained) construction and Type PR (partially restrained) construction. The LRFD Specification permits the evaluation of the flexibility of connections by "rational means".

A large amount of research has been reported on methods of accounting for connection flexibility in the analysis and design of planar frame structures [1, 2, 7, 10, 11, 12, 17, 19, 22, 31, 42, 43, 52, 58, 61, 62, 66, 67, 82]. The connection is usually modeled as a rotational spring to describe the connection flexibility in member stiffness relationships. In this chapter, the formulation of Chen and Lui [22] will be used for the modified incremental force-displacement equations of the beam-column element to account for the presence of end connections.

Connection behavior is represented by its moment-rotation relationship. Extensive experimental work on connections has been performed, and a large body of moment-rotation data collected [21] [32] [46] [68]. With this data base, researchers have developed several connection models including: linear, polynomial, B-spline, power, and exponential. Herein, the three-parameter power model proposed by Kishi and Chen [45] is adopted.

Geometric imperfections should be modeled in frame members when using advanced analysis. Geometric imperfections result from unavoidable error during fabrication or erection. For structural members in building frames, the types of geometric imperfections are out-of-straightness and out-of-plumbness. Explicit modeling and equivalent notional loads have been used to account for geometric imperfections by previous researchers. In this chapter, a new method based on further reduction of the tangent stiffness of members is developed [36] [37]. This method provides a simple means to account for the effect of imperfection without inputting notional loads or explicit geometric imperfections.

The practical advanced analysis method described in this book is limited to two-dimensional braced, unbraced, and semi-rigid frames subject to static loads. The spatial behavior of frames is not considered, and lateral torsional buckling is assumed to be prevented by adequate lateral bracing. A compact W-section is assumed so sections can develop full plastic moment capacity without local buckling. Both strong and weak-axis bending of wide flange sections have been studied using the practical advanced analysis method [36]. The method may be considered an interim analysis/design procedure between the conventional LRFD method widely used now and a more rigorous advanced analysis/design method such as the plastic-zone method to be developed in the future for practical use.

2.2 Key Factors Influencing Steel Frame Behavior

The attributes that affect the behavior of steel members may be grouped into three categories: (1) material, (2) geometric, (3) and connection nonlinearities. Material nonlinearities include gradual yielding associated with flexure and the influence of residual stresses. Geometric nonlinearities include second-order $P - \delta$ and $P - \Delta$ effects as well as geometric imperfections. Connection nonlinearity occurs in the nonlinear moment-rotation relationship of semi-rigid connections.

2.2.1 Gradual Yielding Associated with Flexure

Figure 2.1 shows stress states of section under pure bending. When the bending moment is small, the section remains elastic. If the bending moment increases, the extreme fibers yield and ultimately the section becomes fully plastic. When a wide flange section is subject to pure bending, the moment-curvature relationship may look like that shown

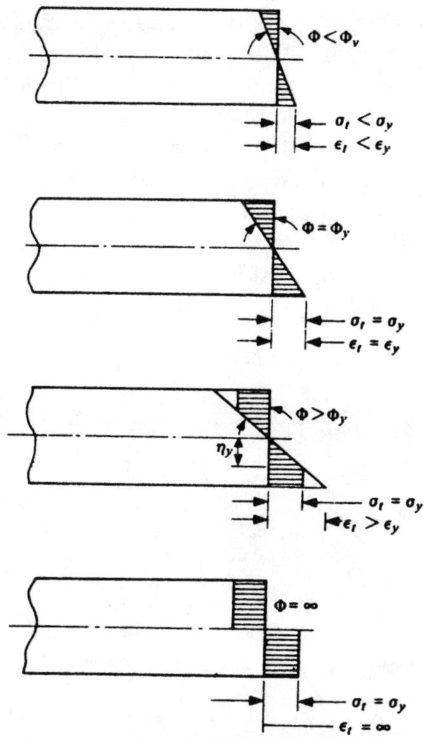

Figure 2.1 States of section under pure bending.

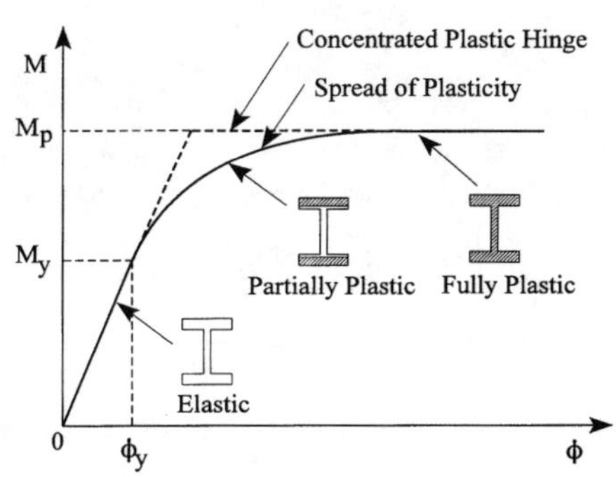

Figure 2.2 Moment-curvature relationships for perfect plastic and work hardening hinges.

in Figure 2.2 [18]. Although the stress-strain relationship of steel is assumed to be elastic-perfectly plastic, the moment-curvature relationship for a section has a smooth transition from elastic to fully plastic. This is because the section starts to yield gradually from extreme fibers which have the highest stresses. The gradual yielding of a section leads to gradual forming of plastic hinges in members in flexure.

The moment-curvature relationship is influenced by the existence of axial thrust P. Figure 2.3 shows a series of strain and stress diagrams that correspond to three stages of loading including elastic (no yielding), primary plastic (yielding in compression zone only), and secondary plastic (yielding in both compression and tension zones). The axial force

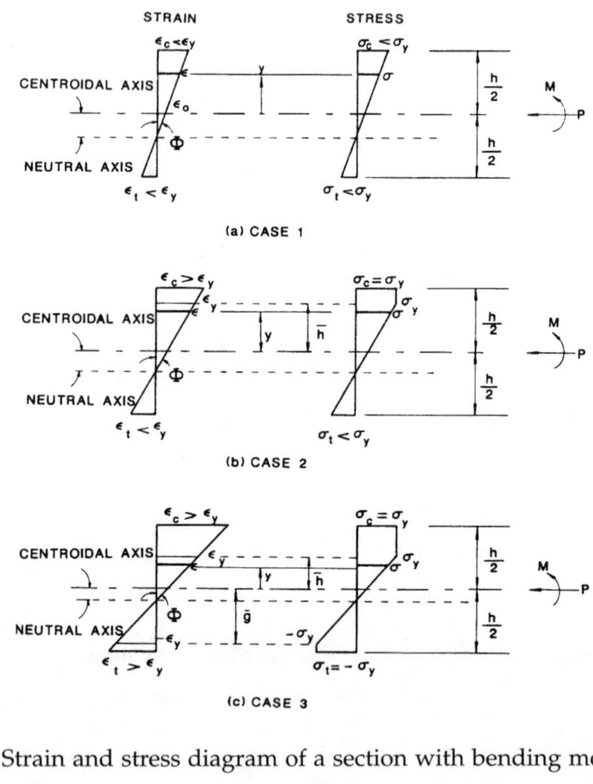

Figure 2.3 Strain and stress diagram of a section with bending moment and thrust.

has no effect on the moment-curvature relationship when the member is elastic (Case 1). However, as soon as the section starts to yield, the moment-curvature relationship will be affected by the axial force (Case 2 and 3). The moment-curvature-thrust relationships of rectangular cross-sections with idealized elastic-plastic, stress-strain behavior is illustrated in Figure 2.4. For other cross-sections, it is often necessary to evaluate the integrals for the axial for P and internal bending moment M numerically. Herein, a simple parabolic stiffness reduction function rather than numerical integration is used to account for the gradual yielding of a wide flange section for practical purposes. This concept is described in detail in Section 2.4.6.

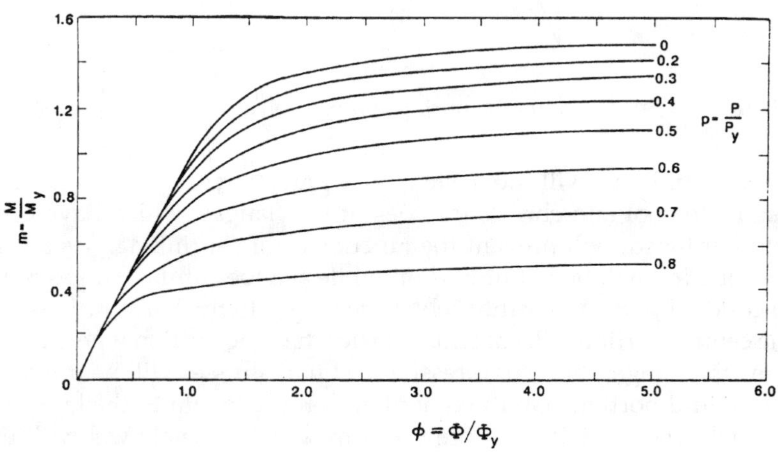

Figure 2.4 Moment-curvature-thrust relationships.

2.2.2 *Gradual Yielding Associated with Residual Stresses*

Material nonlinearities may result from residual stresses. If a simple compression test is carried out on a steel coupon, the stress-strain relationship will be linear elastic-perfectly plastic as shown in Figure 2.5 [23]. However, for a stub column, the stress-strain curve exhibits a smooth transition from elastic to perfectly plastic due to residual stresses in the steel column.

Residual stresses are created in the column in the following way: steel members are usually heated at some stage during the fabrication process. As they cool down, the part of the cross-section for which the surface area to volume ratio is the largest will lose heat more rapidly than the part for which the ratio is small. This uneven cooling creates a set of self-equilibrating stresses in the cross-section. These are "residual" stresses.

For hot-rolled wide-flange shapes, the toes of the flanges have a larger surface area to volume ratio than the region where the web joins the

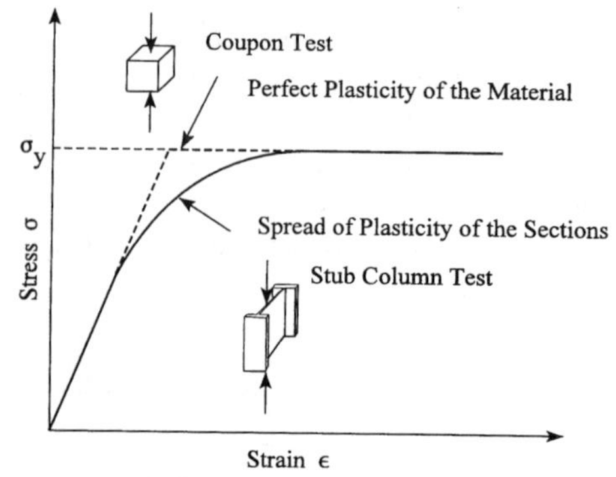

Figure 2.5 Stress-strain relationships for coupon and stub column tests.

flanges, and thus will cool faster. As the junction of the web and flange begins to cool and shrink, the toes of the flanges, which have cooled and hardened, will prevent the junctions from shrinking. As a result, the junction will be left in tension while the toes of the flanges remain in compression. As for the web, if the height to thickness ratio is large, the central portion will cool much faster than the portion where the web joins the flanges, and compressive residual stresses will be induced at the central portion. On the other hand, if the height to thickness ratio is small, cooling will be more uniform, and the whole web will be in tension. Figure 2.6 shows the residual stress pattern widely used in the U.S. where the residual stress is constant in the web; however, for deep sections, the residual stress in the web may not be constant. It may vary parabolically as shown by actual measured data in Figure 2.7.

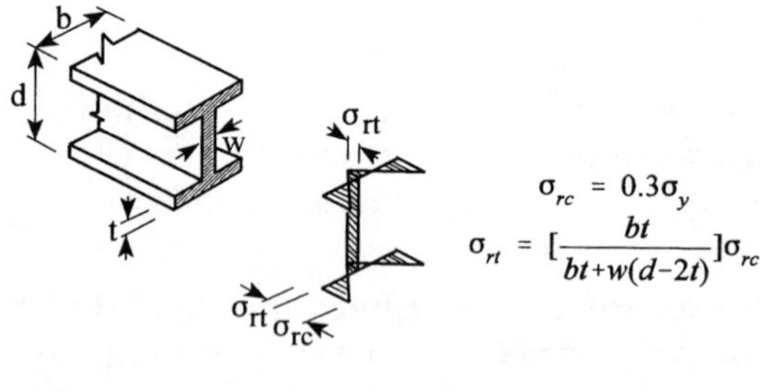

Figure 2.6 Typical residual stress distribution for a hot-rolled wide flange section.

When an axial force is applied to a member, the stress distribution over the cross-section will change. The stages are shown in Figure 2.8. When the stress in any fiber reaches the yield stress, any additional load must be carried by fibers as yet elastic. From the figure, it is clear that

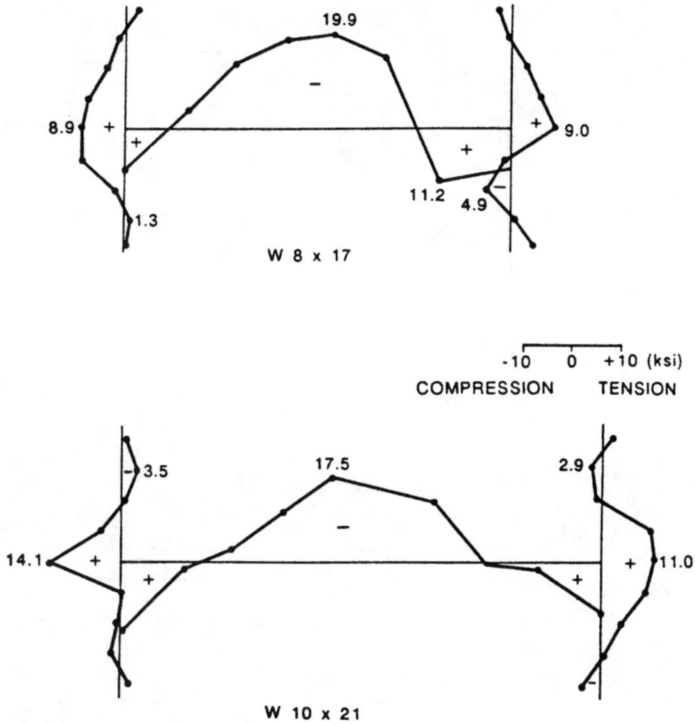

Figure 2.7 Measured residual stresses (Courtesy of T. Kanchanalai, Civil Engineering/Structures Research Laboratory, University of Texas, Austin).

yielding the cross-section is a gradual process. The fibers that have the highest value of compressive residual stress will yield first, and fibers with tensile stress will yield last. Because of this gradual yielding process (plastification) over the cross-section, the stress-strain curve for a stub column is smooth as shown in Figure 2.5. This stiffness degradation effect can be accounted for by adjusting the tangent modulus of the material as described in Section 2.4.5.

2.2.3 *Second-Order Effects*

Beam-columns in frames are subject to both axial compression and bending moment. The bending moment in a beam-column consists of primary and secondary bending moment. Primary bending is caused by the applied end moments and/or transverse loads on members. Secondary bending moments are generated as the axial compressive force acts over the lateral displacements of the member. Secondary bending includes both the $P - \delta$ and the $P - \Delta$ moment.

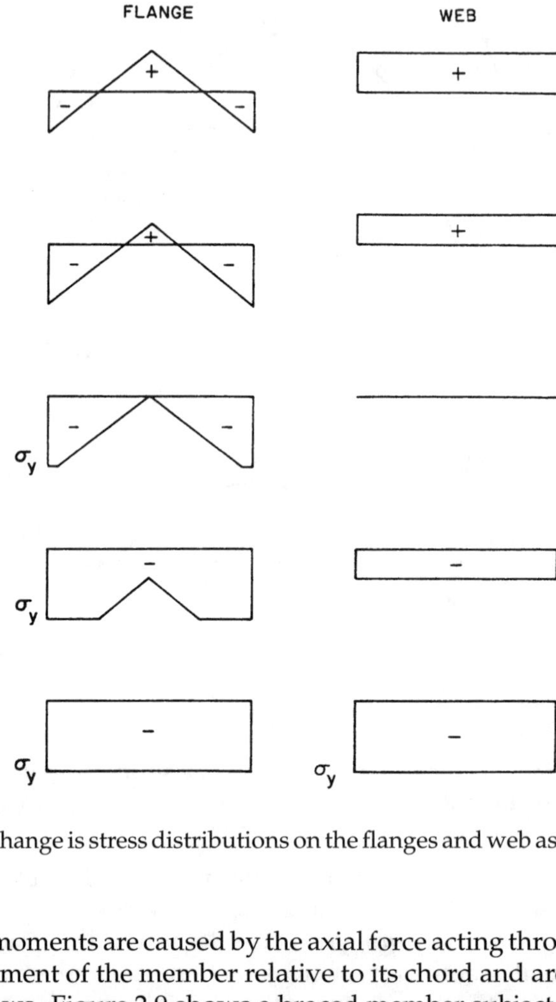

Figure 2.8 Change is stress distributions on the flanges and web as P increases.

The $P-\delta$ moments are caused by the axial force acting through the lateral displacement of the member relative to its chord and are described in what follows. Figure 2.9 shows a braced member subject to end moments causing single-curvature. The primary moment varies linearly, while the secondary moment varies nonlinearly according to the magnitude of deflection along the member. As a result, the primary moment is amplified due to axial compressive force. When a braced member is subject to reverse-curvature bending (Figure 2.10), the maximum primary moment is at one end, and the maximum secondary moment between the ends. According to the magnitude of the compressive force P, the amplified moment can be either greater or smaller than the end moment.

The $P - \Delta$ moment results from the axial force acting through the relative displacement of the ends of the member. Figure 2.11 shows an unbraced frame subject to gravity and lateral forces. The maximum primary moment (due to the lateral load) and the maximum secondary

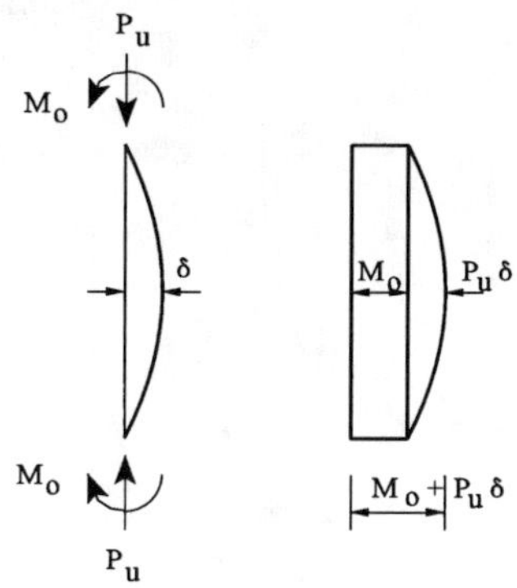

Figure 2.9 $P - \delta$ effect under single-curvature bending.

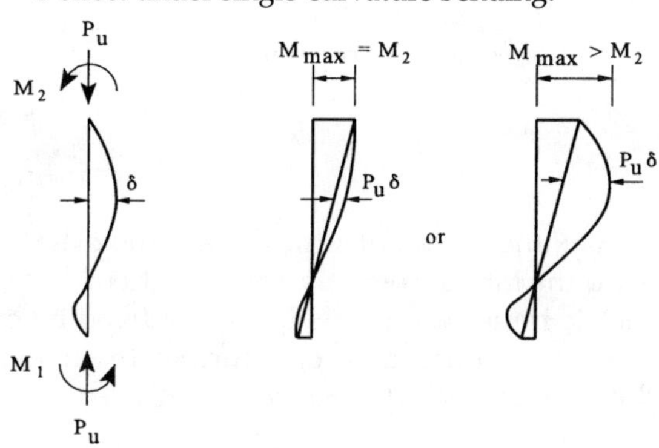

Figure 2.10 $P - \delta$ effect under double-curvature bending.

moment (due to the gravity load) always occur at one end in this sway case. The linear primary moment varies linearly and is amplified by the nonlinear secondary moment. These second-order effects, including $P - \delta$ and $P - \Delta$ effect, can be captured by the stability functions described in Section 2.4.1.

2.2.4 *Geometric Imperfections*

Geometric imperfections result from unavoidable error during fabrication or erection. For structural members, the types of geometric imperfections are out-of-straightness and out-of-plumbness. These imperfections cause additional moments in a column member and further degradation of its bending stiffness. The AISC Code of Standard Practice [5] specifies fabrication and erection tolerances. This code limits maximum out-of-straightness to $L/1000$, where L is distance between braced points, and maximum out-of-plumbness to $L/500$ where L is distance between working points as shown in Figure 2.12. Figure 2.12 also shows the envelope within which all working points must fall. The

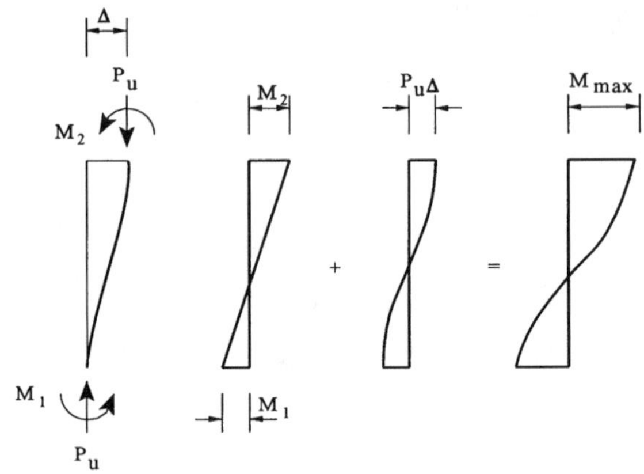

Figure 2.11 $P - \Delta$ effect of unbraced frame.

code allows a maximum lean of 1 in. toward the exterior of the frame and 2 in. toward the interior over the total height of less than 20 stories. Herein, geometric imperfections will be considered by explicit modeling, an equivalent notional load, or a further reduction of members' tangent modulus as presented in Section 2.6.

2.2.5 Connection Nonlinearity

The forces transmitted through beam-column connections consist of axial, shearing, bending, and torsion forces. The effect of axial and shearing forces is negligible since their corresponding deformations are small when compared with the rotational deformation of connections. Torsion is also neglected since this study is limited to plane frames. The deformation behavior of a connection may be described by a moment-rotation relationship, typically nonlinear. When a moment M is applied to a connection, it rotates by an amount θ_r. As depicted in Figure 2.13, the rotation represents the change in angle between the beam and the column from its original configuration. The angle θ_r is a measure of the rotation of the beam relative to the column.

Figure 2.14 shows the moment-rotation ($M - \theta_r$) behavior of a variety of commonly used semi-rigid connections. The single web-angle connection is a very flexible connection and the extended end plate con-

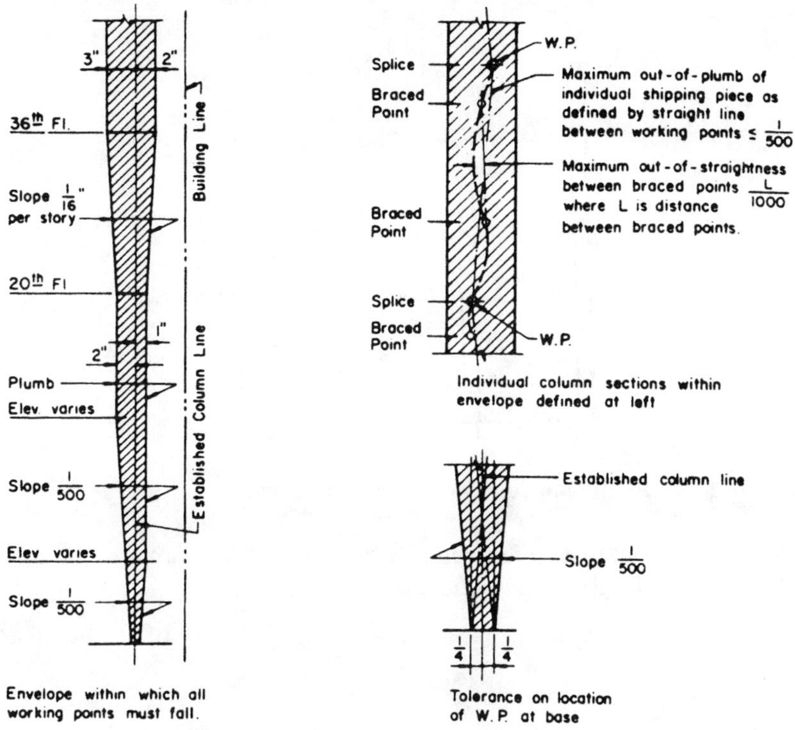

Figure 2.12 Fabrication and erection tolerances [5].

nection is comparatively rigid. Observe that a flexible connection has a smaller ultimate moment capacity and a larger rotation, and vice versa.

The nonlinearity of connection behavior is due to a number of factors [10] [22]. Some of the important ones are:

1. Material discontinuity of the connection assemblage. A connection is usually an arrangement of fasteners (such as welds or bolts) and structural shapes (such as angles and T-stubs). This formation allows irregular slips between components during loading.
2. Local yielding of parts of the connection. This is the primary factor in the nonlinear behavior of connections.
3. Stress and strain concentrations caused by holes, fasteners and bearing contacts of elements used in connection assemblage.

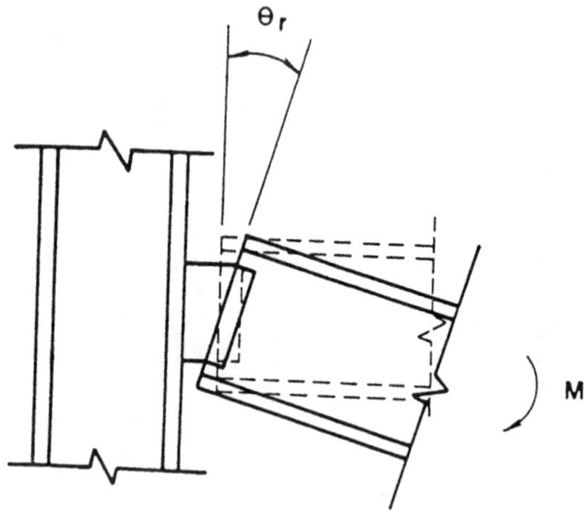

Figure 2.13 Rotational deformation of a connection.

4. Local buckling of flanges and/or web of the beam and the column in the vicinity of a connection.
5. Overall geometric changes under the influence of applied loads.

2.3 *Desirable Attributes For Practical Advanced Analysis*

Referring to the SSRC Task Force report [74], the following are the recommended attributes for plastic-hinge-based elements suitable for practical advanced analysis of two-dimensional frames:

1. The model should be accurate using only one element per member. The element should not be more than 5% unconservative when compared to "exact" solutions for in-plane beam-column strength. Since this is the criterion employed in the development of the AISC-LRFD beam-column interaction equations, an element that meets this requirement satisfies LRFD Specification provisions for in-plane strength.
2. The element relationship should be derived analytically and implemented in explicit form for analysis. Numerically integrated elements do not provide the degree of computational efficiency required for analysis of moderate to large size structural systems.
3. The model should be extensible to three-dimensional analysis. That is, the framework of the model should accommodate the formulation of three-dimensional elements.

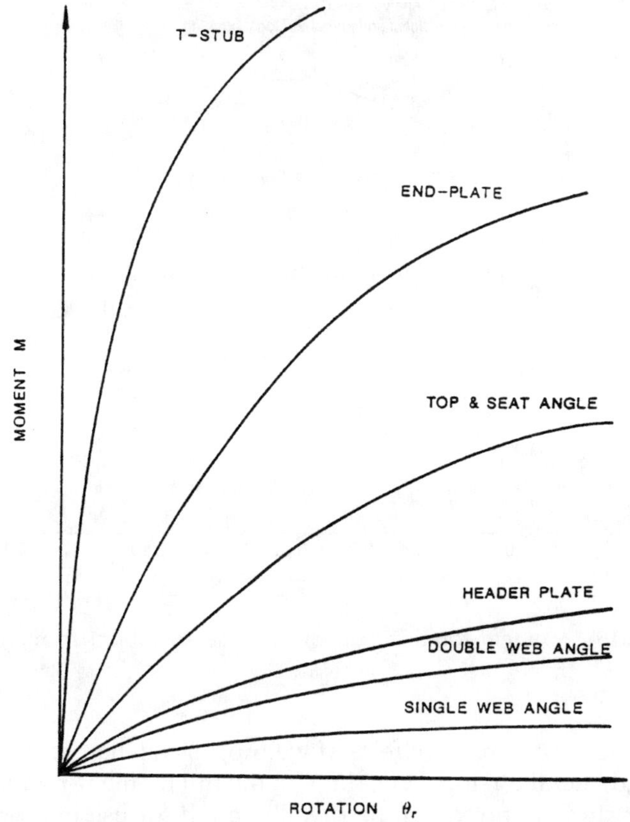

Figure 2.14 Schematic moment-rotation curves of various semi-rigid connections.

4. The effects of inelasticity on axial member deformations should be represented because the column axial stiffness provides a significant portion of the structure's side-sway resistance in many types of frames (particularly tall ones).

5. As the axial load approaches zero, the element behavior should approach that of the elastic-plastic hinge mode because this model provides a good representation of the performance for beam members. The possible benefits of strain-hardening should not be relied upon. In beams with a moment gradient, combined strain-hardening and local/lateral-torsional buckling effects result in the actual moment-rotation curve at a plastic hinge taking a form such as that shown by the solid curve in Figure 2.15. If, however, local and lateral-torsional buckling effects are neglected in a refined inelastic analysis, the dotted curve in Figure 2.15 is typically ob-

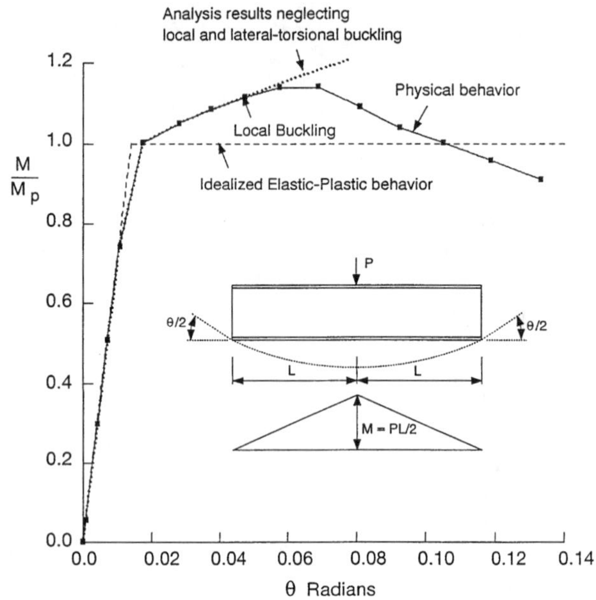

Figure 2.15 Moment-rotation curves for a beam subject to moment gradient [74].

tained. The precise effects of yielding, strain hardening, and local and lateral torsional buckling on the full moment-rotation characteristics are not quantified well enough for use in a practical inelastic analysis model at the present time. Therefore, the moment-rotation behavior of beam members is idealized commonly as the elastic-perfectly plastic curve shown in the figure. This idealization represents the cross-section strength of isolated members conservatively, but also it tends to under-predict the inelastic forces that adjacent beams and beam-columns may actually be subject to.

6. For the case of a member loaded by pure axial load, the element inelastic flexural stiffness should be close to that associated with the inelastic flexural rigidity $E_t I$ implied by the column strength equations of the particular design specification being used (for example, Equation E2-2 of the AISC-LRFD Specification).

7. Member out-of-straightness effect, when important, should be accommodated implicitly within the element model. This would parallel the philosophy behind the development of most modern column strength expressions. That is, the column strength expressions in specifications such as the AISC-LRFD implicitly include the effects of residual stresses and out-of-straightness. In fact, it

can be argued that, to a certain extent, these expressions also include the effect of member out-of-plumbness.

8. For intermediate to high axial loads, the moment gradient along the member length should have a significant effect of the element inelastic stiffness. The reduction in stiffness due to yielding should be largest for single-curvature bending and smallest for full reversed-curvature bending.

9. Once the full plastic strength at a cross-section is reached by member second-order forces, the cross-section forces may vary with continued loading, but these forces should never violate the strength conditions of the fully plastified section. That is, strain hardening effects should be neglected. Thus, if a plastic hinge forms in a beam-column member, the member axial force may still be increased, but this must result in a corresponding decrease in the moment at the hinge. Conversely, the moment may increase at a plastic hinge, but this must be coincident with a decrease in the member axial force.

10. The formation of plastic hinges within the span of a member should be accommodated using one-element per member. This is particularly important for transversely loaded members such as the beams and girders of a frame. Large savings in solution effort may be realized if these members do not need to be discretized into multiple elements to capture internal plastic hinges.

2.4 *Second-Order Refined Plastic Hinge Analysis*

In this section, a method called the refined plastic hinge approach is presented. This method is comparable to the elastic-plastic hinge analysis in efficiency and simplicity, but without its limitations. In this analysis, stability functions are used to predict second-order effects. The benefit of stability functions is that they make the analysis method practical by using only one element per each beam-column. The refined plastic hinge analysis uses a two-surface yield model and an effective tangent modulus to account for stiffness degradation due to distributed plasticity in framed members. The member stiffness is assumed to degrade gradually as the second-order forces at critical locations approach the cross-section plastic strength. Column tangent modulus is used to represent the effective stiffness of the member when it is loaded with a high axial load. Thus, the refined plastic hinge model approximates the effect of distributed plasticity along the element length caused by initial imperfections and large bending and axial force actions. In fact, research by Liew et al. [56] [57], Kim and Chen [37], and Kim [36] have shown that refined plastic hinge analysis captures the interaction of strength and stability of structural systems and that of their component elements. This type of analysis method may, therefore, be classified as an advanced analysis and separate specification member capacity checks are not required.

The general assumptions used in modeling the beam-column elements are:

1. All elements are initially straight and prismatic, and plane cross-section remains plane after deformation.
2. Local buckling and lateral-torsional buckling are not considered. All members are assumed to be fully compact and adequately braced to preclude out-of-plane deformations.
3. Large rigid-body displacements are allowed, but the member deformations and strains are small.
4. The element stiffness formulation is based on conventional beam-column stability functions, including axial and bending deformations, but not those associated with shear. Element bowing effects are neglected.
5. The formulation is limited by its ability to model plastic hinges only at the element ends.
6. Plastic hinges can sustain inelastic rotations only. Strain hardening is not considered.

2.4.1 *Stability Functions Accounting for Second-Order Effect*

The stability functions account for second-order effects using only one element for each beam-column. In the following, the stability functions for an elastic beam-column are derived and discussed. Figure 2.16(a) shows a beam-column subject to end moments M_A and M_B and acted upon by an axial force P. Using the free-body diagram of a segment of beam-column of length x from the left end (Figure 2.16(b)), the external moment acting on the cut section is:

$$M_{ext} = M_A + Py - \frac{M_A + M_B}{L}x \tag{2.1}$$

Equating this to the internal moment of $-EIy''$ and rearranging, we have

$$EIy'' + Py = \frac{M_A + M_B}{L}x - M_A \tag{2.2}$$

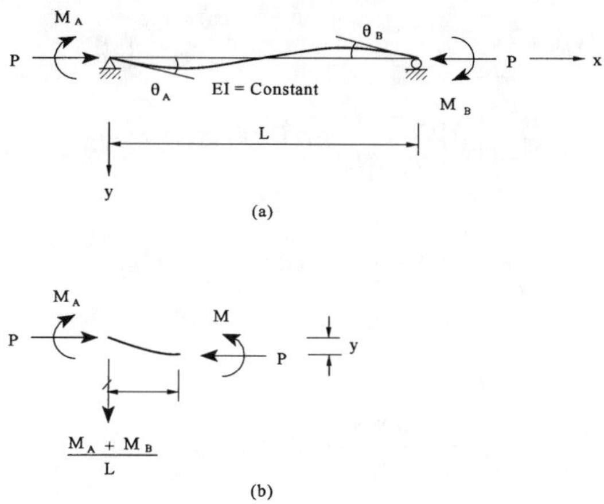

Figure 2.16 Beam-column subject to double-curvature bending.

or, using $k^2 = P/EI$, we can write

$$y'' + k^2 y = \frac{M_A + M_B}{LEI} x - \frac{M_A}{EI} \tag{2.3}$$

The general solution is:

$$y = A \sin kx + B \cos kx + \frac{M_A + M_B}{LEIk^2} x - \frac{M_A}{EIk^2} \tag{2.4}$$

The constants A and B can be evaluated by using the boundary conditions

$$y(0) = 0, \qquad y(L) = 0 \tag{2.5}$$

From the first boundary condition, we obtain:

$$B = \frac{M_A}{EIk^2} \tag{2.6}$$

and from the second boundary condition, we obtain:

$$A = -\frac{1}{EIk^2 \sin kL} (M_A \cos kL + M_B) \tag{2.7}$$

Therefore, Equation (2.4) can be written as:

$$y = -\frac{(M_A \cos kL + M_B)}{EIk^2 \sin kL} \sin kx + \frac{M_A}{EIk^2} \cos kx$$

$$+ \frac{M_A + M_B}{LEIk^2} x - \frac{M_A}{EIk^2} \tag{2.8}$$

Rearranging, we have

$$y = -\frac{1}{EIk^2}\left[\frac{\cos kL}{\sin kL}\sin kx - \cos kx - \frac{x}{L} + 1\right]M_A$$

$$-\frac{1}{EIk^2}\left[\frac{1}{\sin kL}\sin kx - \frac{x}{L}\right]M_B \quad (2.9)$$

from which

$$y' = -\frac{1}{EIk}\left[\frac{\cos kL}{\sin kL}\cos kx + \sin kx - \frac{1}{kL}\right]M_A$$

$$-\frac{1}{EIk}\left[\frac{\cos kx}{\sin kL} - \frac{1}{kL}\right]M_B \quad (2.10)$$

Using Equation (2.10), the end rotation θ_A and θ_B can be obtained:

$$\theta_A = y'(0) = -\frac{1}{EIk}\left[\frac{\cos kL}{\sin kL} - \frac{1}{kL}\right]M_A - \frac{1}{EIk}\left[\frac{1}{\sin kL} - \frac{1}{kL}\right]M_B$$

$$= \frac{L}{EI}\left[\frac{\sin kL - kL\cos kL}{(kL)^2 \sin kL}\right]M_A + \frac{L}{EI}\left[\frac{\sin kL - kL}{(kL)^2\sin kL}\right]M_B \quad (2.11a)$$

$$\theta_B = y'(L) = -\frac{1}{EIk}\left[\frac{1}{\sin kL} - \frac{1}{kL}\right]M_A - \frac{1}{EIk}\left[\frac{\cos kL}{\sin kL} - \frac{1}{kL}\right]M_B$$

$$= \frac{L}{EI}\left[\frac{\sin kL - kL}{(kL)^2\sin kL}\right]M_A + \frac{L}{EI}\left[\frac{\sin kL - kL\cos kL}{(kL)^2 \sin kL}\right]M_B \quad (2.11b)$$

Equations (2.11a) and (2.11b) can be written in matrix form:

$$\begin{bmatrix}\theta_A \\ \theta_B\end{bmatrix} = \begin{bmatrix} f_{11} & f_{12} \\ f_{21} & f_{22}\end{bmatrix}\begin{bmatrix}M_A \\ M_B\end{bmatrix} \quad (2.12)$$

where

$$f_{11} = f_{22} = \frac{L}{EI}\left[\frac{\sin kL - kL\cos kL}{(kL)^2\sin kL}\right] \quad (2.13a)$$

$$f_{12} = f_{21} = \frac{L}{EI}\left[\frac{\sin kL - kL}{(kL)^2\sin kL}\right] \quad (2.13b)$$

From Equation (2.12), we can write:

$$\begin{bmatrix} M_A \\ M_B \end{bmatrix} = \begin{bmatrix} f_{11} & f_{12} \\ f_{21} & f_{22} \end{bmatrix}^{-1} \begin{bmatrix} \theta_A \\ \theta_B \end{bmatrix} \tag{2.14}$$

Equation (2.14) can be written in its expanded form:

$$M_A = \frac{EI}{L}\left(s_{ii}\theta_A + s_{ij}\theta_B\right) \tag{2.15a}$$

$$M_B = \frac{EI}{L}\left(s_{ji}\theta_A + s_{jj}\theta_B\right) \tag{2.15b}$$

where

$$s_{ii} = s_{jj} = \frac{kL \sin kL - (kL)^2 \cos kL}{2 - 2\cos kL - kL \sin kL} \tag{2.16a}$$

$$s_{ij} = s_{ji} = \frac{(kL)^2 - kL(\sin kL)}{2 - 2\cos kL - kL \sin kL} \tag{2.16b}$$

are referred to as stability functions.

Equations (2.15a) and (2.15b) are the slope-deflection equations for a beam-column with no transverse load or relative joint translation (side-sway). Note that when P approaches zero, $kL = \sqrt{P/EIL}$ approaches zero, and by using the L'Hospital's rule, it can be shown that s_{ii} reduces to 4 and s_{ij} reduces to 2, which are the same as conventional slope-deflection equations. Values of s_{ii} and s_{ij} for various values of kL are listed in Table 2.1 and plotted in Figure 2.17. Stability functions account for the effect of an axial force on the bending stiffness of the element.

2.4.2 Incremental Force-Displacement Relationship

From Equations (2.15a) and (2.15b), the incremental force-displacement relationship may be written as:

$$\begin{Bmatrix} \dot{M}_A \\ \dot{M}_B \\ \dot{P} \end{Bmatrix} = \frac{EI}{L} \begin{bmatrix} S_1 & S_2 & 0 \\ S_2 & S_1 & 0 \\ 0 & 0 & A/I \end{bmatrix} \begin{Bmatrix} \dot{\theta}_A \\ \dot{\theta}_B \\ \dot{e} \end{Bmatrix} \tag{2.17}$$

Table 2.1 Values of Stability Functions

			Compression		Tension	
kL	P/P_e	K	S_{ii}	S_{jj}	S_{ii}	S_{jj}
0.00	0.0000	Infinite	4.0000	2.0000	4.0000	2.0000
0.05	0.0003	62.8319	3.9997	2.0001	4.0003	1.9999
0.10	0.0010	31.4159	3.9987	2.0003	4.0013	1.9993
0.15	0.0023	20.9440	3.9970	2.0008	4.0030	1.9993
0.20	0.0041	15.7080	3.9947	2.0013	4.0053	1.9987
0.25	0.0063	12.5664	3.9917	2.0021	4.0083	1.9979
0.30	0.0091	10.4720	3.9880	2.0030	4.0120	1.9970
0.35	0.0124	8.9760	3.9836	2.0041	4.0163	1.9959
0.40	0.0162	7.8540	3.9786	2.0054	4.0213	1.9947
0.45	0.0205	6.9813	3.9729	2.0068	4.0269	1.9933
0.50	0.0253	6.2832	3.9666	2.0084	4.0332	1.9917
0.55	0.0306	5.7120	3.9595	2.0102	4.0402	1.9900
0.60	0.0365	5.2360	3.9518	2.0121	4.0478	1.9881
0.65	0.0428	4.8332	3.9434	2.0143	4.0560	1.9861
0.70	0.0496	4.4880	3.9342	2.0166	4.0649	1.9839
0.75	0.0570	4.1888	3.9244	2.0191	4.0745	1.9816
0.80	0.0648	3.9270	3.9139	2.0218	4.0846	1.9791
0.85	0.0732	3.6960	3.9027	2.0246	4.0954	1.9764
0.90	0.0821	3.4907	3.8908	2.0277	4.1069	1.9737
0.95	0.0914	3.3069	3.8782	2.0309	4.1189	1.9707
1.00	0.1013	3.1416	3.8649	2.0344	4.1316	1.9677
1.05	0.1117	2.9920	3.8508	2.0380	4.1449	1.9645
1.10	0.1226	2.8560	3.8360	2.0419	4.1588	1.9611
1.15	0.1340	2.7318	3.8205	2.0460	4.1734	1.9577
1.20	0.1459	2.6180	3.8043	2.0502	4.1885	1.9541
1.25	0.1583	2.5133	3.7873	2.0547	4.2042	1.9503
1.30	0.1712	2.4166	3.7695	2.0594	4.2205	1.9465
1.35	0.1847	2.3271	3.7510	2.0644	4.2374	1.9425
1.40	0.1986	2.2440	3.7317	2.0695	4.2549	1.9384
1.45	0.2130	2.1666	3.7116	2.0749	4.2729	1.9342
1.50	0.2280	2.0944	3.6907	2.0806	4.2916	1.9299

Table 2.1 Values of Stability Functions (cont.)

			Compression		Tension	
kL	P/P_e	K	S_{ii}	S_{jj}	S_{ii}	S_{jj}
1.55	0.2434	2.0268	3.6690	2.0865	4.3107	1.9255
1.60	0.2594	1.9635	3.6466	2.0926	4.3305	1.9210
1.65	0.2758	1.9040	3.6233	2.0990	4.3508	1.9164
1.70	0.2928	1.8480	3.5991	2.1057	4.3716	1.9116
1.75	0.3103	1.7952	3.5741	2.1127	4.3929	1.9068
1.80	0.3283	1.7453	3.5483	2.1199	4.4148	1.9019
1.85	0.3468	1.6982	3.5216	2.1275	4.4373	1.8969
1.90	0.3658	1.6535	3.4940	2.1353	4.4602	1.8919
1.95	0.3853	1.6111	3.4655	2.1434	4.4836	1.8867
2.00	0.4053	1.5708	3.4361	2.1519	4.5076	1.8815
2.05	0.4258	1.5325	3.4058	2.1607	4.5320	1.8762
2.10	0.4468	1.4960	3.3745	2.1699	4.5569	1.8708
2.15	0.4684	1.4612	3.3422	2.1794	4.5823	1.8654
2.20	0.4904	1.4280	3.3090	2.1893	4.6082	1.8599
2.25	0.5129	1.3963	3.2748	2.1996	4.6345	1.8544
2.30	0.5360	1.3659	3.2395	2.2102	4.6613	1.8488
2.35	0.5595	1.3368	3.2032	2.2213	4.6886	1.8431
2.40	0.5836	1.3090	3.1659	2.2328	4.7163	1.8374
2.45	0.6082	1.2823	3.1274	2.2447	4.7444	1.8317
2.50	0.6333	1.2566	3.0878	2.2572	4.7730	1.8259
2.55	0.6588	1.2320	3.0471	2.2701	4.8020	1.8201
2.60	0.6849	1.2083	3.0052	2.2834	4.8314	1.8142
2.65	0.7115	1.1855	2.9622	2.2974	4.8612	1.8083
2.70	0.7386	1.1636	2.9178	2.3118	4.8915	1.8024
2.75	0.7662	1.1424	2.8723	2.3268	4.9221	1.7965
2.80	0.7944	1.1220	2.8254	2.3425	4.9531	1.7905
2.85	0.8230	1.1023	2.7772	2.3587	4.9845	1.7845
2.90	0.8521	1.0833	2.7276	2.3756	5.0162	1.7785
2.95	0.8817	1.0649	2.6766	2.3932	5.0484	1.7725
3.00	0.9119	1.0472	2.6242	2.4115	5.0809	1.7665

Table 2.1 Values of Stability Functions (cont.)

			Compression		Tension	
kL	P/P_e	K	S_{ii}	S_{jj}	S_{ii}	S_{jj}
3.05	0.9425	1.0300	2.5703	2.4305	5.1137	1.7605
3.10	0.9737	1.0134	2.5148	2.4503	5.1469	1.7544
3.15	1.0054	0.9973	2.4577	2.4709	5.1805	1.7484
3.20	1.0375	0.9817	2.3990	2.4924	5.2143	1.7424
3.25	1.0702	0.9666	2.3385	2.5148	5.2485	1.7363
3.30	1.1034	0.9520	2.2763	2.5382	5.2831	1.7303
3.35	1.1371	0.9378	2.2122	2.5626	5.3179	1.7243
3.40	1.1713	0.9240	2.1463	2.5880	5.3530	1.7183
3.45	1.2060	0.9106	2.0783	2.6146	5.3885	1.7123
3.50	1.2412	0.8976	2.0083	2.6424	5.4242	1.7063
3.55	1.2769	0.8850	1.9362	2.6714	5.4603	1.7003
3.60	1.3131	0.8727	1.8618	2.7017	5.4966	1.6944
3.65	1.3499	0.8607	1.7851	2.7335	5.5332	1.6884
3.70	1.3871	0.8491	1.7060	2.7668	5.5701	1.6825
3.75	1.4248	0.8378	1.6243	2.8016	5.6073	1.6766
3.80	1.4631	0.8267	1.5400	2.8382	5.6447	1.6708
3.85	1.5018	0.8160	1.4528	2.8765	5.6823	1.6649
3.90	1.5411	0.8055	1.3627	2.9168	5.7203	1.6591
3.95	1.5809	0.7953	1.2695	2.9592	5.7584	1.6533
4.00	1.6211	0.7854	1.1731	3.0037	5.7968	1.6476
4.05	1.6619	0.7757	1.0732	3.0507	5.8355	1.6419
4.10	1.7032	0.7662	0.9698	3.1001	5.8744	1.6362
4.15	1.7450	0.7570	0.8624	3.1523	5.9135	1.6305
4.20	1.7873	0.7480	0.7510	3.2074	5.9528	1.6249
4.25	1.8301	0.7392	0.6353	3.2656	5.9923	1.6193
4.30	1.8734	0.7306	0.5149	3.3273	6.0321	1.6138
4.35	1.9173	0.7222	0.3897	3.3926	6.0720	1.6083
4.40	1.9616	0.7140	0.2592	3.4619	6.1122	1.6028
4.45	2.0064	0.7060	0.1231	3.5356	6.1526	1.5974
4.50	2.0518	0.6981	−0.0191	3.6140	6.1931	1.5920
4.55	2.0976	0.6905	−0.1678	3.6975	6.2339	1.5867
4.60	2.1440	0.6830	−0.3234	3.7866	6.2748	1.5814
4.65	2.1908	0.6756	−0.4867	3.8819	6.3159	1.5761

Table 2.1 Values of Stability Functions (cont.)

			Compression		Tension	
kL	P/P_e	K	S_{ii}	S_{jj}	S_{ii}	S_{jj}
4.70	2.2382	0.6684	−0.6582	3.9839	6.3572	1.5709
4.75	2.2861	0.6614	−0.8387	4.0934	6.3987	1.5658
4.80	2.3344	0.6545	−1.0289	4.2112	6.4403	1.5606
4.85	2.3833	0.6478	−1.2299	4.3381	6.4821	1.5556
4.90	2.4327	0.6411	−1.4427	4.4751	6.5241	1.5505
4.95	2.4826	0.6347	−1.6685	4.6235	6.5662	1.5456
5.00	2.5330	0.6283	−1.9087	4.7845	6.6085	1.5406
5.05	2.5839	0.6221	−2.1651	4.9599	6.6509	1.5357
5.10	2.6354	0.6160	−2.4394	5.1514	6.6934	1.5309
5.15	2.6873	0.6100	−2.7340	5.3613	6.7362	1.5261
5.20	2.7397	0.6042	−3.0516	5.5921	6.7790	1.5213
5.25	2.7927	0.5984	−3.3953	5.8470	6.8220	1.5166
5.30	2.8461	0.5928	−3.7688	6.1296	6.8652	1.5120
5.35	2.9001	0.5872	−4.1770	6.4447	6.9084	1.5074
5.40	2.9545	0.5818	−4.6253	6.7977	6.9518	1.5028
5.45	3.0095	0.5764	−5.1210	7.1956	6.9953	1.4983
5.50	3.0650	0.5812	−5.6726	7.6472	7.0390	1.4938
5.55	3.1209	0.5661	−6.2916	8.1635	7.0827	1.4894
5.60	3.1774	0.5610	−6.9922	8.7589	7.1266	1.4851
5.65	3.2344	0.5560	−7.7937	9.4523	7.1706	1.4807
5.70	3.2919	0.5512	−8.7214	10.2692	7.2147	1.4765
5.75	3.3499	0.5464	−9.8105	11.2446	7.2589	1.4722
5.80	3.4084	0.5417	−11.1106	12.4278	7.3033	1.4680
5.85	3.4675	0.5370	−12.6942	13.8914	7.3477	1.4639
5.90	3.5270	0.5325	−14.6715	15.7453	7.3922	1.4598
5.95	3.5870	0.5280	−17.2189	18.1659	7.4369	1.4558
6.00	3.6476	0.5236	−20.6375	21.4540	7.4816	1.4518
6.05	3.7086	0.5193	−25.4863	26.1685	7.5264	1.4478
6.10	3.7702	0.5150	−32.9345	33.4784	7.5714	1.4439
6.15	3.8322	0.5108	−45.9073	46.3087	7.6164	1.4401
6.20	3.8948	0.5067	−74.3621	74.6167	7.6615	1.4363
6.25	3.9579	0.5027	−188.2675	188.3707	7.7067	1.4325
6.2831853	4.0000	0.5000	−Infinite	Infinite	7.7367	1.4300

$P/P_e = \left(\frac{kL}{\pi}\right)^2 \qquad K = \sqrt{P_e/P}$

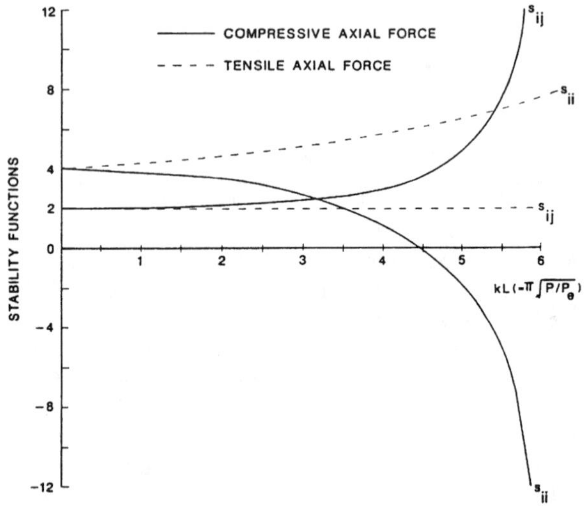

Figure 2.17 Plot of stability functions.

where

$\dot{M}_A, \dot{M}_B, \dot{P}$ = incremental end moments and axial force
$\dot{\theta}_A, \dot{\theta}_B, \dot{e}$ = incremental joint rotations and axial displacement
S_1, S_2 = stability functions, $S_1 = s_{ii} = s_{jj}$, $S_2 = s_{ij} = s_{ji}$

The stability functions given by Equations (2.16a) and (2.16b) may be written as:

$$S_1 = \begin{cases} \dfrac{\pi\sqrt{\rho}\sin(\pi\sqrt{\rho})-\pi^2\rho\cos(\pi\sqrt{\rho})}{2-2\cos(\pi\sqrt{\rho})-\pi\sqrt{\rho}\sin(\pi\sqrt{\rho})} & \text{if } P < 0 \\ \dfrac{\pi^2\rho\cosh(\pi\sqrt{\rho})-\pi\sqrt{\rho}\sinh(\pi\sqrt{\rho})}{2-2\cosh(\pi\sqrt{\rho})+\pi\sqrt{\rho}\sinh(\pi\sqrt{\rho})} & \text{if } P > 0 \end{cases} \quad (2.18a)$$

$$S_2 = \begin{cases} \dfrac{\pi^2\rho-\pi\sqrt{\rho}\sin(\pi\sqrt{\rho})}{2-2\cos(\pi\sqrt{\rho})-\pi\sqrt{\rho}\sin(\pi\sqrt{\rho})} & \text{if } P < 0 \\ \dfrac{\pi\sqrt{\rho}\sinh(\pi\sqrt{\rho})-\pi^2\rho}{2-2\cosh(\pi\sqrt{\rho})+\pi\sqrt{\rho}\sinh(\pi\sqrt{\rho})} & \text{if } P > 0 \end{cases} \quad (2.18b)$$

where

$\rho = P/\left(\pi^2 EI/L^2\right)$, P is positive in tension

The numerical solutions obtained from Equations (2.18a) and (2.18b) are indeterminate when the axial force is zero. To circumvent this problem and to avoid the use of different expressions for S_1 and S_2 for a different sign of axial forces, Goto and Chen (see [22]) have proposed a set of expressions that make use of power-series expansions to approximate the stability functions. The power-series expressions have been shown to converge to a high degree of accuracy within the first 10 terms of the series expansions. Alternatively, if the axial force in the member falls within the range $-2.0 \leq \rho \leq 2.0$, the following simplified expressions may be used to closely approximate the stability functions [61]:

$$S_1 = 4 + \frac{2\pi^2\rho}{15} - \frac{(0.01\rho + 0.543)\rho^2}{4+\rho} - \frac{(0.004\rho + 0.285)\rho^2}{8.183+\rho} \tag{2.19a}$$

$$S_2 = 2 - \frac{\pi^2\rho}{30} + \frac{(0.01\rho + 0.543)\rho^2}{4+\rho} - \frac{(0.004\rho + 0.285)\rho^2}{8.183+\rho} \tag{2.19b}$$

Equations (2.19a) and (2.19b) are applicable for members in tension (positive P) and compression (negative P). For most practical applications, Equations (2.19a) and (2.19b) give an excellent correlation to the "exact" expressions given by Equations (2.18a) and (2.18b). However, for ρ other than the range of $-2.0 \leq \rho \leq 2.0$, the conventional stability functions [Equations (2.18a) and (2.18b)] should be used. The stability function approach uses only one element per member and maintains accuracy in the element stiffness terms and in the recovery of element end forces for all ranges of axial loads. The finite element geometric stiffness approach may lead to some error in the stiffness terms and in the force recovery process; therefore, in using the geometric stiffness approach, its range of applicability must be understood [79].

The element tangent stiffness relationship from Equation (2.17) may be written symbolically:

$$\dot{f}_c = k_c \dot{d}_c \tag{2.20}$$

in which $\dot{f}_c$ and $\dot{d}_c$ are the incremental element end forces and displacements, and k_c is the basic element tangent stiffness matrix. For a plane frame member, three additional degrees of freedom are required to describe the total displacements of the member. If $d_{g1}, d_{g2} \ldots$ and d_{g6} are defined as the global translational and rotational degrees of freedom of a frame member (Figure 2.18), it can be shown that the local displacements are related to the global displacements by

$$d_{c1} = \theta_A = \theta_o + d_{g3} - \tan^{-1}\frac{y_o + d_{g5} - d_{g2}}{x_o + d_{g4} - d_{g1}} \tag{2.21a}$$

$$d_{c2} = \theta_B = \theta_o + d_{g6} - \tan^{-1}\frac{y_o + d_{g5} - d_{g2}}{x_o + d_{g4} - d_{g1}} \tag{2.21b}$$

$$d_{c3} = \frac{(2x_o + d_{g4} - d_{g1})(d_{g4} - d_{g1}) + (2y_o + d_{g5} - d_{g2})(d_{g5} - d_{g2})}{L_f + L} \tag{2.21c}$$

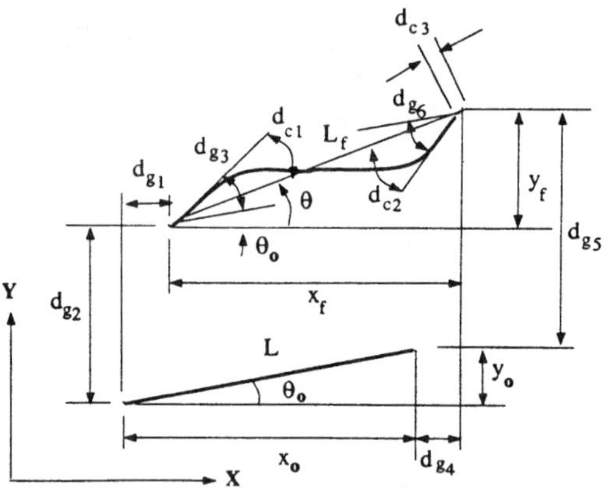

Figure 2.18 Kinematic relationship between local and global displacements of a frame element.

The expression for d_{c3} in Equation (2.21c) is more accurate than the value calculated from $L_f - L$. This is because Equation (2.21c) does not find the small difference between large member lengths [25]. Equation (2.21c) is obtained by writing $d_{c3} = \left(L_f^2 - L^2\right) / \left(L_f + L\right)$, and then solving for d_{c3}. In the denominator, $L_f + L \approx 2L$ may be used, because small displacement theory is presumed for the co-rotational chord element [14].

Upon differentiation of Equations (2.21a) through (2.21c) with respect to time (or pseudo-time), the incremental kinematic relationship relating the two sets of displacement vectors may be written as

$$\dot{d}_c = \frac{\partial d_{cj}}{\partial d_{gk}} \dot{d}_g = T_{cg} \dot{d}_g \tag{2.22}$$

where $j = 1, 2, 3$, $k = 1, 2 \ldots 6$, and

$$\dot{d}_c = \left[\dot{\theta}_A \dot{\theta}_B \dot{e}\right]^T$$

$$\dot{d}_g = \left[\dot{d}_{g1} \dot{d}_{g2} \dot{d}_{g3} \dot{d}_{g4} \dot{d}_{g5} \dot{d}_{g6}\right]^T$$

$$T_{cg} = \begin{bmatrix} -s/L & c/L & 1 & s/L & -c/L & 0 \\ -s/L & c/L & 0 & s/L & -c/L & 1 \\ -c & -s & 0 & c & s & 0 \end{bmatrix} \tag{2.23}$$

in which $c = \cos\theta$, $s = \sin\theta$, and θ is the angle of inclination of the chord of the deformed member.

Based on equilibrium, the forces in the two systems are related by

$$f_g = T_{cg}^T f_c \tag{2.24}$$

Taking derivatives on both sides of Equation (2.24) gives

$$\dot{f}_g = T_{cg}^T \dot{f}_c + \dot{T}_{cg}^T f_c \tag{2.25}$$

In view of Equations (2.20) and (2.22), Equation (2.25) may be further written as

$$\dot{f}_g = T_{cg}^T k_c T_{cg} \dot{d}_g + \dot{T}_{cg}^T f_c \tag{2.26}$$

The matrix $\dot{T}_{cg}^T$ in Equation (2.26) can be evaluated by taking derivative of T_{cg}^T with respect to the pseudo-time as

$$\dot{T}_{cg}^T = \frac{\partial T_{cg}^T}{\partial d_{gk}} \dot{d}_g = \left[\frac{\partial^2 d_{ci}}{\partial d_{gj} \partial d_{gk}} \right]^T \dot{d}_g \tag{2.27}$$

where $i = 1, 2, 3$, and $j = 1, 2, \ldots, 6$. Substituting $\dot{T}_{cg}^T$ from Equation (2.27) into Equation (2.26) gives the following expression for $\dot{f}_g$

$$\dot{f}_g = \left(T_{cg}^T k_c T_{cg} + T_1 M_A + T_2 M_B + T_3 P \right) \dot{d}_g \tag{2.28}$$

where

$$T_1 = T_2 = \frac{1}{L^2} \begin{bmatrix} -2sc & c^2 - s^2 & 0 & 2sc & -(c^2 - s^2) & 0 \\ & 2cs & 0 & -(c^2 - s^2) & -2sc & 0 \\ & & 0 & 0 & 0 & 0 \\ & sym. & & -2sc & c^2 - s^2 & 0 \\ & & & & 2sc & 0 \\ & & & & & 0 \end{bmatrix} \tag{2.29}$$

$$T_3 = \frac{1}{L} \begin{bmatrix} s^2 & -sc & 0 & -s^2 & sc & 0 \\ & c^2 & 0 & sc & -c^2 & 0 \\ & & 0 & 0 & 0 & 0 \\ & sym. & & s^2 & -sc & 0 \\ & & & & c^2 & 0 \\ & & & & & 0 \end{bmatrix} \tag{2.30}$$

Equation (2.28) is the incremental force-displacement relationship of a frame element in the global coordinate system, and it may be written symbolically as:

$$\dot{f}_g = k_g \dot{d}_g \tag{2.31}$$

where k_g represents the tangent stiffness matrix of a frame element. It should be noted that in the derivation of the tangent stiffness matrix (k_g), the joints are assumed to be rigid. If plastic hinges or semi-rigid connections are presented at the element ends, the tangent stiffness matrix needs to be modified. These modifications are discussed in Sections 2.4.4 and 2.5.6.

For structures subject to lateral wind or earthquake loadings, truss diagonal bracings may be used to reduce frame drifts and to enhance the lateral-load resistance of the structure. In design, these braces are usually assumed to carry only axial force. Therefore, it is justifiable to use truss elements to model the bracing members. The truss elements may be used for modeling gravity columns that do not participate in the lateral-force resisting system. These gravity columns (leaning columns) are widely used in many types of low-rise industrial structures and tall office building frames [73]. They can be modeled by the truss element described in this section.

The tangent stiffness relationship for a bracing element can be obtained from the tangent stiffness relationship of a frame element by deleting the appropriate rows and columns in Equation (2.28) that correspond to the rotational degrees of freedom of the element. The resulting tangent stiffness relationship of a truss element is

$$\dot{f}_g = \left(T_{cg}^T k_c T_{cg} + T P\right) \dot{d}_g = k_g \dot{d}_g \tag{2.32}$$

where (refer to Figure 2.19)

$$\dot{f}_g = \left[\dot{f}_{g1} \dot{f}_{g2} \dot{f}_{g3} \dot{f}_{g4}\right]^T \tag{2.33}$$

$$\dot{d}_g = \left[\dot{d}_{g1} \dot{d}_{g2} \dot{d}_{g3} \dot{d}_{g4}\right]^T \tag{2.34}$$

$$T_{cg} = [-c - s\ c\ s] \tag{2.35}$$

$$k_c = \frac{EA}{L} \tag{2.36}$$

$$T = \frac{1}{L} \begin{bmatrix} s^2 & -sc & -s^2 & sc \\ & c^2 & sc & -c^2 \\ sym. & & s^2 & -sc \\ & & & c^2 \end{bmatrix} \tag{2.37}$$

in which $s = \sin\theta$ and $c = \cos\theta$. θ is the inclination angle of the displaced element chord.

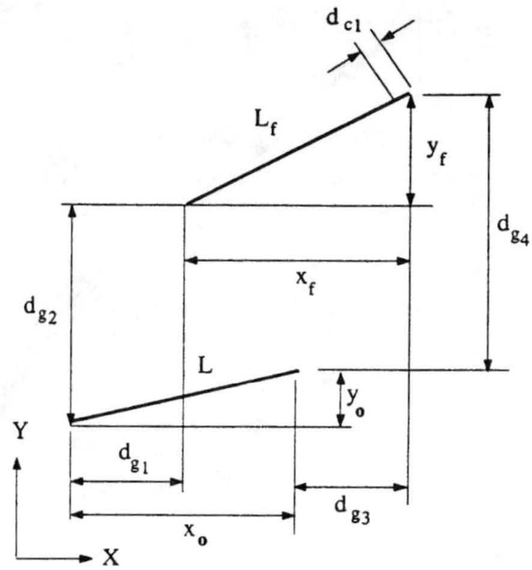

Figure 2.19 Kinematic relationship between local and global displacements of a truss element.

2.4.3 *Cross-Section Plastic Strength*

Based on the AISC-LRFD bilinear interaction equations [5] the cross-section plastic strength may be expressed

$$\frac{P}{P_y} + \frac{8}{9}\frac{M}{M_p} = 1.0 \quad \text{for} \quad \frac{P}{P_y} \geq 0.2 \tag{2.38a}$$

$$\frac{P}{2P_y} + \frac{M}{M_p} = 1.0 \quad \text{for} \quad \frac{P}{P_y} < 0.2 \tag{2.38b}$$

where

P_y = squash load
M_p = plastic moment capacity
P = second-order axial force
M = second-order bending moment

These AISC-LRFD cross-section plastic strength curves are used for both strong- and weak-axis bendings. The strength curves are plotted in Figure 2.20 for comparison with "exact" analytical solutions. While the equations provide a reasonable lower-bound to strong-axis strength, they are conservative to some degree for the weak-axis strength. Regardless, the identical plastic strength curves are used for strong and weak-axis strengths in this method because the bending stiffness of the

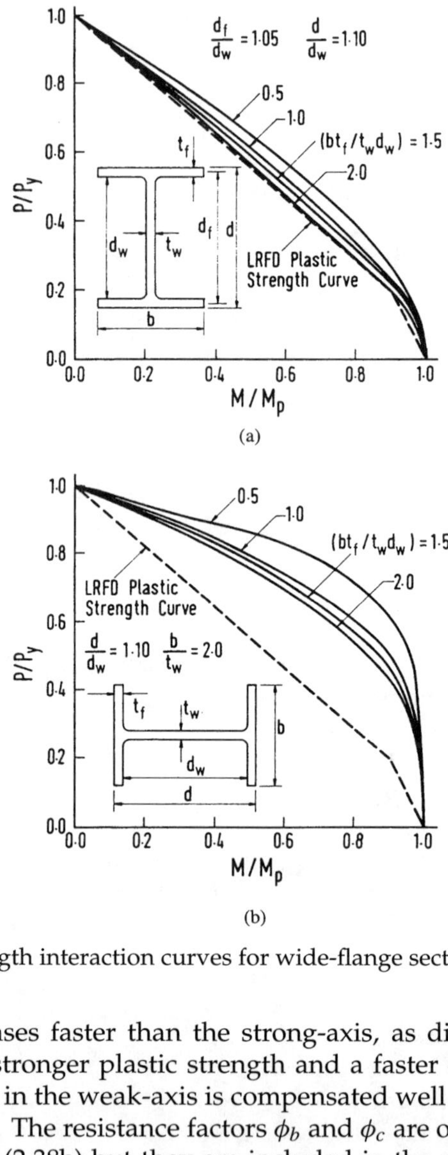

Figure 2.20 Strength interaction curves for wide-flange sections.

weak-axis decreases faster than the strong-axis, as discussed in Section 2.4.5. This stronger plastic strength and a faster deterioration of bending stiffness in the weak-axis is compensated well and this is verified in Chapter 3. The resistance factors ϕ_b and ϕ_c are omitted in Equations (2.38a) and (2.38b) but they are included in the complete design procedures given in Chapter 4 and 6.

2.4.4 *Modification of Element Stiffness for the Presence of Plastic Hinges*

If the state of forces at any cross-section equals or exceeds the plastic section capacity, a plastic hinge is formed, and a slope discontinuity at this location occurs. Therefore, the element force-displacement relationships need to be modified to reflect the change in element behavior due to formation of plastic hinges at the element ends.

If a plastic hinge forms in an element at end A, the incremental force-displacement relationship from Equation (2.17) may be written as

$$\begin{Bmatrix} \Delta M_{pcA} \\ \dot{M}_B \\ \dot{P} \end{Bmatrix} = \frac{EI}{L} \begin{bmatrix} S_1 & S_2 & 0 \\ S_2 & S_1 & 0 \\ 0 & 0 & A/I \end{bmatrix} \begin{Bmatrix} \dot{\theta}_A \\ \dot{\theta}_B \\ \dot{e} \end{Bmatrix} \tag{2.39}$$

where ΔM_{pcA} is the change in plastic moment capacity at the end A as P changes. $\dot{\theta}_A$ can be solved from the first row of Equation (2.39) as:

$$\dot{\theta}_A = \frac{L \Delta M_{pcA}/EI - S_2 \dot{\theta}_B}{S_1} \tag{2.40}$$

Back-substituting Equation (2.40) into the second and third row of Equation (2.39), the modified element force-displacement relationship can be obtained:

$$\begin{Bmatrix} \dot{M}_A \\ \dot{M}_B \\ \dot{P} \end{Bmatrix} = \frac{EI}{L} \begin{bmatrix} 0 & 0 & 0 \\ 0 & (S_1 - S_2^2/S_1) & 0 \\ 0 & 0 & A/I \end{bmatrix} \begin{Bmatrix} \dot{\theta}_A \\ \dot{\theta}_B \\ \dot{e} \end{Bmatrix} + \begin{Bmatrix} 1 \\ S_2/S_1 \\ 0 \end{Bmatrix} \Delta M_{pcA} \tag{2.41}$$

Equation (2.41) represents the modified incremental force-displacement relationship of a frame element with a plastic hinge formed at end A. If a plastic hinge is formed at end B, a similar approach can be followed. The corresponding basic incremental force-displacement relationship is

$$\begin{Bmatrix} \dot{M}_A \\ \dot{M}_B \\ \dot{e} \end{Bmatrix} = \frac{EI}{L} \begin{bmatrix} (S_1 - S_2^2/S_1) & 0 & 0 \\ 0 & 0 & 0 \\ 0 & 0 & A/I \end{bmatrix} \begin{Bmatrix} \dot{\theta}_A \\ \dot{\theta}_B \\ \dot{P} \end{Bmatrix} + \begin{Bmatrix} S_2/S_1 \\ 1 \\ 0 \end{Bmatrix} \Delta M_{pcB} \tag{2.42}$$

If plastic hinges are formed at both ends, θ_A and θ_B can be written in terms of the change in moment at the respective end of the element. The resulting matrix equation is

$$\begin{Bmatrix} \dot{M}_A \\ \dot{M}_B \\ \dot{P} \end{Bmatrix} = \frac{EI}{L} \begin{bmatrix} 0 & 0 & 0 \\ 0 & 0 & 0 \\ 0 & 0 & A/I \end{bmatrix} \begin{Bmatrix} \dot{\theta}_A \\ \dot{\theta}_B \\ \dot{e} \end{Bmatrix} + \begin{Bmatrix} \Delta M_{pcA} \\ \Delta M_{pcB} \\ 0 \end{Bmatrix} \tag{2.43}$$

where ΔM_{pcA} and ΔM_{pcB} are the change in the plastic moment capacity at the respective end of the members as P changes.

Equations (2.41) through (2.43) account for the presence of plastic hinge(s) at the element end(s). They may be written symbolically as

$$\dot{f}_c = k_{ch}\dot{d}_c + \dot{f}_{cp} \tag{2.44}$$

where k_{ch} is the modified basic tangent stiffness matrix due to the presence of plastic hinge(s). $\dot{f}_{cp}$ is an equilibrium force correction vector that comes from the change in moment capacity as P changes.

If Equation (2.20) is replaced by Equation (2.44) and the procedure in Section 2.4.2 is followed, the modified element force-displacement relationship in global coordinates may be written as

$$\dot{f}_g = \left(T_{cg}^T k_{ch} T_{cg} + T_1 M_A + T_2 M_B + T_3 P\right)\dot{d}_g + T_{cg}^T \dot{f}_{cp} \tag{2.45}$$

or

$$\dot{f}_g = K_{gh}\dot{d}_g + \dot{f}_{gp} \tag{2.46}$$

where

$$K_{gh} = T_{cg}^T k_{ch} T_{cg} + T_1 M_A + T_2 M_B + T_3 P \tag{2.47}$$

is the modified element tangent stiffness matrix, and

$$\dot{f}_{gp} = T_{cg}^T \dot{f}_{cp} \tag{2.48}$$

is the global equilibrium force correction vector.

Since the force-point movement keeps on the plastic strength surface of a member, the plastic strength of the section is not violated by the change of member forces after the full plastic strength of a cross-section is reached. Referring to Figure 2.21, after the force point reaches point Q on the plastic strength surface and the member axial force is increased, a corresponding decrease of the moment at the member results. The force point then moves from Q to R. The moment at a plastic hinge may increase from point R, but this causes the axial forces to decrease and the force point goes to Q.

2.4.5 *Tangent Modulus Model Associated with Residual Stresses*

When a concentric axial force is applied to a column, a part of cross-section will yield and the yield portion cannot carry any additional loads, leaving the elastic portion to provide additional flexural rigidity. When residual stresses are assumed to be $F_y/3$, the bending stiffness reduction in strong- and weak-axis can be written as [23]:

$$\frac{EI_e}{EI} = \sqrt{3\left(1 - \frac{P}{P_y}\right)} \quad \text{for strong axis} \tag{2.49a}$$

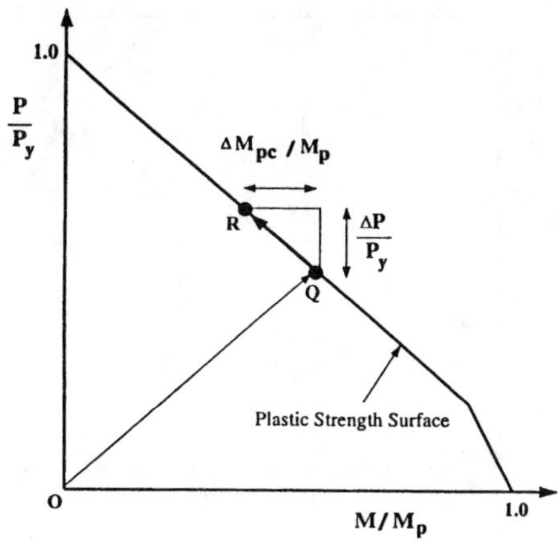

Figure 2.21 Equilibrium correction for force-point movement on the plastic strength surface.

$$\frac{EI_e}{EI} = \left[3\left(1 - \frac{P}{P_y}\right)\right]^{3/2} \quad \text{for weak axis} \tag{2.49b}$$

where

I_e = effective moment of inertia for the elastic portion of cross-section

The bending stiffness in the weak-axis degrades faster than in the strong-axis as shown in Figure 2.22. This difference results from the fact that yielding of the flange tips due to the highest compressive residual stresses causes a larger deterioration in the weak-axis strength than the strong-axis.

Herein, the tangent modulus concept is used to account for the gradual yielding due to residual stresses along the length of members under axial loads between two plastic hinges. The elastic modulus E (instead of moment of inertia I) is reduced to account for the reduction of the elastic portion of the cross-section since the reduction of elastic modulus is easier to implement than a new moment of inertia for every different section. The stiffness reduction rate in the weak- and strong-directions are different, but that is not considered and the same tangent modulus is used because rapid degradation in weak-axis strength is compensated by the stronger weak-axis plastic strength. This simplification makes the proposed methods practical. The CRC tangent modulus is a better choice than the LRFD E_t. When this model incorporates appropriate geometrical imperfections, it can approximate very well the plastic-zone solutions. From Chen and Lui [22], the CRC E_t can be written as (Fig-

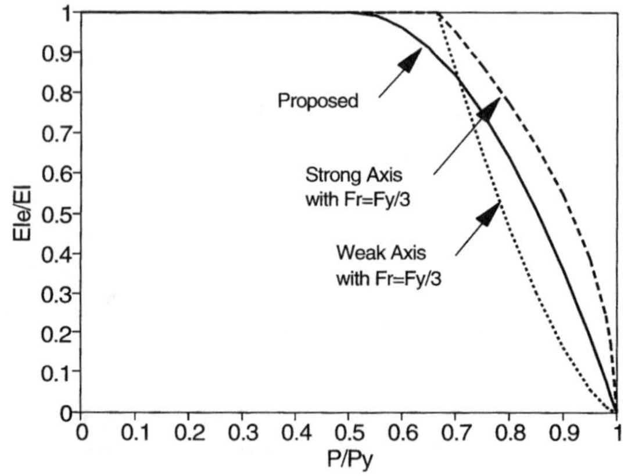

Figure 2.22 Comparison of bending stiffness degradations for strong and weak-axis, and the proposed method.

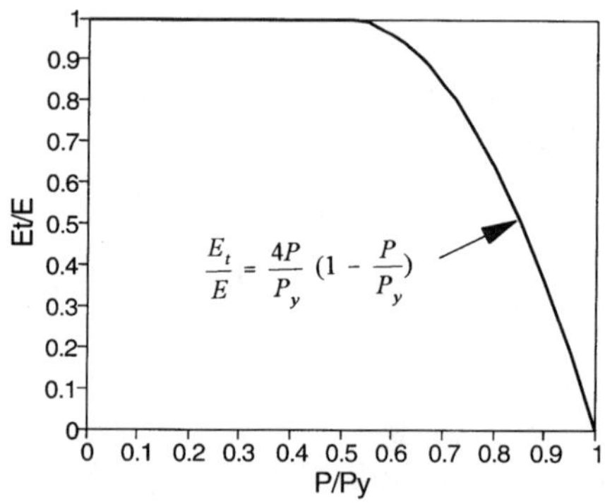

Figure 2.23 Member tangent stiffness degradation derived from the CRC column curve.

ure 2.23):

$$E_t = 1.0E \quad \text{for } P \leq 0.5P_y \tag{2.50a}$$

$$E_t = 4\frac{P}{P_y}E\left(1 - \frac{P}{P_y}\right) \quad \text{for } P > 0.5P_y \tag{2.50b}$$

The total axial displacement can be derived from Equations (2.50a) and (2.50b) as follows:

$$e = \frac{PL}{EA} \quad \text{for } P \leq 0.5P_y \tag{2.51a}$$

$$e = \frac{0.5P_yL}{EA} + \int_{0.5P_y}^{P} \frac{L}{AE_t} dP \quad \text{for } P > 0.5P_y \tag{2.51b}$$

By integrating Equations (2.51a) and (2.51b), the axial force-deformation relationship can be derived as:

$$P = P_y \frac{\varepsilon}{\varepsilon_y} \quad \text{for } P \leq 0.5P_y \tag{2.52a}$$

$$P = \frac{P_y}{1 + \exp\left(2 - \frac{4\varepsilon}{\varepsilon_y}\right)} \quad \text{for } P > 0.5P_y \tag{2.52b}$$

where $\varepsilon/\varepsilon_y = E_e/F_yL$ is the total normalized axial strain in the member. The normalized axial stress-strain relationship is plotted in Figure 2.24.

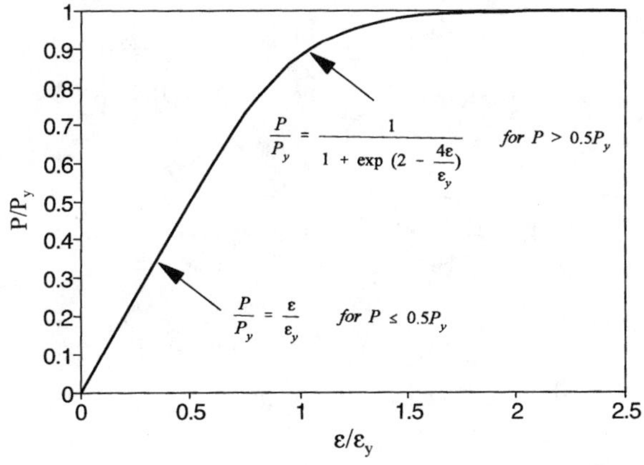

Figure 2.24 Normalized axial force-strain relationship of the CRC tangent modulus.

2.4.6 *Two-Surface Stiffness Degradation Model Associated with Flexure*

In the refined plastic hinge approach, the element stiffness is assumed to degrade parabolically after the element end forces exceed a defined initial yield function. When the plastic strength is reached, the plastic hinge is modeled as a true hinge with an applied moment, M_{pc}. The element tangent stiffness is then adjusted to account for the presence of the plastic hinge.

The force-displacement relationships, with and without plastic hinges at their ends, have been derived in Section 2.4.4, and are summarized in Figure 2.25. M_{pcA} and M_{pcB} in Figures 2.25(b)-(c) are the plastic moment

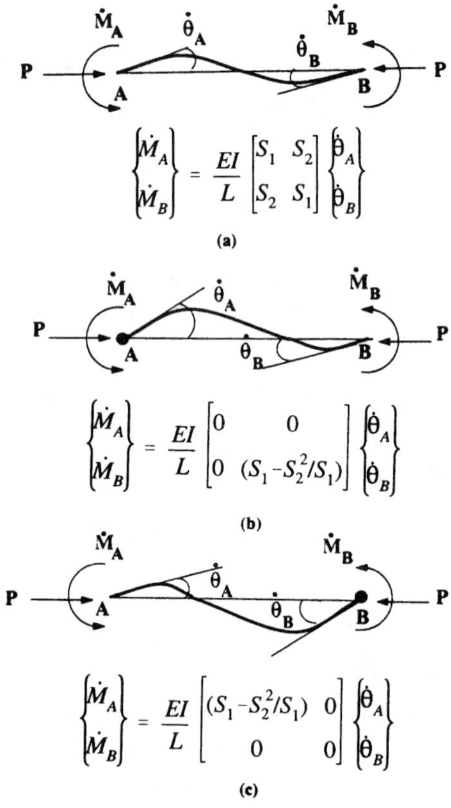

Figure 2.25 Incremental force displacement relationships of frame element.

capacity at element ends A and B. These plastic moment capacities are reduced by the effect of axial force, and their terms computed by solving Equations (2.38a) and (2.38b) for M based on the value of P at the plastic-hinge location.

Effect of Plastification at End A Only

Consider the frame elements shown in Figure 2.25(a)-(b). To represent a gradual transition from the elastic stiffness at the onset of yielding [Figure 2.25(a)] to the stiffness associated with a full plastic hinge at end A [Figure 2.23(b)], the element incremental force-displacement relationship may be written as

$$\begin{Bmatrix} \dot{M}_A \\ \dot{M}_B \end{Bmatrix} = \frac{E_t I}{L} \begin{bmatrix} \eta_A S_1 & \eta_A S_2 \\ \eta_A S_2 & \left[S_1 - \frac{S_2^2}{S_1}(1-\eta_A) \right] \end{bmatrix} \begin{Bmatrix} \dot{\theta}_A \\ \dot{\theta}_B \end{Bmatrix} \tag{2.53}$$

The terms S_1 and S_2 in this equation are the conventional beam-column stability functions, but with E_t used in place of the elastic modulus. The term η_A is a scalar parameter that allows for gradual inelastic stiffness reduction of the element associated with plastification at end A. This term is equal to 1.0 when the element is elastic (Figure 2.25(a)), and zero when a plastic hinge is formed at end A (Figure 2.25(b)). The parameter η is assumed to vary according to a prescribed function

$$\eta = 1 \quad \text{for } \alpha \leq 0.5 \tag{2.54a}$$

$$\eta = 4\,\alpha(1-\alpha) \quad \text{for } \alpha > 0.5 \tag{2.54b}$$

where α is a force-state parameter that measures the magnitude of axial force (P) and bending moment (M) at the element end. The term α is expressed as:

$$\alpha = \frac{P}{P_y} + \frac{8}{9}\frac{M}{M_p} \quad \text{for } \frac{P}{P_y} \geq \frac{2}{9}\frac{M}{M_p} \tag{2.55a}$$

$$\alpha = \frac{P}{2P_y} + \frac{M}{M_p} \quad \text{for } \frac{P}{P_y} < \frac{2}{9}\frac{M}{M_p} \tag{2.55b}$$

Initial yielding is assumed to occur based on a yield surface that has the same shape as the plastic-strength surface and with the force-state parameter denoted as $\alpha_0 = 0.5$. The plastic-strength surface and initial yield surface with $\alpha_0 = 0.5$ are shown in Figure 2.26. If the forces change so the force point moves inside or along the initial yield surface, the element is assumed to remain fully elastic with no stiffness reduction. If the force point moves beyond the initial yield surface, the element stiffness is reduced to account for the effect of plastification at the element end.

Figure 2.27 shows the relationship between the stiffness reduction factor η, and the force state parameter α, modeled by Equations (2.54a) and (2.54b). A gradual stiffness degradation is observed when the force state moves from the initial yield surface (α_0) toward the plastic strength surface ($\alpha = 1.0$). The expression of η in Equations (2.54a) and (2.54b) is obtained based on a calibration with "exact" plastic-zone solutions

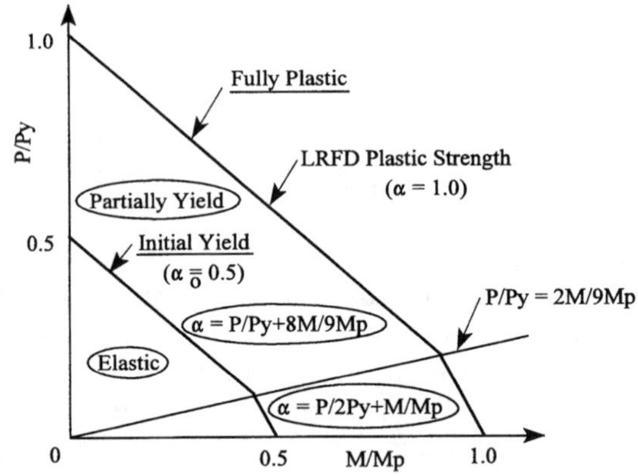

Figure 2.26 Smooth stiffness degradation for a softening plastic hinge based on LRFD sectional strength curve.

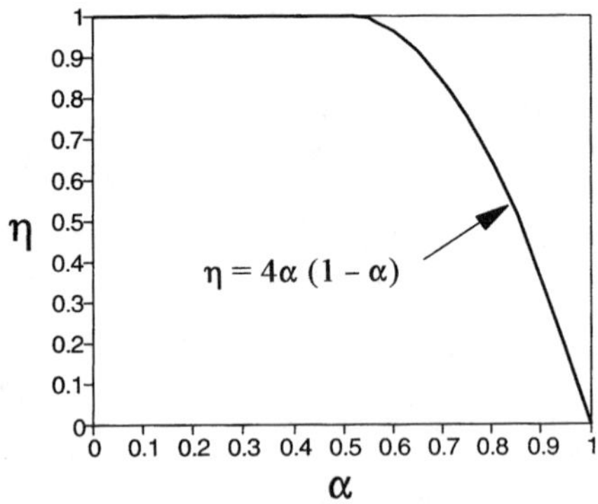

Figure 2.27 Parabolic plastic hinge stiffness degradation function with $\alpha_0 = 0.5$ based on LRFD sectional strength equation.

of simple portal frames and beam-column subassemblages [54] [58]. It is important to note that only a simple relationship for η is required to "adequately" describe the degradation in stiffness associated with distributed plasticity effects. Although more complicated expressions for η can be developed and proposed, a simple expression for η is needed for keeping the analysis model simple and straightforward.

Effect of Plastification at End B Only

To represent a gradual transition from the elastic beam-column stiffness [Figure 2.25(a)] to the stiffness associated with a plastic hinge at end *B*

(Figure 2.23(c)), the incremental stiffness equations may be written as

$$\begin{Bmatrix} \dot{M}_A \\ \dot{M}_B \end{Bmatrix} = \frac{E_t I}{L} \begin{bmatrix} \left[S_1 - \frac{S_2^2}{S_1}(1-\eta_B) \right] & \eta_B S_2 \\ \eta_B S_2 & \eta_B S_1 \end{bmatrix} \begin{Bmatrix} \dot{\theta}_A \\ \dot{\theta}_B \end{Bmatrix} \tag{2.56}$$

Equations (2.53) and (2.56) represent the gradual plastification effects at one end of the beam-column element only.

Effect of Plastification at Both Ends

For the general case where plastic hinges may form at both ends of an element, the incremental force-displacement relationship may be written as

$$\begin{Bmatrix} \dot{M}_A \\ \dot{M}_B \end{Bmatrix} = \frac{E_t I}{L} \begin{bmatrix} \eta_A \left[S_1 - \frac{S_2^2}{S_1}(1-\eta_B) \right] & \eta_A \eta_B S_2 \\ \eta_A \eta_B S_2 & \eta_B \left[S_1 - \frac{S_2^2}{S_1}(1-\eta_A) \right] \end{bmatrix} \begin{Bmatrix} \dot{\theta}_A \\ \dot{\theta}_B \end{Bmatrix} \tag{2.57}$$

It should be noted that when:

1. $1 > \eta_A > 0$ and $1 > \eta_B > 0$, Equation (2.57) accounts for the effect of partial plastification at both ends of the element.
2. $\eta_A = \eta_B = 1$, both ends of the beam-column element are fully elastic. Equation (2.57) reduces to the stiffness equations shown in Figure 2.25(a).
3. $\eta_A = 1$ and $1 > \eta_B > 0$, end A is elastic and end B is partially yielded. When $\eta_B = 0$, Equation (2.57) then reduces to the stiffness equations shown in Figure 2.25(c).
4. $\eta_B = 1$ and $1 > \eta_A > 0$, end B is elastic and end A is partially yielded. When $\eta_A = 0$, Equation (2.57) then reduces to the form shown in Figure 2.25(b).

In a similar formulation proposed by King et al. [44], it is assumed that once a plastic hinge forms, the moment at the hinge will not change as the axial force is increased. However, as discussed in Section 2.4.4, neglecting force point movement on the plastic strength surface may violate the cross-section plastic strength, which is not acceptable for maximum strength analysis of members. In the refined plastic hinge formulation, the force-point movement on the plastic strength surface is handled in the same way as in the elastic-plastic hinge formulation in Section 2.4.4.

2.5 Analysis of Semi-Rigid Frames

In conventional analysis and design of a steel frame, the beams to columns connections are assumed fully rigid or ideally pinned. The rigid joint assumption implies that slope continuity exists between adjoining members, and that full gravity moment is transferred from the beam to the column. On the other hand, the assumption of ideally pinned connections implies that beams will behave as simply supported members and that the columns will carry no gravity moments from the beams. Although the assumption of fully rigid and ideally pinned connections drastically simplifies analysis and design procedures, the validity may be questionable for cases in which connections are between rigid and pinned cases. In fact, as is evident from experimental observations, all connections used in practice possess stiffnesses that fall between the extreme cases of fully rigid and ideally pinned.

The LRFD Specifications [4] [5] designate two types of constructions in their provisions: Type FR (fully restrained) and Type PR (partially restrained). Type FR corresponds to ASD Type 1. Type PR includes ASD Types 2 and 3. If Type PR construction is used, the effect of connection flexibility must be taken into account in the analysis and design procedures.

The influence of partially restrained connections on structural response not only changes the moment distribution in the beams and columns, but also increases the frame drift and thus the $P - \Delta$ effect. In this section, the real behavior of beam-column connections, as distinct from the ideal assumptions, is discussed. The role flexible connections play in the analysis of steel frames is explained. Common connection types are presented with their general behavior and characteristics. Several empirical equations based on curve-fit techniques that model the behavior of connections have been developed. Herein, the "three-parameter power model" proposed by Kishi and Chen [45] is adopted for practical use. The practical procedure for determining the three-parameters is presented for various connections, and the moment-rotation relationship is given.

When a moment is applied to a connection in Figure 2.13, the connection rotates according to the curves shown in Figure 2.14; however, if the direction of moment is reversed, the connection will unload and follow a different path which is almost linear with a slope equal to the initial slope of the $M - \theta_r$ curve. This phenomenon is depicted in Figure 2.28. Because of this behavior, identical connections at the ends of a beam may not always behave identically. This can best be illustrated by means of the simple portal frame shown in Figure 2.29, subject only to gravity loads. In this case, the connections experience a moment of M_g on the $M - \theta_r$ curve (Figure 2.29). If a lateral force is applied to the frame,

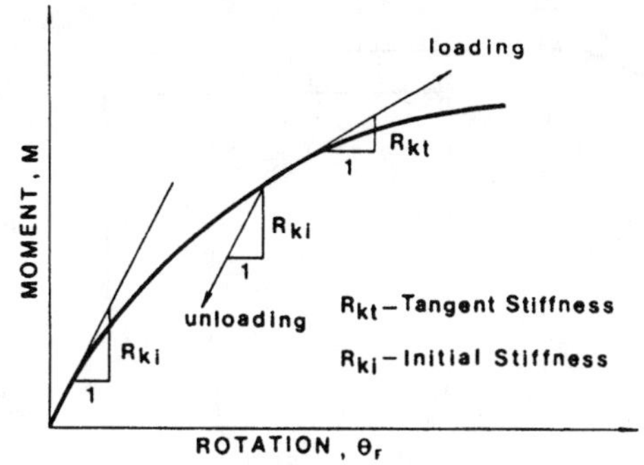

Figure 2.28 Typical moment-rotation behavior of semi-rigid connections.

the leeward connection will continue to hold higher loads but the windward connection will unload (Figure 2.29). As a result, the apparent stiffnesses of the connections under the action of this lateral force will be different. Depending on the magnitude of M_g and the characteristic of the $M-\theta_r$ curve of a particular connection, the difference in stiffnesses of the two connections may be substantial. For the extreme case when the gravity load moment equals the ultimate moment capacity of the connections, the leeward connection will behave virtually like a pinned connection with a nearly constant ultimate moment capacity, whereas the windward connection will respond like a linear elastic connection with a stiffness equal to its initial stiffness. This loading/unloading phenomenon of the connection must be properly modeled to predict reliably the response of a frame.

2.5.1 *Types of Semi-Rigid Connections*

Single Web-Angle and Single Plate Connections

A single web-angle connection consists of an angle either bolted or welded to both the column flange and the beam web [Figure 2.30(a)]. The most popular design is an angle shop welded to the column with the beam bolted to the angle on the site. A single plate connection, where a plate is used in place of the clip angle, requires less material than a single angle connection and minimizes the influence of the eccentricity. These connections have little moment-rotation rigidity and are very flexible.

Double Web-Angle Connections

Double web-angle connections consist of two angles either welded or bolted to both the column and the beam web as shown in Figure 2.30(b). Experiments have indicated that these connections are capable of developing as much as 20% of the full fixed-end moment of the beam at working loads especially for deep connections [53]. On the other hand,

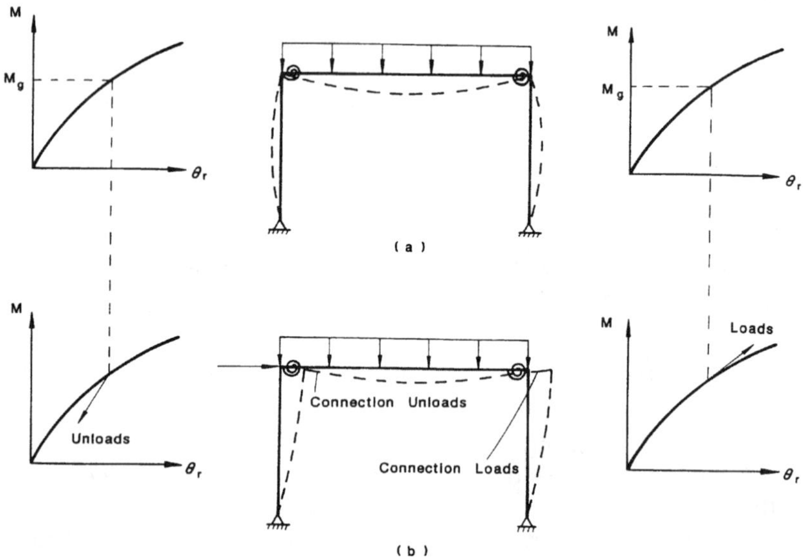

Figure 2.29 Illustrative example of loading/unloading characteristics of connections in a frame.

the AISC-ASD Specifications [6] consider this type of connection as a Type 2 construction connection (simple connection or shear connection).

Header Plate Connection

A header plate connection consists of an end plate (whose length is less than the depth of the beam), welded to the beam web and bolted to the column as shown in Figure 2.30(c). The moment-rotation characteristics of these connections are similar to those of double web-angle connections and belong to Type 2 framing of the AISC-ASD Specifications [6]. A header plate connection is used mainly to transfer the reaction of the beam to the column.

Top- and Seat-Angle Connection

A typical top- and seat-angle connection is shown in Figure 2.30(d). The AISC-ASD Specifications [6] describe the top- and seat-angle connection as follows:

1. The seat-angle transfers only vertical reaction and should not give significantly restraining moments on the end of the beam.
2. The top-angle is merely for lateral stability and is not considered to carry any gravity loads.

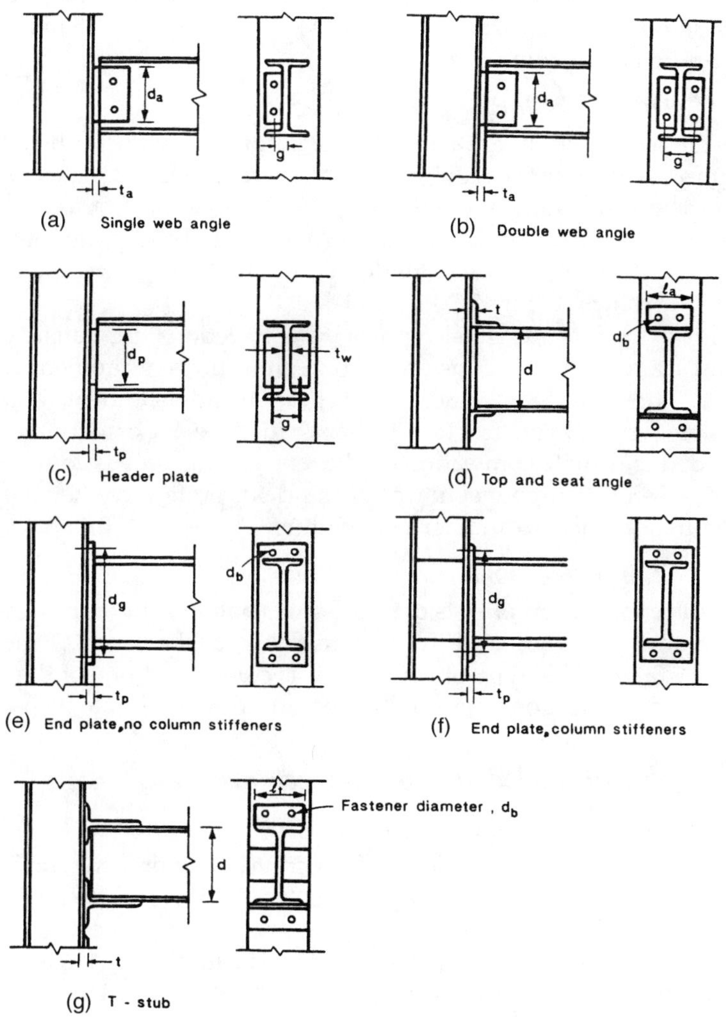

Figure 2.30 Common types of beam-column connections.

However, according to the experimental results, these connections will be able to resist some end moment of the beam.

Top- and Seat-Angle with Double Web-Angle

This type of connection is a combination of a top- and seat-angle connection and a double web-angle connection. This type of connection is considered as Type 3 framing of the AISC-ASD Specifications [6] (i.e., semi-rigid connection).

Extended/Flush End-Plate Connections

End-plate connections are a popular means of connecting the beam to the column when a moment connection is required. The end plate is welded to the beam end along both the flanges and web in the fabricator's shop and bolted to the column in the field. An extended end-plate connection may take 1 of 2 forms: an extended end-plate only on the tension side or on both sides [Figure 2.30(e and f)].

Both extended and flush end-plate connections are considered as Type 1 framing connections (i.e., rigid connections or moment connections). AISC-LRFD Specifications [4] [5] considers end-plate connections as Type FR construction rather than Type PR construction. An extended end-plate connection on the tension side only is common. An extended end-plate connection on both sides is preferred when the frame structure is subject to an alternate loading.

T-Stub Connections

A T-stub connection consists of two T-stubs bolted to both the beam and the column at the top and the bottom flange of the beam as shown in Figure 2.30(g). This type of connection is considered one of the stiffest semi-rigid connections, especially when used with double web-angles.

2.5.2 Practical Modeling of Connections

Various Connection Modelings

A number of analytical studies of connection behavior using the finite element method have been reported thus far [26] [50] [51] [69]. However, these approaches are unacceptable for practical use because cumbersome calculations are required to consider the material and geometrical nonlinearities. The most common way to describe the $M - \theta_r$ curve is to curve-fit the experimental data with simple expressions. Several analytical models have been developed to represent connection flexibility using available experimental test data.

Early models used the initial connection stiffness as the key parameter in a linear $M - \theta_r$ model [60] [65] [72]. Although the linear model is very easy to use, it has one serious disadvantage. It is suitable for only a small range of the initial relative rotation. A closer approximation of true connection behavior can be obtained by using either a bilinear model [73] or a piecewise linear model. The piecewise linear model is composed of a series of straight line segments. In these models, the abrupt change in connection stiffness at transition points makes their practical use difficult. Jones et al. [34] [35] proposed a cubic-B-spline model to obtain a more suitable function; however, this model requires a large number of sampling data during the formulation process. Frye and Morris [30] have reported a polynomial model to evaluate the behavior of several types of connections. In this model, the $M - \theta_r$ behavior is represented by an odd power polynomial.

Lui and Chen (see [22]) used an exponential function to curve-fit the experimental $M - \theta_r$ data. This model is a good representation of the monotonic nonlinear connection behavior; however, if there are some sharp changes in slopes in the $M - \theta_r$ curve, this model does not adequately represent it [77]. Kishi and Chen [46] refined the Lui-Chen exponential model to accommodate any sharp changes in slopes in the $M - \theta_r$ (modified exponential model). The Kishi-Chen model can be used in lieu of the $M - \theta_r$ of experimental data. Other exponential models (Yee-Melchers model [81] and Wu-Chen model [78]) have been reported. The former is a four-parameter model using the initial connection stiffness, the strain-hardening connection stiffness, plastic moment capacity, and constant. The latter is the three-parameter model which is composed of initial connection stiffness, ultimate moment capacity, and shape parameter.

Models using power functions, namely, "power models", have also been reported. Colson and Louveau [24] and Kishi and Chen [45] proposed similar models independently using three-parameters: (1) initial connection stiffness, (2) ultimate moment capacity, and (3) shape parameter. Since these models use only three-parameters, they are not as accurate as the cubic B-spline and/or the modified exponential models. And, they cannot estimate the $M - \theta_r$ curve which does not flatten out near the final loading. However, amount of data required for this model is drastically reduced. Ang and Morris [7] used a standardized Ramberg-Osgood function [70] in the power form. The Ang-Morris model is composed of four parameters. Richard et al. [72] analytically described the $M - \theta_r$ curve of single plate connections with a nondimensional power model. The procedure used to establish this model was based on a nonlinear finite element method in which nonlinear behavior of the bolts and connected plates was modeled by using a force-deformation relationship obtained from single shear bolt tests.

Herein, the three-parameter power model proposed by Kishi and Chen [45] is adopted as practical connection modeling among various connection modelings.

Three-Parameter Power Model

The Kishi-Chen power model contains three-parameters: (1) initial connection stiffness R_{ki}, (2) ultimate connection moment capacity M_u, and (3) shape parameter n. The power model may be written as (Figure 2.31):

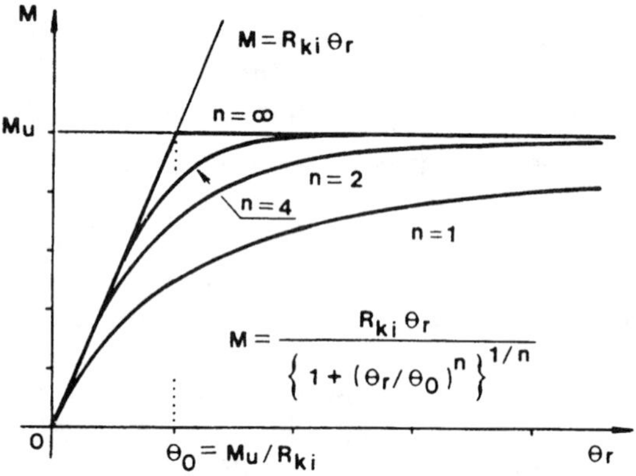

Figure 2.31 Moment-rotation behavior of three-parameter model.

$$m = \frac{\theta}{(1+\theta^n)^{1/n}} \quad \text{for } \theta > 0,\ m > 0 \tag{2.58}$$

where

m = M/M_u
θ = θ_r/θ_o
θ_o = reference plastic rotation, ($= M_u/R_{ki}$)
M_u = ultimate moment capacity of the connection
R_{ki} = initial connection stiffness
n = shape parameter

When the connection is loaded, the connection tangent stiffness R_{kt} at an arbitrary rotation θ_r can be derived by simply differentiating Equation (2.58) as:

$$R_{kt} = \frac{dM}{d\,|\theta_r|} = \frac{M_u}{\theta_o\,(1+\theta^n)^{1+1/n}} \tag{2.59}$$

When the connection is unloaded, the tangent stiffness is equal to the initial stiffness:

$$R_{kt} = \frac{dM}{d\,|\theta_r|_{\theta_r=0}} = \frac{M_u}{\theta_o} = R_{ki} \tag{2.60}$$

It is observed that a small value of the power index n makes a smooth transition curve from the initial stiffness R_{kt} to the ultimate moment

M_u. On the contrary, a large value of the index n makes the transition more abrupt. In the extreme case, when n is infinity, the curve becomes bilinear consisting of the initial stiffness R_{ki} and the ultimate moment capacity M_u.

This power model is an effective tool for designers to execute the second-order nonlinear structural analysis quickly and accurately. This is because the tangent connection stiffness R_{kt} and relative rotation θ_r can be determined directly from Equation (2.59) without iteration.

2.5.3 *Formulation of Initial Stiffness and Ultimate Moment Capacity*

An important task for a practical use of the power model is to determine the three-parameters for a given connection configuration. Herein, the practical procedures for determining the three-parameters are presented for the four types of connections with angles: single/double web-angle connections and top- and seat-angle with/without double web-angle connections. The initial stiffness and the ultimate connection moment capacity are evaluated analytically, while the shape parameter is obtained by a curve-fit technique in what follows.

Single/Double Web-Angle Connections

Initial Connection Stiffness Based on experimental results reported by Bell et al. [13] and Lewitt et al. [53] of double web-angle connections, the web-angle in the initial elastic region is assumed to behave in the following manner [47] [48]:

1. The effect of shear force on connection deformation is ignored.
2. The part of the angle connected to the column behaves in a linearly elastic manner, while the part of the angle connected to the beam behaves as a rigid body.
3. The connection deformation is small.
4. The part of the angle fastened to the column flange acts as a moderately thick plate in which the fixed support is assumed to be at the fastener-nut edge close to the beam web and the concentrated torsional moment is in equilibrium with the connection moment acting at the free edge (Figures 2.32 and 2.33).

With these assumptions and by applying simple bending torsion theory to the part of the angle connected to the column, the governing

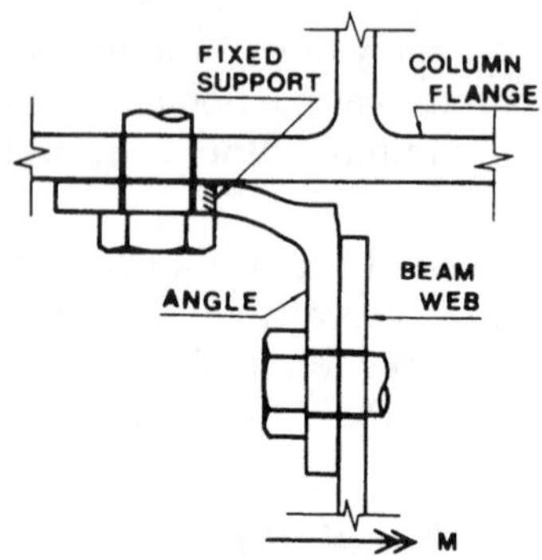

Figure 2.32 Deformation of the angle during the initial loading.

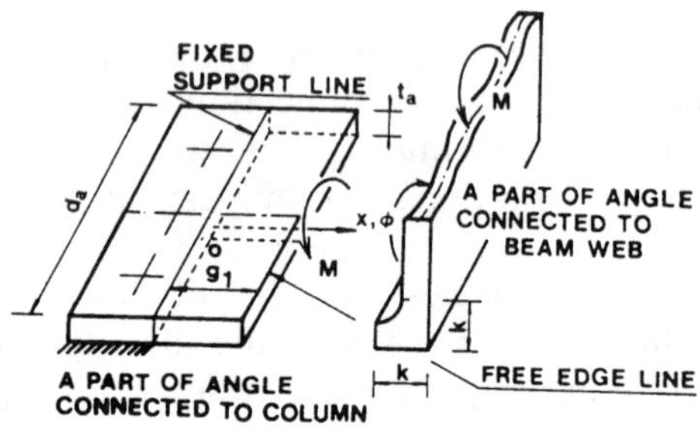

Figure 2.33 Moderate thick plate modeling of web-angle.

differential equation for the angle of rotation ϕ can be written:

$$GJ\frac{d^2\phi}{dx^2} - EC_w\frac{d^4\phi}{dx^4} = 0 \tag{2.61}$$

in which the variables for a single web-angle are the uniform torsion constant, J, the warping constant, C_w, and the Young's and shear moduli, E and G, respectively. The boundary conditions are as follows:

At the free edge ($x = g_1$)

$$\frac{d^2\phi}{dx^2} = 0 \tag{2.62}$$

$$GJ\frac{d\phi}{dx} - EC_w\frac{d^3\phi}{dx^3} = M \tag{2.63}$$

At the fixed edge ($x = 0$)

$$\frac{d\phi}{dx} = 0 \tag{2.64a}$$

$$\phi = 0 \tag{2.64b}$$

in which g_1 is the gage distance from the fixed support line to the free edge line as shown in Figure 2.33.

Solving Equation (2.61) using the above boundary conditions and setting ϕ equals θ_r at $x = g_1$, the relationship between the connection moment M and the rotation θ_r is obtained as

$$M = R_{ki}\theta_r \tag{2.65}$$

where the initial connection stiffness R_{ki} is given by

$$R_{ki} = G\frac{t_a^3}{3}\frac{\alpha \cosh(\alpha\beta)}{(\alpha\beta)\cosh(\alpha\beta) - \sinh(\alpha\beta)} \tag{2.66}$$

and α and β are defined as

$$\alpha = 4.2967 \quad \text{when Poisson's ratio is 0.3} \tag{2.67a}$$

$$\beta = g_1/d_a \;\; (d_a \text{ is the height of the web angle}) \tag{2.67b}$$

Equation (2.66) is the initial connection stiffness of a single web-angle connection. For a double web-angle connection, R_{ki} is twice the value given by Equation (2.66).

Ultimate Connection Moment Capacity Figure 2.34 shows the assumed global collapse mechanism for the web-angle when the connection reaches its full capacity. Since the distance between the two plastic hinges is of the same order of magnitude when compared with the angle thickness, we must therefore consider the effect of flexural shear interaction on the yielding of the material.

The work equation for the mechanism at an arbitrary section (y) in Figure 2.34, with the plastic moment capacity M_{py} and the shear force V_{py} per unit length, is given by

$$2M_{py}\theta = V_{py}g_y\theta \tag{2.68}$$

If Drucker's yield criterion [27] for the combined bending moment M_{py} and shear force V_{py} is used, we have

$$\left(\frac{M_{py}}{M_0}\right) + \left(\frac{V_{py}}{V_0}\right)^4 = 1 \tag{2.69}$$

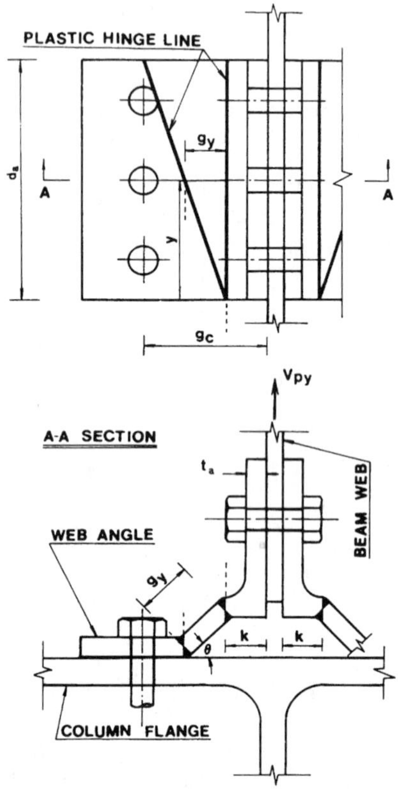

Figure 2.34 Mechanism of the web-angle at the ultimate condition.

in which M_0 and V_0 are the pure plastic bending moment capacity in the absence of shear and the shear force capacity per unit length in the absence of bending of the web-angle.

Using Equation (2.68) and the relationship between M_0 and V_0 based on Tresca's yield criterion, Equation (2.69) can be written as

$$\left(\frac{V_{py}}{V_0}\right)^4 + \frac{g_y}{t_a}\left(\frac{V_{py}}{V_0}\right) - 1 = 0 \tag{2.70}$$

From the above equation, it can be seen that V_{py} varies nonlinearly along the height of the web-angle. To simplify the analytical procedure, the variation of V_{py} is assumed to be linear. The minimum value is obtained when V_{py} is evaluated at $y = d_a$ (i.e., at the upper edge of the angle) and the maximum value is obtained when V_{py} is evaluated at $y = 0$ (i.e., at the lower edge of the angle). For subsequent discussions,

V_{pu} will designate the minimum value of V_{py} and V_0 will designate the maximum value.

With the assumption that V_{py} is linearly distributed along the angle height, the ultimate connection moment capacity can be obtained by summing moments about the lower edge of the angle.

$$M_u = \left(\frac{2V_{pu} + V_0}{6} \right) d_a^2 \tag{2.71}$$

The above equation gives the ultimate connection moment capacity for a single web-angle connection. For a double web-angle connection, the value M_u is twice that of Equation (2.71).

Top- and Seat-Angle Connections

Initial Connection Stiffness According to the experimental results reported by Hechtman and Johnston [33], Altman et al. [8], and Azizinamini et al. [9], it has been demonstrated that, at failure, this connection type rotates about a critical section of the seat-angle (point C of Figure 2.35), while the top-angle provides resistance to the bending forces at the end of the beam (Figure 2.35).

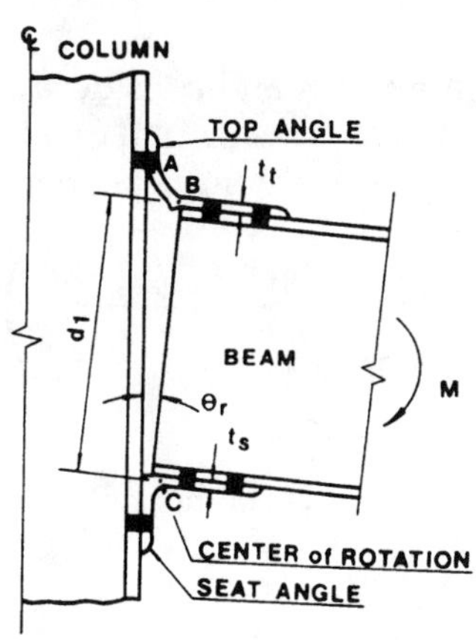

Figure 2.35 Deflected configuration of top- and seat-angles at the elastic condition.

In order to determine the initial elastic stiffness R_{ki} based on the above observed conditions, it is assumed that the connection behaves in the following manner:

1. The center of rotation for the connection is located at the leg adjacent to the compression beam flange at the end of the beam (point C in Figure 2.35).
2. The top-angle acts as a cantilever beam in which the fixed support is assumed to be at the fastener-hold edge near the beam flange in the leg adjacent to the column face as shown in Figure 2.36.
3. The resisting moment at the center of rotation is so small that it can be neglected.

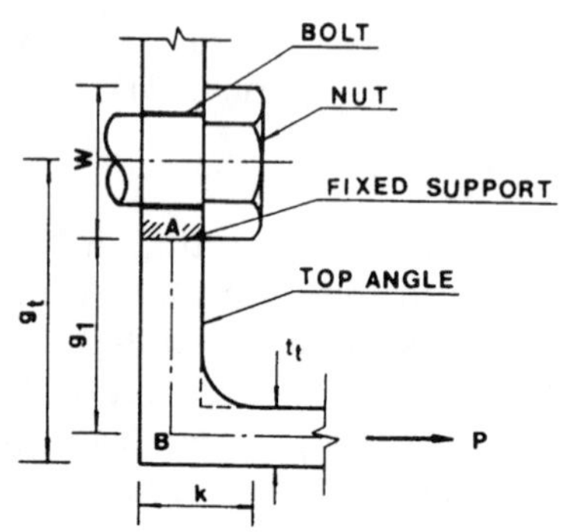

Figure 2.36 Cantilever beam model of top-angle.

Based on these assumptions and considering the shear deformation in the leg of the top-angle, the horizontal displacement of Δ of the heel of the top-angle corresponding to the beam flange force P can be shown to be [21]

$$\Delta = \frac{P g_1^3}{3(EI)} \left(1 + \frac{0.78 t_t^2}{g_1^2} \right) \tag{2.72}$$

where

EI = the bending stiffness of the angle's leg adjacent to the column face
g_1 = $g_t - D/2 - t_t/2$
D = d_b, if rivets are used as fasteners
D = W, if bolts are used as fasteners
d_b = the diameter of the fastener
W = the diameter of the nut
t_t = the thickness of the top-angle
g_t = the gage distance from the top-angle's heel to the center of the fastener holes in the leg adjacent to the column face

In deriving Equation (2.72), the coefficient of shear deformation is taken as $\kappa = 6/5$.

From kinematics and equilibrium, the relationship between the horizontal displacement Δ and the end rotation θ_r, and the connection moment M and between the beam force P acting at the top-angle, are

$$\Delta = d_1 \theta_r \tag{2.73}$$

$$M = d_1 P \tag{2.74}$$

where d_1 is the distance between the centers of the legs of the top-and bottom-angles as shown in Figure 2.35 ($= d + t_t/2 + t_s/2$, where t_t and t_s are the thickness of the top- and bottom-angles and d is the total depth of the beam section). Eliminating Δ and P from Equations (2.72) through (2.74), it can be shown that the initial connection stiffness is given by

$$R_{ki} = \frac{M}{\theta_r} = \frac{3\,(EI)}{1 + \left(0.78 t_t^2 / g_1^2\right)} \left(\frac{d_1^2}{g_1^3}\right) \tag{2.75}$$

Ultimate Connection Moment Capacity Based on experimental results reported by Altman et al. [8], it is assumed that the collapse mechanism for the top- and seat-angle connection is as shown in Figure 2.37. Since the distance between the two plastic hinges at H_1 and H_2 is short compared to the top-angle's thickness, the effect of shear force on yielding of the material is taken into account in the derivation of the ultimate connection moment capacity.

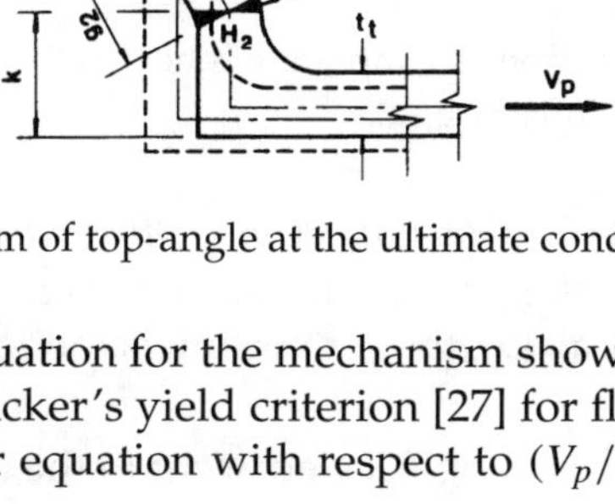

Figure 2.37 Mechanism of top-angle at the ultimate condition.

Using the work equation for the mechanism shown in Figure 2.37 in conjunction with Drucker's yield criterion [27] for flexural-shear interaction, a fourth-order equation with respect to (V_p/V_0) can be written as:

$$\left(\frac{V_p}{V_0}\right)^4 + \frac{g_2}{t_t}\left(\frac{V_p}{V_0}\right) - 1 = 0 \tag{2.76}$$

where

$V_0 = \sigma_y l_t t_t / 2$
l_t = the width of the top-angle
t_t = the thickness of the top-angle
g_2 = the distance between the plastic hinges at H_1 and H_2

Once V_p is known, the ultimate moment capacity M_u is obtained by summing moment about the center of rotation in the leg adjacent to the compression beam flange (point C in Figure 2.35).

$$M_u = M_{0s} + M_p + V_p d_2 \tag{2.77}$$

where M_{0s} is the plastic moment capacity at point C of the seat-angle shown in Figure 2.35.

$$M_{0s} = \sigma_y l_s t_s^2 / 4 \tag{2.78}$$

where

l_s = the width of the seat-angle
t_s = the thickness of the seat-angle
M_p = the plastic moment capacity at point H_2 of the top-angle (Figure 2.37)

$$M_p = V_p g_2 / 2 \tag{2.79}$$

and d_2 is given by

$$d_2 = d + t_s/2 + k \tag{2.80}$$

where

d = beam depth
k = distance from the top-angle's heel to the toe of the fillet as shown in Figure 2.37

Top- and Seat-Angle Connections with Double Web-Angles

Initial Connection Stiffness Based on the results of Altman et al. [8] and Azizinamini et al. [9], the connection is modeled as follows.

1. The beam and the angle leg adjacent to the beam act rigidly.

2. The center of rotation for the connection is located at the angle leg adjacent to the compression beam flange.
3. The top-angle acts as a cantilever beam similar to that of the top- and seat-angle connection (Figure 2.36).
4. The web-angle acts as a cantilever beam similar to the behavior of the top-angle shown in Figure 2.38.
5. The resisting moment at the center of rotation is neglected.

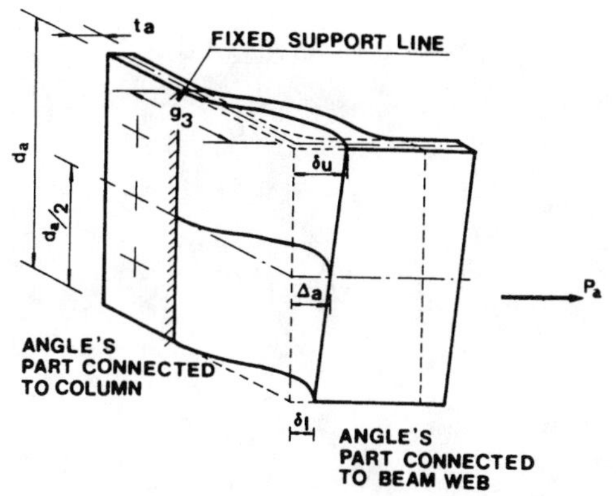

Figure 2.38 Deflected configuration for cantilever model for web-angle.

With these assumptions, the horizontal displacement Δ_t of the heel of the top-angle due to a beam flange force of P_t (similar to P in Figure 2.36) and the mean value of Δ_a of the horizontal displacement of the angle's heel corresponding to a single web-angle force of P_a (Figure 2.38) can be derived as [47]:

$$\Delta_t = \frac{P_t g_1^3}{3\,(EI_t)} \left(1 + \frac{0.78 t_t^2}{g_1^2}\right) \tag{2.81}$$

and

$$\Delta_a = \frac{P_a g_3^3}{3\,(EI_a)} \left(1 + \frac{0.78 t_a^2}{g_3^2}\right) \tag{2.82}$$

where

EI_t, EI_a = the bending stiffnesses of the legs adjacent to the column face of the top-angle and the web-angle
g_1 = as defined in Equation (2.72)
t_t = the thickness of the top-angle
g_3 = $g_c - W/2 - t_a/2$
W = the diameter of the nut
t_a = the thickness of the top-angle

Assuming that P_a acts at midheight of the web-angle and referring to the deflected configuration shown in Figure 2.39, kinematic relationships for Δ_t and Δ_a expressed in terms of the connection rotation θ_r can be written as:

$$\Delta_t = d_1\theta_r \tag{2.83}$$

$$\Delta_a = d_3\theta_r \tag{2.84}$$

where

$$d_3 = d/2 + t_s/2 \tag{2.85}$$

The variables d_1, d_3, d, and t_s are shown in Figure 2.39.

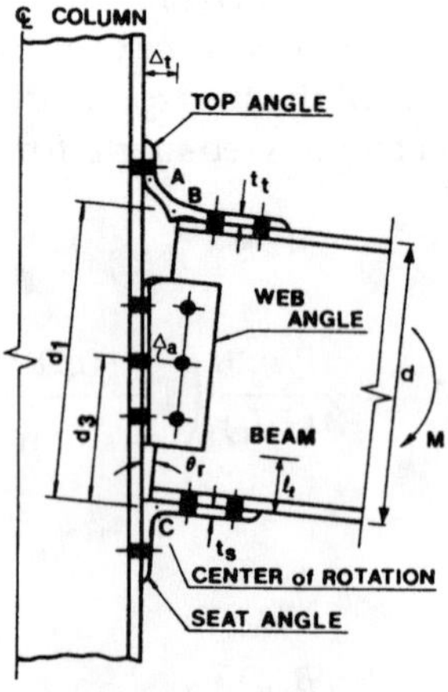

Figure 2.39 Deflected configuration of the angles connected to beam and column at the elastic condition.

Summing moments about point C of the seat-angle shown in Figure 2.39, the moment in the connection assemblage is:

$$M = P_t d_1 + 2P_a d_3 \tag{2.86}$$

Eliminating Δ_t, Δ_a, P_t, and P_a from Equations [(2.81) through (2.84), and (2.86)] it can be shown that the initial connection stiffness is:

$$R_{ki} = \frac{M}{\theta_r} = \frac{3\,(EI_t)\,d_1^2}{g_1\left(g_1^2 + 0.78t_t^2\right)} + \frac{6\,(EI_a)\,d_3^2}{g_3\left(g_3^2 + 0.78t_a^2\right)} \tag{2.87}$$

Ultimate Connection Moment Capacity Based on experiments by Altman et al. [8] and Azizinamini et al. [9], the collapse mechanism of a top- and seat-angle connection with double web-angles is assumed to be a combination of the collapse mechanisms for a double web-angle connection and a top- and seat-angle connection described above.

First, we shall consider the ultimate moment capacity of the top-angle. Referring to Figure 2.37, and assuming M_{pt} and V_{pt} are, respectively, the bending moment and shearing force acting on the plastic hinges H_1 and H_2, we can write Drucker's yielding criterion as:

$$\left(\frac{V_{pt}}{V_{0t}}\right)^4 + \frac{g_2}{t_t}\left(\frac{V_{pt}}{V_{0t}}\right) - 1 = 0 \tag{2.88}$$

where

V_{0t} = the plastic shear capacity of the top-angle leg when the effect of bending is ignored ($= \sigma_y l_t t_t/2$)
l_t = the width of the top-angle

t_t = the thickness of the top-angle
g_2 = the distance between the plastic hinges at H_1 and H_2 as shown in Figure 2.37.

Once V_{pt} is calculated from Equation (2.88), the ultimate moment capacity for the top-angle M_{pt} can be obtained from

$$M_{pt} = V_{pt}\,g_2/2 \tag{2.89}$$

On the other hand, the ultimate moment capacity of the seat-angle M_{0s} is given by

$$M_{0s} = \sigma_y l_s t_s^2/4 \tag{2.90}$$

where

l_s = the width of the seat-angle
t_s = the thickness of the seat-angle

In addition to the contribution at the top- and seat-angles, the web-angles also contribute to the ultimate moment capacity of the connection. For web-angles, if the same failure mechanism as described previously is used, it can be shown that the resultant plastic shear force V_{pa} acting on one web-angle is expressed by

$$V_{pa} = \left(\frac{V_{pu} + V_{0a}}{2} \right) d_a \tag{2.91}$$

where

V_{pu} = the value of shear at the upper edge
V_{0a} = the value of shear at the lower edge of the web-angle

Since the value of V_{pa} is for one web-angle, the value for a double web-angle connection is twice that of Equation (2.91).

Figure 2.40 shows the forces acting on the connection at its ultimate state. Using the ultimate strength equations [Equations (2.89) through (2.91)] and summing moments about the center of rotation (point C),

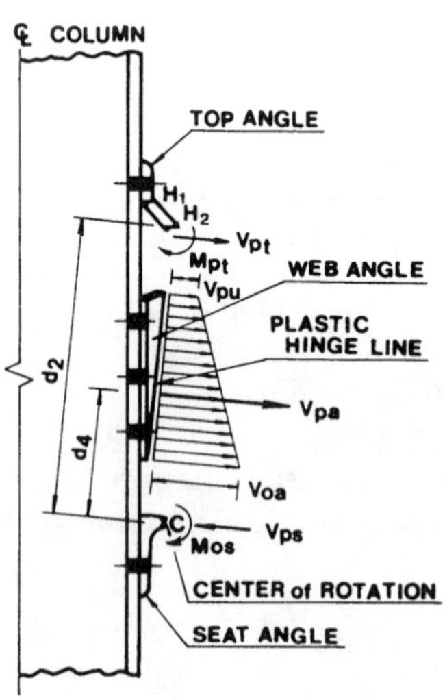

Figure 2.40 Applied forces in the ultimate state of the condition.

the ultimate connection moment capacity M_u can be written as:

$$M_u = M_{0s} + M_{pt} + V_{pt}d_2 + 2V_{pa}d_4 \tag{2.92}$$

where

$$d_4 = \frac{(2V_{pu} + V_{0a})}{3\left(V_{pu} + V_{0a}\right)}d_a + l_i + t_s/2 \tag{2.93}$$

where

l_i = the distance measured from the outer edge of the bottom beam flange to the lower edge of the web-angle

t_s = the thickness of the seat-angle as shown in Figure 2.39.

2.5.4 *Empirical Equation for Shape Parameter*

One disadvantage of the Kishi-Chen power model is that the shape parameter n is a curve-fitting parameter. Test data must be provided for its evaluation, and these data are not readily available in design process. To overcome this difficulty, Wu and Chen [78] proposed empirical formulas obtained from a statistical regression analysis of experimental curves for connections. The formulas are linear functions of the reference angle $\theta_0 = (M_u/R_{ki})$. Kishi and Chen et al. [49] proposed another empirical equation for shape parameter. While Wu and Chen's formulas deal with top- and seat-angle connections with/without double web-angles, Kishi and Chen's equations cover not only top- and seat-angle connections but also single and double web-angle connections. Herein, the Kishi and Chen equations will be used to estimate the shape parameter; its equation for each connection type is listed in Table 2.2. Figure 2.41 compares the distribution of n values of the empirical equation with the experimental test data installed in the program SCDB [46].

2.5.5 *Practical Estimation of Three-Parameters Using Computer Program*

Among several connection types, single web/double web-angle and header plate connections may be considered as hinged since they are very flexible as shown in Figure 2.14. On the contrary, extended end plate and flush end plate connections may be considered rigid connections. For practical purposes, any top and seat-angle with/without web-angles may be regarded as a semi-rigid connection.

One difficulty in determining the three-parameters is the need for numerical iteration, especially to estimate the ultimate moment, M_u. A set of nomographs was proposed by Kishi et al. [49] to overcome the

Table 2.2 Empirical Equations for Shape Parameter n [49]

Connection type	n	
Single web-angle connection	$0.520 \log_{10} \theta_o + 2.291$	for $\log_{10} \theta_o > -3.073$
	0.695	for $\log_{10} \theta_o < -3.073$
Double web-angle connection	$1.322 \log_{10} \theta_o + 3.952$	for $\log_{10} \theta_o > -2.582$
	0.573	for $log_{10}\theta_o < -2.582$
Top- and seat-angle connection	$2.003 \log_{10} \theta_o + 6.070$	for $\log_{10} \theta_o > -2.880$
	0.302	for $\log_{10} \theta_o < -2.880$
Top- and seat-angle connection with double web-angle	$1.398 \log_{10} \theta_o + 4.631$	for $\log_{10} \theta_o > -2.721$
	0.827	for $\log_{10} \theta_o < -2.721$

difficulty. Even though the purpose of these nomographs is to allow the engineer to rapidly determine the three-parameters for a given connection configuration, the nomographs require other efforts for engineers to know how to use those, and they depend on approximate values between lines on the nomographs.

Herein, one simple way to avoid difficulties described above is presented. A direct and easy estimation of the three-parameters may be achieved by use of a simple computer program listed in Table 2.3 and provided in the form of a disk with this book.

The operating procedure of the program is shown in Figure 2.42. The input data CONN.DAT may be easily generated corresponding to the input format listed in Table 2.4.

Figure 2.43 shows an example of a top- and seat-angle connection. Input data may be generated as listed in Table 2.5. The computer output may be obtained as listed in Table 2.6 by the execution of the program 3PARA.f. The three-parameters for the given configuration are determined as $R_{ki} = 128,414$ k-in/rad, $M_u = 383.5$ k-in, and $n = 1.013$. The units of the three-parameters are consistent with the units of the input data provided by users.

Figure 2.44 shows a top- and seat-angle with web-angle connection. Input data for this connection are written in Table 2.7. The computer output may be obtained as listed in Table 2.8 by the execution of the program 3PARA.f. The three-parameters for the given configuration are estimated as $R_{ki} = 146,942$ k-in/rad, $M_u = 651.8$ k-in, and $n = 1.341$. The units of the three-parameters are consistent with the units of the input data given by users.

Table 2.3 Computer Program Estimating Three-Parameters, 3PARA.f

```
C
C*******************************************************************
C
C      PROGRAM 3PARA (THREE PARAMETERS)
C
C      THIS PROGRAM COMPUTES THE THREE PARAMETERS:
C
C         1) INITIAL STIFFNESS, Rki
C         2) ULTIMATE MOMENT, Mu
C         3) SHAPE PARAMETER, n
C
C      FOR FOLLOWING SEMI-RIGID CONNECTIONS:
C
C         1) TOP- AND SEAT-ANGLE CONNECTION
C         2) TOP- AND SEAT-ANGLE WITH WEB-ANGLE CONNECTION
C
C      PROGRAMMED BY KIM, SEUNG-EOCK
C                    STRUCTURES AREA, SCHOOL OF CIVIL ENGINEERING
C                    PURDUE UNIVERSITY
C
C      LAST CHANGED ON JUNE 1996
C
C******************************************************************
C
      IMPLICIT REAL (A-H,K-Z)
      COMMON VOT,G2,TT,VOA,GY,TA
      EXTERNAL FUNC1,FUNC2
      OPEN (UNIT=5, FILE='CONN.DAT')
      OPEN (UNIT=6, FILE='CONN.OUT')
C
C     DATA INPUT AND WRITE
C
      DA=1.0
      TA=1.0
      KA=1.0
      GA=1.0
      READ (5,*) ITYPE,FY,E
      READ (5,*) LT,TT,KT,GT,W,BD
      IF (ITYPE.EQ.2) READ (5,*) DA,TA,KA,GA
      LS=LT
      TS=TT
      KS=KT
      GS=GT
      WRITE (6,10)
      WRITE (6,*) ('            CONNECTION TYPE:')
      WRITE (6,*) ('            ---------------')
      IF (ITYPE.EQ.1) WRITE (6,20)
      IF (ITYPE.EQ.2) WRITE (6,30)
      WRITE (6,*) ('            MATERIAL PROPERTY:')
      WRITE (6,*) ('            -----------------')
      WRITE (6,40) FY,E
      WRITE (6,*) ('            DIMENSIONS RELATED TO TOP ANGLE:')
      WRITE (6,*) ('            -------------------------------')
      WRITE (6,50) LT,TT,KT,GT,W,BD
      WRITE (6,*) ('            DIMENSIONS RELATED TO BOTTOM ANGLE:')
      WRITE (6,*) ('            ----------------------------------')
      WRITE (6,60) LS,TS,KS,GS,W
      IF (ITYPE.EQ.2) THEN
```

Table 2.3 Computer Program Estimating Three-Parameters, 3PARA.f (cont.)

```
         WRITE (6,*) ('          DIMENSIONS RELATED TO WEB ANGLE:')
         WRITE (6,*) ('          --------------------------------')
         WRITE (6,70) DA,TA,KA,GA
      ENDIF
      WRITE (6,*) ('       OUTPUT OF THREE PARAMETERS')
      WRITE (6,*) ('       ==========================')
C
C     CALCULATION OF INITIAL STIFFNESS, Rki
C
      LL=(BD-DA)/2.0
      EIT=E*LT*TT**3/12.0
      EIA=E*DA*TA**3/12.0
      D1=BD+TS/2.0+TT/2.0
      D3=BD/2.0+TS/2.0
      G1=GT-W/2.0-TT/2.0
      G3=GA-W/2.0-TA/2.0
      RKIT=3.0*EIT*D1**2/(G1*(G1**2+0.78*TT**2))
      RKI=RKIT
      RKIA=6.0*EIA*D3**2/(G3*(G3**2+0.78*TA**2))
      IF(ITYPE.EQ.2) RKI=RKI+RKIA
C
C     CALCULATION OF ULTIMATE MOMENT, Mu
C
      MOS=FY*LS*TS**2/4.0
      VOT=FY*LT*TT/2.0
      G2=GT-KT-W/2.0-TT/2.0
      VPTMIN=0.0
      VPTMAX=FY*LT*TT
      VPT=RTBIS(FUNC1,VPTMIN,VPTMAX)
      MPT=FY*LT*TT**2/4.0
      VOA=FY*TA/2.0
      GY=GA-KA
      VPYMIN=0.0
      VPYMAX=FY*DA*TA
      VPU=RTBIS(FUNC2,VPYMIN,VPYMAX)
      VPA=0.5*(VPU+VOA)*DA
      D2=BD+TS/2.0+KT
      D4=DA*(2*VPU+VOA)/(3.0*(VPU+VOA))+LL+TS/2.0
      MU=MOS+MPT+VPT*D2
      IF (ITYPE.EQ.2) MU=MU+2.0*VPA*D4
C
C     CALCULATION OF SHAPE PARAMETER, n
C
      THETA0=MU/RKI
      IF (ITYPE.EQ.1) N=MAX(2.003*LOG10(THETA0)+6.070,0.302)
      IF (ITYPE.EQ.2) N=MAX(1.398*LOG10(THETA0)+4.631,0.827)
C
C     WRITE THREE PARAMETERS
C
      WRITE (6,80) RKI
      WRITE (6,90) MU
      WRITE (6,100) N
10    FORMAT (///,'        INPUT DATA',/,
     &         '        ==========',/)
20    FORMAT (/,14X,'TOP- AND SEAT- ANGLE CONNECTION',//)
30    FORMAT (/,14X,'TOP- AND SEAT- ANGLE WITH WEB-ANGLE ',
     &        'CONNECTION',//)
40    FORMAT (/,14X,'Fy =',F8.4,4X,'E  =',F8.1,//,)
50    FORMAT (/,14X,'lt =',F8.4,4X,'tt =',F8.4,4X,'kt =',F8.4,//,
     &        14X,'gt =',F8.4,4X,'W  =',F8.4,4X,'d  =',F8.4,///)
60    FORMAT (/,14X,'ls =',F8.4,4X,'ts =',F8.4,4X,'ks =',F8.4,//,
     &        14X,'gs =',F8.4,4X,'W  =',F8.4,///)
70    FORMAT (/,14X,'da =',F8.4,4X,'ta =',F8.4,4X,'ka =',F8.4,//,
     &        14X,'ga =',F8.4,///)
80    FORMAT (/,14X,'INITIAL STIFFNESS, Rki = ',F10.2,/)
90    FORMAT (/,14X,'ULTIMATE MOMENT,   Mu  = ',F10.2,/)
```

Table 2.3 Computer Program Estimating Three-Parameters, 3PARA.f (cont.)

```
100     FORMAT (/,14X,'SHAPE PARAMETER,    n    = ',F10.3,/)
        STOP
        END
C
C       FUNCTIONS
C
        FUNCTION RTBIS (FUNC,X1,X2)
        COMMON VOT,G2,TT,VOA,GY,TA
        PARAMETER (JMAX=10000)
        FMID=FUNC(X2)
        F=FUNC(X1)
        IF (F.LT.0.0) THEN
          RTBIS=X1
          DX=X2-X1
        ELSE
          RTBIS=X2
          DX=X1-X2
        ENDIF
        DO 110 J=1,JMAX
          DX=0.5*DX
          XMID=RTBIS+DX
          FMID=FUNC(XMID)
          IF(FMID.LE.0.0) RTBIS=XMID
          IF(ABS(DX).LT.0.00001.OR.FMID.EQ.0.0) RETURN
110     CONTINUE
        END
C
        FUNCTION FUNC1(VPT)
          COMMON VOT,G2,TT
          FUNC1=(VPT/VOT)**4+(G2/TT)*(VPT/VOT)-1
        RETURN
        END
C
        FUNCTION FUNC2(VPY)
          COMMON VOT,G2,TT,VOA,GY,TA
          FUNC2=(VPY/VOA)**4+(GY/TA)*(VPY/VOA)-1
        RETURN
        END
```

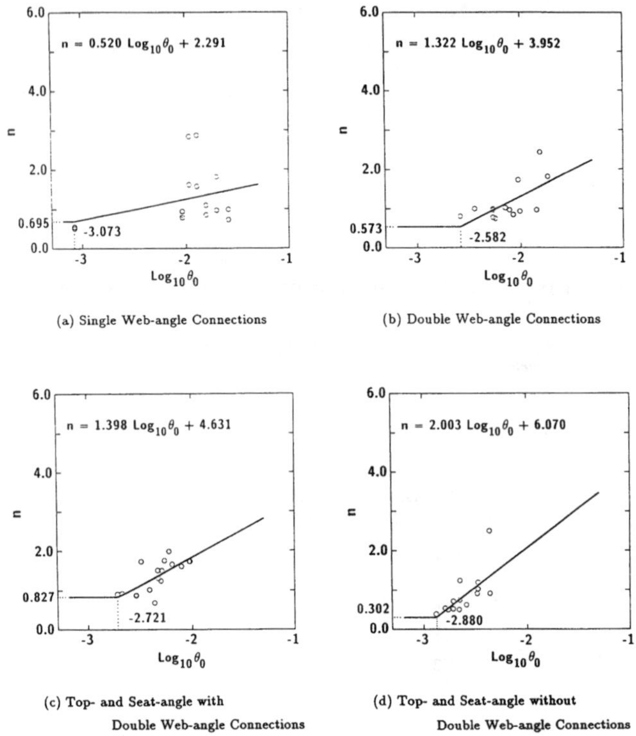

Figure 2.41 Comparison of the shape parameter n of the empirical equation with experimental test data [49].

2.5.6 *Incremental Force-Displacement Relationship Accounting for Semi-Rigid Connections*

To incorporate the effect of connection flexibility into the member stiffness relationships, the connection is modeled as a rotational spring with the moment-rotation relationship described by Equation (2.58). Consider a beam-column element subject to end moments (M_A and M_B) and axial force (P) with the connections attached at both ends as shown in Figure 2.45. The presence of the connections introduces relative incremental rotations of θ_{rA} and θ_{rB} at the A-th and B-th end of the member. Denoting R_{ktA} and R_{ktB} as the tangent stiffness of connections A and B, the relative incremental rotation between the joint and the beam end (i.e., rotational deformation of the connection) can be expressed as

$$\dot{\theta}_{rA} = \frac{\dot{M}_A}{R_{ktA}}; \qquad \dot{\theta}_{rB} = \frac{\dot{M}_B}{R_{ktB}} \tag{2.94}$$

Table 2.4 Input Format

Line	Input data	Remark
1	ITYPE F_y E	Connection type and material properties
2	l_t t_t k_t g_t W d	Top/seat-angle data
3	l_a t_a k_a g_a	Web-angle data

Description

ITYPE = Connection type (1 = top-and-seat-angle connection, 2 = with web-angle connection)
F_y = yield strength of angle
E = Young's modulus (= $29,000 ksi$)
l_t = length of top-angle
t_t = thickness of top-angle
k_t = k value of top-angle
g_t = gauge of top-angle (= 2.5 in., typical)
W = width of nut (W = 1.25 in. for 3/4D bolt, W = 1.4375 in. for 7/8D bolt)
d = depth of beam
l_a = length of web-angle
t_a = thickness of web-angle
k_a = k value of web-angle
g_a = gauge of web-angle

(1) Top- and seat-angle connections need line 1 and 2 for input data, and top- and seat-angle with web-angle connections need line 1, 2, and 3.
(2) All input data are in free format.
(3) Top- and seat-angle sizes are assumed to be the same.
(4) Bolt sizes of top-angle, seat-angle, and web-angle are assumed to be the same.

If $\dot{\theta}_A$ and $\dot{\theta}_B$ are the incremental joint rotations at A-th and B-th end of the beam-column, the incremental force-displacement equations for the beam-column element modified for the presence of end connections can be expressed as [22]:

$$\dot{M}_A = \frac{E_t I}{L}\left[S_{ii}\left(\dot{\theta}_A - \frac{\dot{M}_A}{R_{kt\,A}}\right) + S_{ij}\left(\dot{\theta}_B - \frac{\dot{M}_B}{R_{kt\,B}}\right)\right] \qquad (2.95a)$$

Table 2.5 Input Data of the Top- and Seat-Angle Connection

1	36.0	29000.0			
8.0	0.375	0.875	2.5	1.25	14.1

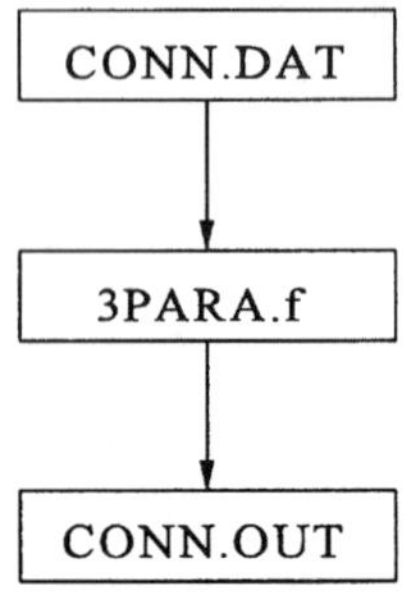

Figure 2.42 Operating procedure of computer program estimating the three-parameters.

$$\dot{M}_B = \frac{E_t I}{L}\left[S_{ij}\left(\dot{\theta}_A - \frac{\dot{M}_A}{R_{ktA}}\right) + S_{jj}\left(\dot{\theta}_B - \frac{\dot{M}_B}{R_{ktB}}\right)\right] \tag{2.95b}$$

S_{ii}, S_{ij}, and S_{jj} are defined in the following manner.

1. If elastic-plastic hinge assumption is used, then

$$S_{ii} = S_{jj} = S_1 \text{ and } S_{ij} = S_2 \tag{2.96}$$

   where S_1 and S_2 are the stability functions as defined in Section 2.4.2.

2. If gradual stiffness degradation due to plasticity effects in the elements is considered, then

$$S_{ii} = \left[S_1 - \frac{S_2^2}{S_1}(1 - \eta_B)\right]\eta_A \tag{2.97a}$$

$$S_{jj} = \left[S_1 - \frac{S_2^2}{S_1}(1 - \eta_A)\right]\eta_B \tag{2.97b}$$

$$S_{ij} = \eta_A \eta_B S_2 \tag{2.97c}$$

The elastic modulus, E, in Equations (2.95a) and (2.95b) is replaced by the tangent modulus E_t, and solving Equations (2.95a) and (2.95b) for $\dot{M}_A$ and $\dot{M}_B$ gives

$$\dot{M}_A = \frac{E_t I}{L}\left[S_{ii}^* \dot{\theta}_A + S_{ij}^* \dot{\theta}_B\right] \tag{2.98a}$$

$$\dot{M}_B = \frac{E_t I}{L}\left[S_{ij}^* \dot{\theta}_A + S_{jj}^* \dot{\theta}_B\right] \tag{2.98b}$$

Table 2.6 Output of the Top- and Seat-Angle Connection

```
INPUT DATA
==========

    CONNECTION TYPE:
    ----------------

       TOP- AND SEAT- ANGLE CONNECTION

    MATERIAL PROPERTY:
    ------------------

       Fy = 36.0000    E  = 29000.0

    DIMENSIONS RELATED TO TOP ANGLE:
    --------------------------------

       lt =  8.0000    tt =  0.3750    kt =  0.8750

       gt =  2.5000    W  =  1.2500    d  = 14.1000

    DIMENSIONS RELATED TO BOTTOM ANGLE:
    -----------------------------------

       ls =  8.0000    ts =  0.3750    ks =  0.8750

       gs =  2.5000    W  =  1.2500

  OUTPUT OF THREE PARAMETERS
  ==========================

       INITIAL STIFFNESS, Rki =  128414.30
```

where

$$S_{ii}^* = \left(S_{ii} + \frac{E_t I S_{ii} S_{jj}}{L R_{ktB}} - \frac{E_t I S_{ij}^2}{L R_{ktB}} \right) / R^* \tag{2.99a}$$

$$S_{jj}^* = \left(S_{jj} + \frac{E_t I S_{ii} S_{jj}}{L R_{ktA}} - \frac{E_t I S_{ij}^2}{L R_{ktA}} \right) / R^* \tag{2.99b}$$

$$S_{ij}^* = S_{ij} / R^* \tag{2.99c}$$

Table 2.7 Input Data of the Top- and Seat-Angle with Web-Angle Connection

2	36.0	29000.0			
8.0	0.375	0.875	2.5	1.25	14.1
8.5	0.25	0.6875	2.5		

Table 2.8 Output of the Top- and Seat-Angle with Web-Angle Connection

```
INPUT DATA
==========

    CONNECTION TYPE:
    ----------------

       TOP- AND SEAT- ANGLE WITH WEB-ANGLE CONNECTION

    MATERIAL PROPERTY:
    ------------------

       Fy = 36.0000     E  = 29000.0

    DIMENSIONS RELATED TO TOP ANGLE:
    --------------------------------

       lt =  8.0000     tt =  0.3750     kt =  0.8750

       gt =  2.5000     W  =  1.2500     d  = 14.1000


    DIMENSIONS RELATED TO BOTTOM ANGLE:
    -----------------------------------

       ls =  8.0000     ts =  0.3750     ks =  0.8750

       gs =  2.5000     W  =  1.2500


    DIMENSIONS RELATED TO WEB ANGLE:
    --------------------------------

       da =  8.5000     ta =  0.2500     ka =  0.6875

       ga =  2.5000


  OUTPUT OF THREE PARAMETERS
  ==========================

       INITIAL STIFFNESS, Rki =  146941.56

       ULTIMATE MOMENT,    Mu  =     651.79

       SHAPE PARAMETER,    n   =      1.341
```

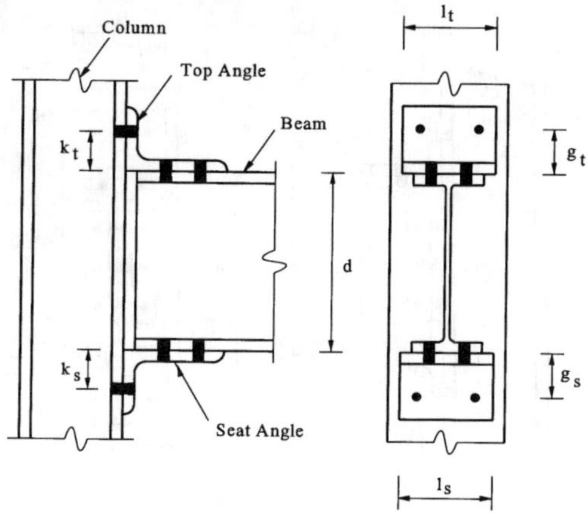

Figure 2.43 An example of top- and seat-angle connections.

and

$$R^* = \left(1 + \frac{E_t I S_{ii}}{L R_{ktA}}\right)\left(1 + \frac{E_t I S_{jj}}{L R_{ktB}}\right) - \left(\frac{E_t I}{L}\right)^2 \frac{S_{ij}^2}{R_{ktA} R_{ktB}} \tag{2.100}$$

The incremental axial force-displacement equation is

$$\dot{P} = \frac{E_t A}{L}\dot{e} \tag{2.101}$$

Finally, the incremental element force-displacement relationship can be written as:

$$\begin{Bmatrix} \dot{M}_A \\ \dot{M}_B \\ \dot{P} \end{Bmatrix} = \frac{E_t I}{L}\begin{bmatrix} S_{ii}^* & S_{ij}^* & 0 \\ S_{ij}^* & S_{jj}^* & 0 \\ 0 & 0 & A/I \end{bmatrix}\begin{Bmatrix} \dot{\theta}_A \\ \dot{\theta}_B \\ \dot{e} \end{Bmatrix} \tag{2.102}$$

The element tangent stiffness matrix expressed in Equation (2.102) accounts for both the effect of distributed plasticity and connection flexibility on the element. In addition, the connection stiffnesses at both ends of the element need not be equal. In other words, this formulation allows consideration of cases where the axial force (P) in the element is appreciable and the connections at the ends of the element may behave differently. For members subject to end loading only, Equation (2.102), formulated based on the use of one frame element per member, is suf-

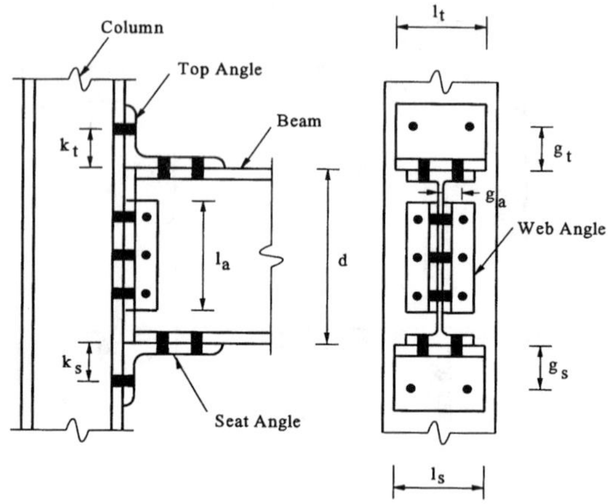

$l_t = l_s = 8$ in.
$t_t = t_s = 0.375$ in.
$k_t = k_s = 0.875$ in.
$W = 1.25$ in.
$d = 14.1$ in.
$l_a = 8.5$ in.
$t_a = 0.25$ in.
$k_a = 0.6875$ in.
$g_a = 2.5$ in.

$F_y = 36$ ksi
$E = 29{,}000$ ksi

Figure 2.44 An example of top- and seat-angle with web-angle connections.

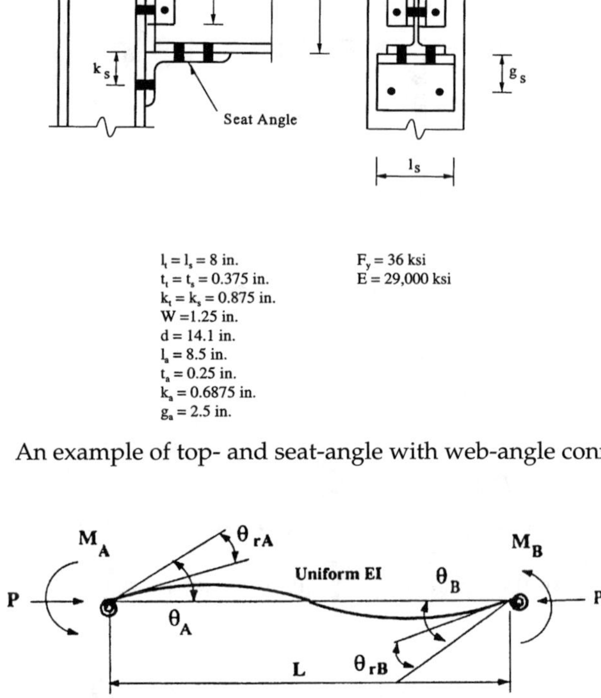

Figure 2.45 Beam-column-element with semi-rigid connections.

ficiently accurate to capture the inelasticity and connection nonlinear effects on member performance.

2.6 Geometric Imperfection Methods

Geometric imperfection modeling combined with the CRC tangent modulus model is discussed in what follows. There are three methods: (1) the explicit imperfection modeling method, (2) the equivalent notional load method, and (3) the further reduced tangent modulus method.

2.6.1 Explicit Imperfection Modeling Method

Braced Frame

The refined plastic-hinge analysis implicitly accounts for the effects of both residual stresses and spread of yielded zones. To this end, refined plastic-hinge analysis may be regarded as equivalent to the plastic-zone analysis. As a result, geometric imperfections are necessary only to consider fabrication error. For braced frames, member out-of-straightness, rather than frame out-of-plumbness, needs to be used for geometric imperfections. This is because the $P - \Delta$ effect due to the frame out-of-plumbness is diminished by braces. The ECCS [28] [29], AS [75], and CSA [15], [16] Specifications recommend an initial crookedness of column equal to $1/1{,}000$ times the column length. The AISC Code recommends the same maximum fabrication tolerance of $L_c/1,000$ for member out-of-straightness. In this study, a geometric imperfection of $L_c/1,000$ is adopted.

The ECCS [28] [29], AS [75], and CSA [15] [16] Specifications recommend the out-of-straightness varying parabolically with a maximum in-plane deflection at the mid-height. They do not, however, describe how the parabolic imperfection should be modeled in analysis. Ideally, many elements are needed to model the parabolic out-of-straightness of a beam-column member, but it is not practical. In this study, two elements with a maximum initial deflection at the mid-height of a member are found adequate for capturing the imperfection. Figure 2.46 shows the out-of-straightness modeling for a braced beam-column member. It

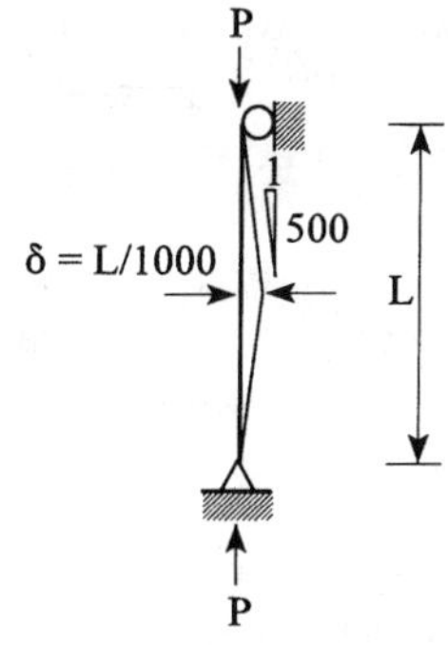

Figure 2.46 Explicit imperfection modeling of braced member.

may be observed that the out-of-plumbness is equal to $1/500$ when the half segment of the member is considered. This value is identical to that of sway frames as discussed in recent papers by Kim and Chen [37] [39]. Thus, it may be stated that the imperfection values are essentially identical for both sway and braced frames. It is noted that this explicit modeling method in braced frames requires the inconvenient imperfection modeling at the center of columns although the inconvenience is much lighter than that of the conventional LRFD method for frame design.

Unbraced Frame

Referring to the European Convention for Constructional Steelwork [28] [29], an out-of-plumbness of a column equal to 1/200 times the column height is recommended for the elastic plastic-hinge analysis. For multi-story and multi-bay frames, the geometric imperfections may be reduced to 1/200 $k_c k_s$ since all columns in buildings may not lean in the same direction. The coefficients k_c and k_s account for the situation where a large number of columns in a story and stories in a frame would reduce the total magnitude of geometric imperfections. According to the ECCS [28] [29], the member initial out-of-straightness should be modeled at the same time with the initial out-of-plumbness if the column parameter $L\sqrt{P_u/EI}$ is larger than 1.6. This may be necessary to consider residual stresses and possible member instability effects for highly compressed slender columns; however, the magnitude of the imperfection is not specified in the ECCS [28] [29].

Since plastic-zone analysis accounts for both residual stresses and the spread of yielding, only geometric imperfections for erection tolerances need be included in the analysis. The ECCS recommends the out-of-plumbness of columns equal to 1/300 $r_1 r_2$ times the column height shown in Figure 2.47 where r_1 and r_2 are factors which account for the length and number of columns, respectively. For the plastic-zone analysis, the ECCS does not specify the requirement of the initial out-of-straightness to be modeled in addition to the out-of-plumbness when the column parameter $L\sqrt{P_u/EI}$ is larger than 1.6, since the plastic-zone analysis already includes residual stresses and spread of yielding in its formulation.

Since refined-plastic hinge analysis implicitly accounts for both residual stresses and the spread of yielding, it may be considered equivalent to the plastic-zone analysis. Thus, modeling the out-of-plumbness for erection tolerances is used here without the out-of-straightness for the column, regardless of the value of the column parameter, so that the same ultimate strength can be predicted for mathematically identical braced and unbraced members verified in Section 3.4. This simplification enables us to use the proposed methods easily with consistent imperfection modeling. The Canadian Standard [15] [16] and the AISC Code of Standard Practice [5] set the limit of erection out-of-plumbness to $L_c/500$. The maximum erection tolerances in the AISC are limited to 1 in. toward the exterior of buildings and 2 in. toward the interior of buildings less than 20 stories. Considering the maximum permitted

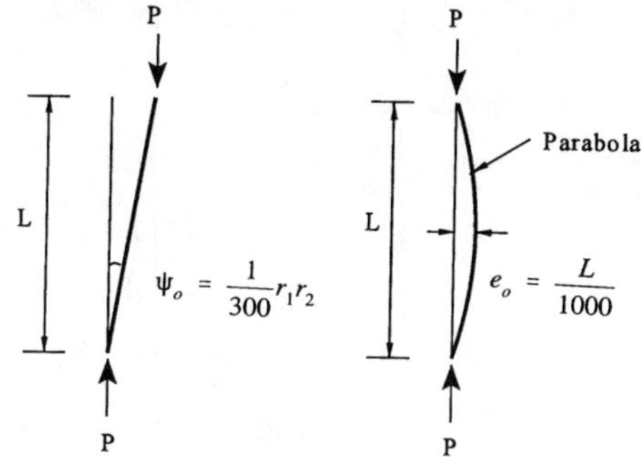

$r_1 = \sqrt{\frac{5}{L}}$ if L > 5 m (L must be stated in meters)

$r_1 = 1$ if $L \leq 5$ m

$r_2 = \frac{1}{2}(1 + \frac{1}{n})$ (n: number of columns in one story in the plane)

Figure 2.47 Geometric imperfection for plastic-zone analysis recommended by ECCS [29].

average lean of 1.5 in. in the same direction of a story, the geometric imperfection of $L_c/500$ can be used for buildings up to 6 stories with each story approximately 10 ft high. For taller buildings, this imperfection value of $L_c/500$ is conservative because the accumulated geometric imperfection calculated by 1/500 times building height is greater than the maximum permitted erection tolerance.

In this study, we shall use $L_c/500$ for the out-of-plumbness without any modification because the system strength is often governed by a weak story which has an out-of-plumbness equal to $L_c/500$ [64] and a constant imperfection has the benefit of simplicity in practical design. The explicit geometric imperfection modeling for an unbraced frame is illustrated in Figure 2.48.

2.6.2 *Equivalent Notional Load Method*

Braced Frame

The ECCS [28] [29] and the Canadian Standard [15] and [16] introduced the equivalent load concept which accounted for the geometric imperfections in an unbraced frame, but not in braced frames. The notional load approach for braced frames is also necessary to use the proposed methods for braced frames.

For braced frames, an equivalent notional load may be applied at mid-height of a column since the ends of the column are braced. An

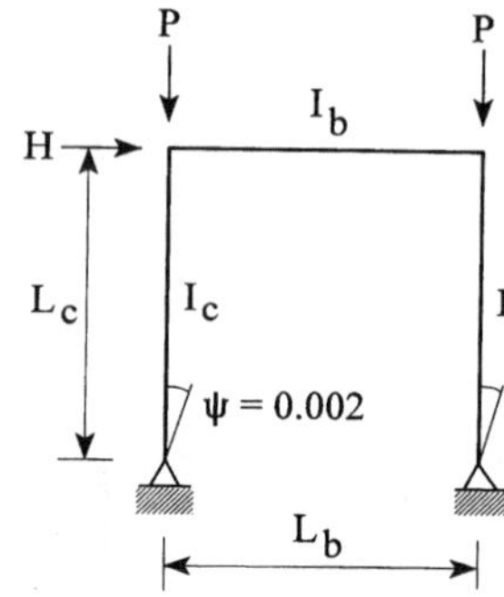

Figure 2.48 Explicit imperfection modeling of unbraced frame.

equivalent notional load factor equal to 0.004 is proposed here, and it is equivalent to the out-of-straightness of $L_c/1,000$. When the free body of the column shown in Figure 2.49 is considered, the notional load factor α results in 0.002 with respect to one half of the member length. Here, as in explicit imperfection modeling, the equivalent notional load factor is the same in concept for both sway and braced frames.

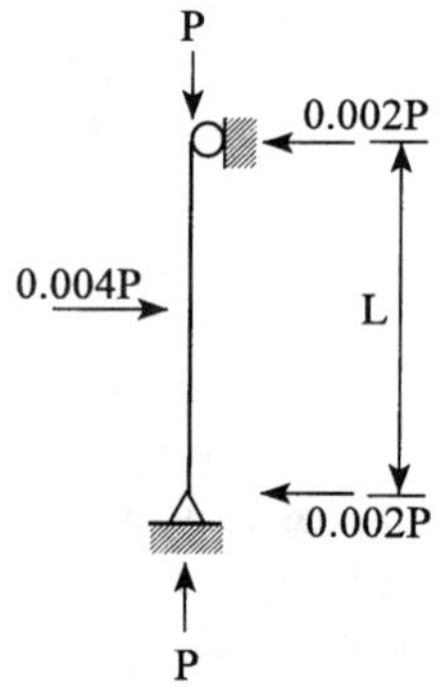

Figure 2.49 Equivalent notional load modeling for geometric imperfection of braced member.

One drawback of this method for braced frames is that it requires tedious input of notional loads at the center of each column. Another is the axial force in the columns must be known in advance to determine the notional loads before analysis, but these are often difficult to calculate for large structures subject to lateral wind loads. To avoid this difficulty, it is recommended that either the explicit imperfection modeling method or the further reduced tangent modulus method be used (see Section 2.6.3).

Unbraced Frame

The geometric imperfections of a frame may be replaced by the equivalent notional lateral loads expressed as a fraction of the gravity loads acting on the story. The ECCS [28] [29] and the new Canadian Standard [15] recommended a notional lateral load of $0.005P_u$ equivalent to the out-of-plumbness of $L/200$. Liew [54] also adopted the notional load factor of 0.005 in the second-order plastic-hinge analysis.

Herein, the equivalent notional load factor of 0.002 is used instead of 0.005 since the proposed method accounts for the residual stresses and the plastification associated with flexure. A cantilever column (shown in Figure 2.50) is used here to discuss the idea of the equivalent notional load method. The moment M_{IM} caused by the geometric imperfection is equal to $PL_c/500$ at the base of the column shown in Figure 2.50(a). The moment M_{NL} caused by equivalent notional load is equal to αPL_c in Figure 2.50(b), where α is an equivalent notional load factor. M_{IM} and M_{NL} should be identical, and thus the notional load factor α results in the value of 0.002.

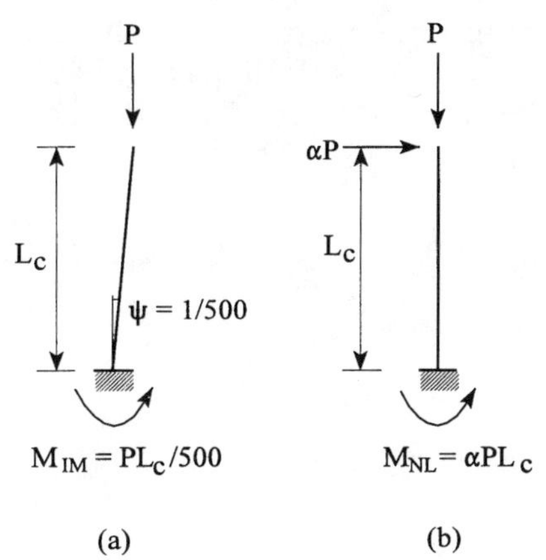

Figure 2.50 Equivalent notional load concept for modeling geometric imperfection of unbraced member.

The notional load should be applied laterally at the top of each story. For sway frames subject to combined gravity and lateral loads, the notional loads should be added to the lateral loads. Figure 2.51 shows an illustration of the equivalent notional load for a portal frame.

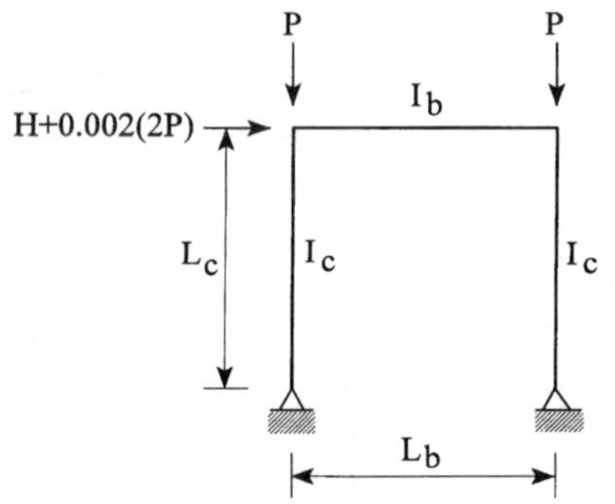

Figure 2.51 Equivalent notional load for modeling geometric imperfection of unbraced frame.

2.6.3 *Further Reduced Tangent Modulus Method*

Braced Frame

The idea of using the reduced tangent modulus concept is to further reduce the tangent modulus E_t to account for further stiffness degradation due to geometrical imperfections. The degradation of member stiffness due to geometric imperfections may be simulated by an equivalent reduction of member stiffness. This may be achieved by a further reduction of tangent modulus as [36] [37]:

$$E_t' = 4\frac{P}{P_y}\left(1 - \frac{P}{P_y}\right) E\xi_i \quad \text{for} \quad P > 0.5P_y \tag{2.103a}$$

$$E_t' = E\xi_i \quad \text{for} \quad P \leq 0.5P_y \tag{2.103b}$$

where

E_t' = reduced E_t
ξ_i = reduction factor for geometric imperfection

Herein, the reduction factor of 0.85 is used, and the further reduced tangent modulus curves for the CRC E_t with geometric imperfections are shown in Figure 2.52.

The further reduced tangent modulus concept satisfies one of the requirements for advanced analysis recommended by the SSRC task force report [74], that is: "The geometric imperfections should be accommodated implicitly within the element model. This would parallel the philosophy behind the development of most modern column strength expressions. That is, the column strength expressions in specifications such as the AISC-LRFD implicitly include the effects of residual stresses and out-of-straightness."

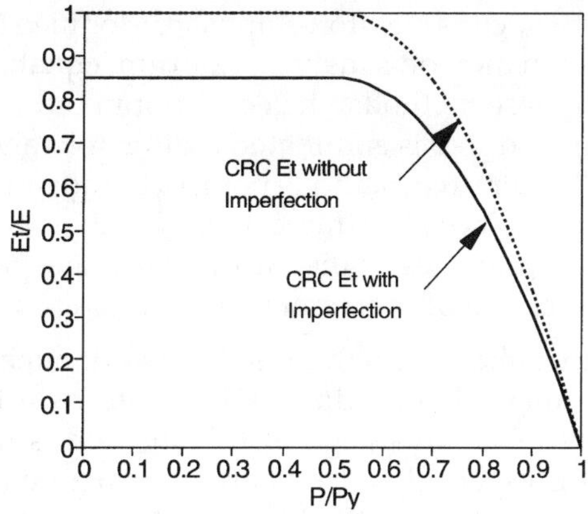

Figure 2.52 Further reduced CRC tangent modulus for members with geometric imperfections.

The advantage of this method over the other two methods is its convenience for design use because it eliminates the inconvenience of explicit imperfection modeling or equivalent notional loads. Another benefit of this method is that it does not require the determination of the direction of geometric imperfections, often difficult to determine in a large system. On the other hand, in the other two methods, the direction of geometric imperfections must be taken correctly in coincidence with the deflection direction caused by bending moments, otherwise the wrong direction of geometric imperfection in braced frames may help the bending stiffness of columns rather than reduce it.

Unbraced Frame

The idea of the further reduced tangent modulus concept may also be used in the analysis of unbraced frames. Herein, as in braced frame case, an appropriate reduction factor 0.85 to E_t can be used [39] [41]. The advantage of this approach over the other two methods is its convenience and simplicity because it completely eliminates the inconvenience of explicit imperfection modeling or the notional load input.

2.7 Numerical Implementation

The nonlinear global solution method may be divided into two subgroups:

1. Iterative methods
2. Simple incremental method

Iterative methods such as Newton-Raphson, Modified Newton-Raphson, and Quasi-Newton method satisfy equilibrium equations at specific external loads. In these methods, the equilibrium out-of-balance present following linear load step is eliminated (within tolerance) by taking corrective steps. The iterative methods possess the advantage of providing the "exact" load-displacement frame; however, they are inefficient, especially for practical purposes, in the trace of the hinge-by-hinge formation due to the requirement of the numerical iteration process.

The simple incremental method is a direct nonlinear solution technique. This numerical procedure is straightforward in concept and implementation. The advantage of this method is its computational efficiency. This is especially true when the structure is loaded into the inelastic region since tracing the hinge-by-hinge formation is required in the element stiffness formulation. For a finite increment size, this approach only approximates the nonlinear structural response, and equilibrium between the external applied loads and the internal element forces is not satisfied. To avoid this, an improved incremental method is used in this program. The applied load increment is automatically reduced to minimize the error when the change in the element stiffness parameter ($\Delta\eta$) exceeds a defined tolerance. To prevent plastic hinges from forming within a constant-stiffness load increment, load step sizes less than or equal to the specified increment magnitude are internally computed so plastic hinges form only after the load increment. Subsequent element stiffness formations account for the stiffness reduction due to the presence of the plastic hinges. For elements partially yielded at their ends, a limit is placed on the magnitude of the increment in the element end forces.

The applied load increment in the above solution procedure may be reduced for any of the following reasons:

1. Formation of new plastic hinge(s) prior to the full application of incremental loads.
2. The increment in the element nodal forces at plastic hinges is excessive.
3. Nonpositive definiteness of the structural stiffness matrix.

As the stability limit point is approached in the analysis, large step increments may overstep a limit point. Therefore, a smaller step size is used near the limit point to obtain accurate collapse displacements and second-order forces.

2.8 Summary

This chapter has described the refined-plastic hinge method for the direct design of steel frames without separate member capacity checks. The refined-plastic hinge method is developed and refined by simple modifications to the conventional elastic-plastic hinge method. The method uses the stability functions to account for second-order effects. The advantage of stability functions is that they enable designers to use only one element per beam-column. The tangent modulus concept is employed to account for gradual yielding due to residual stresses along the length of members under axial load between two plastic hinges. The two-surface stiffness degradation model is used to describe gradual plastification of cross-section due to flexure.

The flexibility of connections influences the distribution of moment in beams and columns, and the magnitude of frame drift. To incorporate this flexibility into the member stiffness relationship, the connection is modeled as a rotational spring. The three-parameter power model is employed to account for the nonlinear behavior of these connections. Practical procedures for determining these three-parameters have been presented.

The geometric imperfection models described in this chapter include: the explicit imperfection model, the equivalent notional load model, and the further reduced tangent modulus model. The further reduced tangent modulus model is the most convenient and is recommended for general use because it eliminates the inconvenience input of the explicit imperfection or the equivalent notional load models.

References

[1] Ackroyd, M. H. (1985) Design of flexibility connected steel frames, Final research report to the American Iron and Steel Institute for Project No. 333, Rensselaer Polytechnic Institute, Troy, NY.

[2] Ackroyd, M. H. and Gerstle, K. H. (1983) Strength of flexibly connected steel frames, *Eng. Struct.*, 5, 31-37.

[3] Al-Mashary, F. and Chen, W. F. (1991) Simplified second-order inelastic analysis for steel frames, *J. Inst. Struct. Eng.*, 69(23), 395-399.

[4] American Institute of Steel Construction (AISC, 1986) *Load and Resistance Factor Design, Manual of Steel Construction,* 1st ed., Chicago, IL.

[5] American Institute of Steel Construction (AISC, 1994) *Load and Resistance Factor Design, Manual of Steel Construction,* Vol. 1 and 2, 2nd ed., Chicago, IL.

[6] American Institute of Steel Construction (1989) *Allowable Stress Design and Plastic Design Specifications for Structural Steel Buildings,* Chicago, IL, 329 pp.

[7] Ang, K. M. and Morris, G. A. (1984) Analysis of three dimensional frames with flexible beam-column connections, *Can. J. Civ. Eng.,* 11, 245-254.

[8] Altman, W.G., Azizinamini, A., Bradburn, J.H., and Radziminski, J.B. (1982) Moment-Rotation Characteristics of Semi-Rigid Steel Beam-to-Column Connections, Civil Engineering Department, University of South Carolina, Columbia, SC.

[9] Azizinamini, A., Bradburn, J.H., and Radziminski, J.B. (1985) Static and Cyclic Behavior of Semi-Rigid Steel Beam-Column Connections, Structural Research Studies, Department of Civil Engineering, University of South Carolina, Columbia, SC.

[10] Barakat, M. (1988) Simplified Design Analysis of Frames with Semi-Rigid Connections, Ph.D. dissertation, School of Civil Engineering, Purdue University, West Lafayette, IN.

[11] Barakat, M. and Chen, W. F. (1991) Design analysis of semi-rigid frames, evaluation and implementation, *Eng. J.,* AISC, 28, 55-64.

[12] Barakat, M. and Chen, W. F. (1990) Practical analysis of semi-rigid frames, *Eng. J.,* AISC, 27(2), 54-68.

[13] Bell, W.G., Chesson, E., Jr., and Munse, W.H. (1958) Static Tests of Standard Riveted and Bolted Beam-to-Column Connections, University of Illinois Engineering Experiment Station, Urbana, IL.

[14] Belytschko, T. and Hsieh, B.J. (1973) Non-linear transient finite element analysis with convected coordinates, *Int. J. Num. Meth. Eng.,* 7(3), 255-271.

[15] Canadian Standard Association (1994) *Limit States Design of Steel Structures,* CAN/CSA-S16.1-M94.

[16] Canadian Standard Association (1989) *Limit States Design of Steel Structures,* CAN/CSA-S16.1-M89.

[17] Chen, W. F. (1987) *Joint Flexibility in Steel Frames,* Elsevier Science Publication Co., NY, 290pp.

[18] Chen, W. F. and Atsuta, T. (1976) *Theory of Beam-Columns, Vol. 1 In-Plane Behavior and Design,* McGraw-Hill, New York, 513 pp.

[19] Chen, W. F. and Chan, S. L. (1994) Second-order inelastic analysis of steel frames by personal computers, *J. Struc. Eng.,* 21(2), 99-106.

[20] Chen, W.F. and Kishi, N. (1989a) Semirigid steel beam-to-column connections data base and modeling. *J. Struct. Eng.,* ASCE, 115(1), 105-119.

[21] Chen, W.F. and Kishi, N. (1989b) Moment-rotation relation of top- and seat-angle connections, *Proceedings of the International Colloquium on Bolted and Special Connections,* May 15-20, Moscow,

USSR National Committee of the International Association for Bridge and Structural Engineering, Moscow.

[22] Chen, W. F. and Lui, E. M. (1991) *Stability Design of Steel Frames,* CRC Press, Boca Raton, FL, 380 pp.

[23] Chen, W. F. and Lui, E. M. (1986) *Structural Stability - Theory and Implementation,* Elsevier Science Publication Co., NY, 490 pp.

[24] Colson, A. and Louveau, J.M. (1983) Connections incidence on the inelastic behavior of steel structures, *Euromech Coll.,* 174, October.

[25] Cook, R.D., Malkus, D.S., and Plesha, M.E. (1989) *Concepts and Applications of Finite Element Analysis,* 3rd ed., John Wiley & Sons, New York, 630 pp.

[26] Driscoll, G.C. (1987) Elastic-plastic analysis of top- and seat-angle connections, *J. Constr. Steel Res.,* 8, 119-136.

[27] Drucker, D.C. (1956) The effect of shear on the plastic bending of beams, *J. Appl. Mech.,* ASME, 23(4) 509-514.

[28] ECCS (1991) *Essentials of Eurocode 3 Design Manual for Steel Structures in Building,* ECCS-Advisory Committee 5, No. 65, 60 pp.

[29] ECCS (1984) *Ultimate Limit State Calculation of Sway Frames with Rigid Joints,* Technical Committee 8 - Structural stability technical working group 8.2 - system, Publication No. 33, 20 pp.

[30] Frye, M.J. and Morris, G.A. (1975) Analysis of flexibly connected steel frames, *Can. J. Civ. Eng.,* 2, 280-291.

[31] Goto, Y., Suzuki, S., and Chen, W. F. (1991) Analysis of critical behavior of semi-rigid frames with or without load history in connections, *Int. J. Sol. Struct.,* 27(4), 467-483.

[32] Goverdhan, A. V. (1983) A Collection of Experimental Moment-rotation Curves and Evaluation of Prediction Equations for Semi-rigid Connections, Masters thesis, Vanderbilt University, Nashville, TN, 490 pp.

[33] Hetchman, R. A. and Johnston, B. G. (1947) Riveted Semi-Rigid Beam-to-Column Building Connections, Progress report No. 1, AISC Research at Lehigh University, Bethlehem, PA.

[34] Jones, S. W., Kirby, P. A., and Nethercot, D. A. (1980) Effect of semi-rigid connection on steel column strength, *J. Const. Steel Res.,* 1(1), 38-46.

[35] Jones, S. W., Kirby, P. A., and Nethercot, D. A. (1981) Modeling of semi-rigid connection behavior and its influence on steel column behavior, in *Joints in Structural Steel Work,* Howlett, J. H., Jenkins, W. M., and Stainsby, R., Eds., Pentech Press, United Kingdom, 5.73-5.78.

[36] Kim, S.E. (1996) Practical Advanced Analysis for Steel Frame Design, Ph.D. thesis, School of Civil Engineering, Purdue University, West Lafayette, IN, 271 pp.

[37] Kim, S. E. and Chen, W. F. (1996a) Practical advanced analysis for steel frame design, *ASCE Structural Congress XIV,* April 1996, Chicago, Special Proceeding Volume on "Analysis and Computation," 19-30.
[38] Kim, S. E. and Chen, W. F. (1996b) Practical advanced analysis for braced steel frame design, *J. Struct. Eng.,* ASCE 122(11).
[39] Kim, S. E. and Chen, W. F. (1996c) Practical advanced analysis for unbraced steel frame design, *J. Struct. Eng.,* ASCE 122(11).
[40] Kim, S. E. and Chen, W. F. (1996d) Practical advanced analysis for semi-rigid frame design, *AISC Eng. J.* (Accepted for publication).
[41] Kim, S. E. and Chen, W. F. (1995) Practical advanced analysis for frame design—Case study, *SSSS J.* 6(1), 61-73.
[42] King, W. S. and Chen, W. F. (1994) Practical second-order inelastic analysis of semirigid frames. *J. Struct. Eng.,* ASCE, 120(7), pp. 2156-2175.
[43] King, W. S. and Chen, W. F. (1993) LRFD analysis for semi-rigid frame design, *AISC Eng. J.,* 4th quarter, pp. 130-140.
[44] King, W. S., White, D. W., and Chen, W. F. (1991) On second-order inelastic methods for steel frame design, *J. Struct. Eng.,* ASCE 118(2), 408-428.
[45] Kish, N. and Chen, W. F. (1990) Moment-rotation relations of semi-rigid connections with angles, *J. Struct. Eng.,* ASCE 116(7), 1813-1834.
[46] Kishi, N. and Chen, W. F. (1986) Data Base of Steel Beam-to-Column Connections, Structural Engineering Report No. CE-STR-86-26, School of Civil Engineering, Purdue University, West Lafayette, IN, 653 pp.
[47] Kishi, N., Chen, W.F., Matsuoka, K.G., and Nomachi, S.G. (1987a) Moment-rotation relation of top-and seat-angle with double web-angle connections, *Proc. Of the State-of-the-Art Workshop on Connections and the Behavior, Strength and Design of Steel Structures,* Bjorhovde, R., Brozzetti, J., and Colson, A., Eds., Ecole Normale Superieure de Cachan, France, May 25-27, 1987.
[48] Kishi, N., Chen, W.F., Matsuoka, K.G., and Nomachi, S.G. (1987b) Moment-rotational relation of single/double web-angle connections, *Proc. Of the State-of-the-Art Workshop on Connections and the Behavior, Strength and Design of Steel Structures,* Bjorhovde, R., Brozzetti, J., and Colson, A., Eds., Ecole Normale Superieure de Cachan, France, May 25-27, 1987.
[49] Kishi, N., Goto, Y., Chen, W.F., and Matsuoka, K. G. (1993) Design aid of semi-rigid connections for frame analysis, *Eng. J.,* AISC, 4th quarter, pp. 90-107.

[50] Krishnamurthy, N., Huang, H.T., Jefferey, P.K., and Avery, L.K. (1979) Analytical $M - \theta$ curves for end-plate connections, *ASCE*, 105(ST1), 133-145.

[51] Kukreti, A.R., Murray, T.M., and Abolmaali, A. (1987) End-plate connection moment-rotation relationship, *J. Constr. Steel Res.*, 8, 137-157.

[52] Lei, X., Archie, N., Sherbourne, and Donald, E. G. (1995) Optimal cost design of semi-rigid, low-rise industrial frames, *Eng. J.*, 3rd quarter, pp. 87-97.

[53] Lewitt, C.W., Chesson, E., and Munse, W.H. (1966) Restraint Characteristics of Flexible Riveted and Bolted Beam-to-Column Connections, Bulletin No. 500, Engineering Experiment Station, University of Illinois, Urbana, IL.

[54] Liew, J. Y. R. (1992) Advanced Analysis for Frame Design, Ph.D. thesis, School of Civil Engineering, Purdue University, West Lafayette, IN, 392 pp.

[55] Liew, J. Y. R. and Chen, W. F. (1991) Refining the plastic hinge concept for advanced analysis/design of steel frames, *J. Singapore Struct. Steel Soc., Steel Struct.*, 2(1), 13-28.

[56] Liew, J.Y.R., White, D.W., and Chen, W.F. (1993a) Second-order refined plastic hinge analysis of frame design: Part I, *J. Struct. Eng., ASCE*, 119(11), 3196-3216.

[57] Liew, J. Y. R., White, D. W., and Chen, W. F. (1993b) Second-order refined plastic-hinge analysis for frame design: Part 2, *J. Struct. Eng., ASCE*, 119(11), 3217-3237.

[58] Liew, J. Y. R., White, D. W., and Chen, W. F. (1993c) Limit states design of semi-rigid frames using advanced analysis: Part 2: Analysis and design, *J. Construct. Steel Res.*, 26(1), 29-57.

[59] Liew, J.Y.R., White, D.W., and Chen, W.F. (1993d) Limit states design of semi-rigid frames using advanced analysis: Part 1: Connection modeling and classification, *J. Construct. Steel Res.*, 26(1), 1-28.

[60] Lightfoot, E. and LeMessurier, A.P. (1974) Elastic analysis of frameworks with elastic connections, *ASCE* , 100(ST6), 1297-1309.

[61] Lui, E.M. (1985) Effects of Connection Flexibility and Panel Zone Deformation on the Behavior of Plane Steel Frames, Ph.D. dissertation, School of Civil Engineering, Purdue University, West Lafayette, IN, 440 pp.

[62] Lui, E. M. and Chen, W. F. (1988) Behavior of braced and unbraced semi-rigid frames, *Int. J. Solids Struct.*, 24(9), 893-913.

[63] Lu, L. W. (1985) Note on a beam-column moment amplification factor, *AISC Eng. J.*, 1st quarter, 47-49.

[64] Maleck, A. E., White, D. W., and Chen, W. F. (1995) Practical application of advanced analysis in steel design, *Proceedings the 4th Pacific Structural Steel Conference*, Vol. 1, Steel Structures, pp. 119-126.

[65] Monforton, A.R. and Wu, T.S. (1963) Matrix analysis of semi-rigidly connected frames, *J. Struct. Div., ASCE*, 87, ST6; 13-42.

[66] Narayanan, R. (1989) (Ed.) *Structural Connections - Stability and Strength*, Elsevier Applied Science, London, 452 pp.

[67] Nethercot, D. A. (1989) Frame analysis and the link between connection behavior and frame performance, in *Frame and Steel Structures*, Armer G. G. T. and Moore D. B., Eds., Butterworths, London, pp. 57-74.

[68] Nethercot, D. A. (1985) Steel Beam-to-Column Connections — A Review of Test Data and its Applicability to the Evaluation of Joint Behavior in the Performance of Steel Frames, CIRIA Project Record, RP 338.

[69] Patel, K.V. and Chen, W.F. (1984) Nonlinear analysis of steel moment connections, *ASCE*, 110(ST8), 1861-1874.

[70] Ramberg, W. and Osgood, W.R. (1943) Description of Stress-Strain Curves by Three-Parameters, Technical Note, 902, National Advisory Committee for Aeronautics, Washington, D.C.

[71] Rathbun, J.C. (1936) Elastic properties of riveted connections, *Trans. ASCE*, 101, 524-563.

[72] Richard, R.M., Gillet P.E., Kriegh, J.D., and Lewis, B.A. (1980) The analysis and design of single plate framing connections, *AISC Eng. J.*, 2nd quarter, 38-52.

[73] Springfield, J. (1991) Limits on second-order elastic analysis, *SSRC Proceedings*, 1991 Annual Technical Session on Inelastic Behavior and Design of Frames, SSRC, April 25-27, Chicago, IL, 89-99.

[74] SSRC (1993) Plastic-Hinge-Based Methods for Advanced Analysis and Design of Steel Frames, an Assessment of the State-of-the-art, White, D. W. and Chen, W. F., Eds., SSRC, Lehigh University, Bethlehem, PA, 299 pp.

[75] Standards Australia (1990) *AS4100-1990, Steel Structures*, Sydney, Australia.

[76] Tarpy, T.S. and Cardinal, J.W. (1981) Behavior of semi-rigid beam-to-column end plate connections, in *Joints in Structural Steelwork*, Howlett, J.H., Jenkins, W.M., and Stainsby, R., Eds., Pentech Press, United Kingdom, 2.3-2.25.

[77] Wu, F. H. (1989) Semi-Rigid Connections in Steel Frames, Ph.D. dissertation, School of Civil Engineering, Purdue University, West Lafayette, IN.

[78] Wu, F.S. and Chen, W.F. (1990) A design model for semi-rigid connections, *Eng. Struct.*, 12(2), 88-97.

[79] White, D.W. and Hajjar, J.F. (1991) Application of second-order elastic analysis in LRFD: research to practice, *Eng. J.*, 28(4), 133-148.

[80] White, D.W., Liew, J.Y.R., and Chen, W.F. (1991) Second-Order Inelastic Analysis for Frame Design: A Report to SSRC Task Group 29 on Recent Research and Perceived State-of-the-Art, Structural Engineering Report, CE-STR-91-12, Purdue University, West Lafayette, IN, 116 pp.

[81] Yee, Y.L. and Melchers, R.E. (1986) Moment-rotation curves for bolted connections, *ASCE*, 112(ST3), 615-635.

[82] Yu, C. H., and Shanmugam, N. E. (1986) Stability of frames with semi-rigid joints, *Comp. Struct.*, 23(5), 639-648.

chapter three

Verifications

3.1 Introduction

In the previous chapter, a practical advanced analysis method was presented for a direct two-dimensional frame design. The practical approach of geometric imperfections and of semi-rigid connections was also discussed together with the advanced analysis method. The practical advanced analysis method was developed using simple modifications to the conventional elastic-plastic hinge analysis.

In this chapter, the practical advanced analysis method will be verified by the use of several benchmark problems available in the literature. Verification studies are carried out by comparing with the plastic-zone solutions as well as the conventional LRFD solutions. The strength predictions and the load-displacement relationships are checked for a wide range of steel frames including axially loaded columns, isolated beam-columns, mathematically identical columns, a rigidly jointed truss, braced frames, sway frames, special frames, and semi-rigid frames [11]. The three imperfection modelings including explicit imperfection modeling, equivalent notional load modeling, and further reduced tangent modulus modeling are also verified for a wide range of steel frames [11].

3.2 Axially Loaded Columns

The AISC-LRFD column strength curve is used for the calibration because it properly accounts for second-order effects, residual stresses, and geometric imperfections in a practical manner. In this study, the column strength of proposed methods is evaluated for columns with slenderness parameters $\left[\lambda_c = \frac{KL}{r}\sqrt{F_y/\left(\pi^2 E\right)}\right]$ varying from 0 to 2, which is equivalent to slenderness ratios (L/r) from 0 to 180 when the yield stress is equal to 36 ksi.

In explicit imperfection modeling, the two-element column is assumed to have an initial geometric imperfection equal to $L_c/1,000$ at column mid-height. The predicted column strengths are summarized in Table 3.1 and compared with the LRFD curve in Figure 3.1. The errors are found to be less than 5% for slenderness ratios up to 140 (or λ_c up to 1.57). This range includes most columns used in engineering practice.

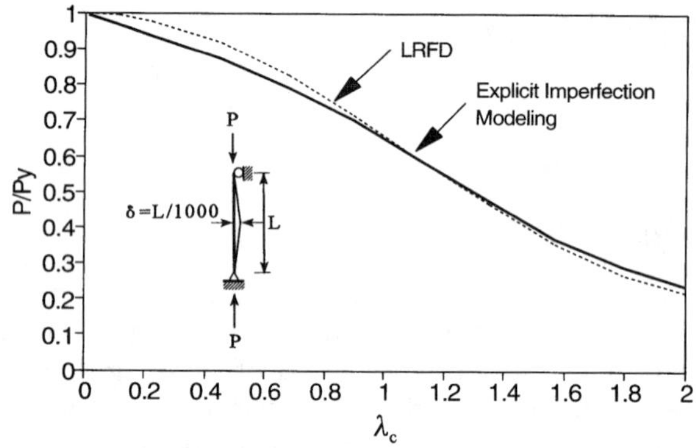

Figure 3.1 Comparison of strength curves for axially loaded pin-ended column (explicit imperfection modeling method).

In the equivalent notional load method, notional loads equal to 0.004 times the gravity loads are applied at mid-height to the column. The strength predictions are the same as those of the explicit imperfection model (Figure 3.2).

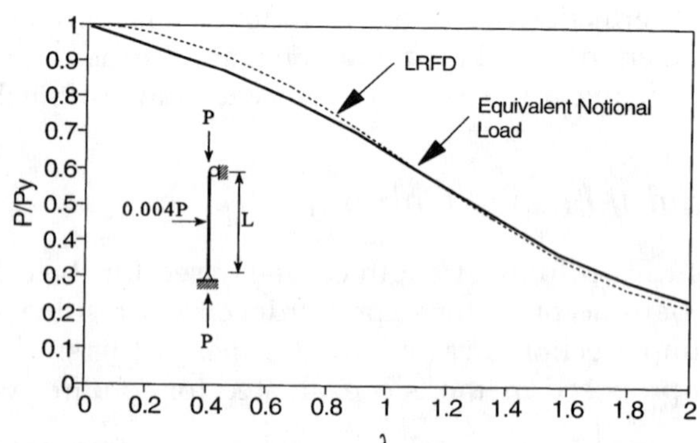

Figure 3.2 Comparison of strength curves for axially loaded pin-ended column (equivalent notional load method).

Table 3.1 Comparison of Strength for the Axially Loaded Pin-Ended Column

L/r	λ_c	LRFD	Proposed method		Error	
		P_L/P_y	P_{IM}/P_y P_{NL}/P_y	P_{RE}/P_y	P_{IM}/P_L P_{NL}/P_L	P_{RE}/P_L
1	0.0112	1.000	0.994	0.997	0.994	0.997
10	0.1122	0.995	0.969	0.971	0.974	0.976
20	0.2243	0.979	0.938	0.942	0.958	0.962
40	0.4486	0.919	0.871	0.882	0.948	0.960
60	0.6729	0.827	0.792	0.811	0.958	0.981
80	0.8972	0.714	0.700	0.726	0.980	1.017
100	1.1215	0.590	0.590	0.618	1.000	1.047
120	1.3458	0.468	0.475	0.470	1.015	1.004
140	1.5701	0.356	0.369	0.345	1.037	0.969
160	1.7944	0.272	0.293	0.264	1.077	0.971
180	2.0187	0.215	0.235	0.209	1.093	0.972

P_{IM} = ultimate strength by the explicit imperfection modeling;
P_{NL} = ultimate strength by the equivalent notional load method;
P_{RE} = ultimate strength by the further reduced tangent modulus method;
P_L = ultimate strength by the LRFD method;
P_y = squash load.

In the further reduced tangent modulus method, the reduced tangent modulus factor equal to 0.85 results in an excellent fit to the LRFD column strengths. The errors are less than 5% for columns of all slenderness ratios. These comparisons are summarized in Table 3.1 and Figure 3.3.

3.3 Isolated Beam-Columns

Galambos and Ketter [7] [10] developed interaction curves for pin-ended beam-columns subject to axial thrust and bending moments using plastic-zone theory. The stress-strain relationship was assumed to be elastic-perfectly plastic with a 33 ksi yield stress and an elastic modulus of 30,000 ksi. Galambos and Ketter's interaction curves only account for residual stresses of $0.3F_y$ (Figure 2.6), not geometric imperfections. An adjustment can be made to these interaction curves to account for these imperfections as described in Section 1.3.3 [16] [17].

In addition, the LRFD interaction curves can be used for strength comparisons without resistance factors ϕ_c and ϕ_b. The two-dimensional LRFD interaction equations take the simple form:

$$\frac{P_u}{P_n} + \frac{8}{9}\left(\frac{M_u}{M_p}\right) \leq 1.0 \quad \text{for} \quad \frac{P_u}{P_n} \geq 0.2 \tag{3.1a}$$

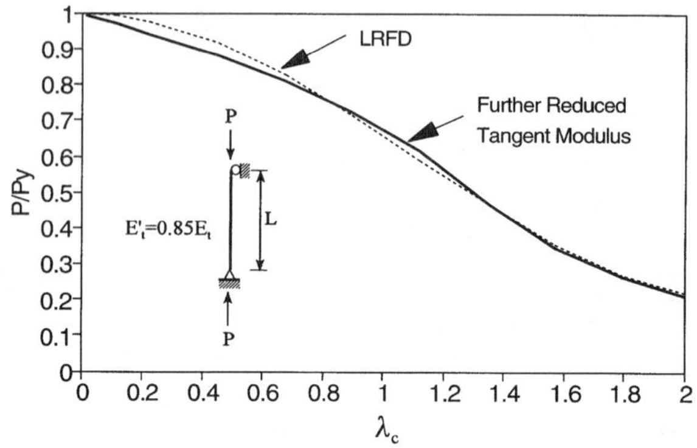

Figure 3.3 Comparison of strength curves for axially loaded pin-ended column (further reduced tangent modulus method).

$$\frac{P_u}{2P_n} + \frac{M_u}{M_p} \leq 1.0 \quad \text{for} \quad \frac{P_u}{P_n} < 0.2 \tag{3.1b}$$

where

P_n = nominal axial compressive strength
P_u, M_u = second-order axial force and bending moment

When a geometric imperfection of $L_c/1,000$ at mid-height is used, the strength curves compare well with LRFD interaction curves and adjusted plastic-zone curves. The errors do not exceed 5% (Figures 3.4 and 3.5). When equivalent notional loads equal to 0.004 times gravity loads are applied at mid-height of the column, the results match those of the explicit imperfection model (Figures 3.6 and 3.7). When the tangent modulus E_t is reduced by 0.85, the predicted beam-column strength curves compared well with LRFD and adjusted plastic-zone curves. The errors are less than 5% as shown in Figures 3.8 and 3.9.

3.4 Mathematically Identical Columns

When the LRFD interaction equations were developed, one of the six guidelines outlined in Section 1.3.3 was that the same ultimate strength should be predicted for mathematically identical members (Figure 1.7) in both elastic and inelastic regions. The KL of the fixed end column in Figure 1.7 is $1L$ since the effective length K and the column length L are 0.5 and $2L$, respectively. The KL of the hinged-ended column and the guided-ended column is also equal to $1L$ obtained by multiplying the K of 1 by the $1L$. For the cantilever column, the K of 2 and the length of $0.5L$ result in the KL equal to $1L$. As a result, the four columns should

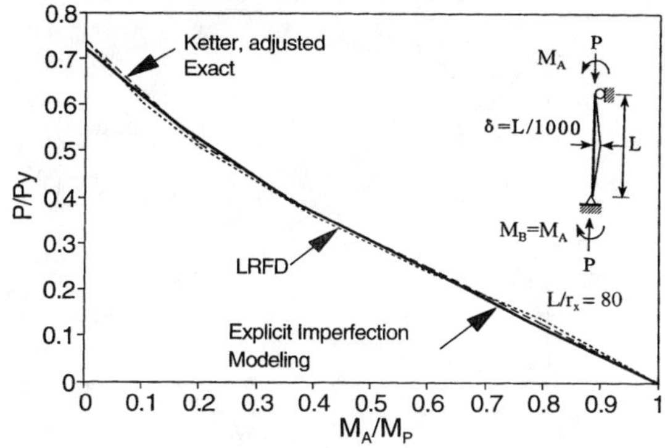

Figure 3.4 Comparison of strength curves for beam-column of $L/r = 80$ (explicit imperfection modeling method).

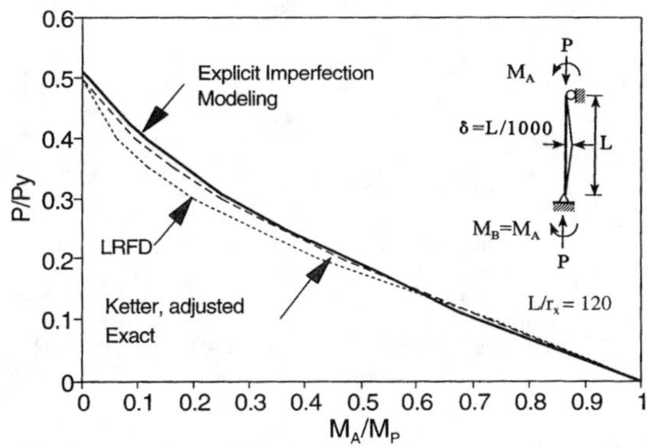

Figure 3.5 Comparison of strength curves for beam-column of $L/r = 120$ (explicit imperfection modeling method).

have mathematically identical strengths. Columns with a slenderness ratio (L/r) equal to 80 are used to verify that indeed this is the case.

In explicit imperfection modeling of the fixed-ended column and the hinged-ended column, the two-element column is assumed to have an initial out-of-straightness equal to $L_c/1,000$ at column mid-height. For the guided-ended column and the cantilever column, the one-element column is used with an initial out-of-plumbness equal to $L_c/500$ at the top of the columns where L_c is the total column length. Advanced anal-

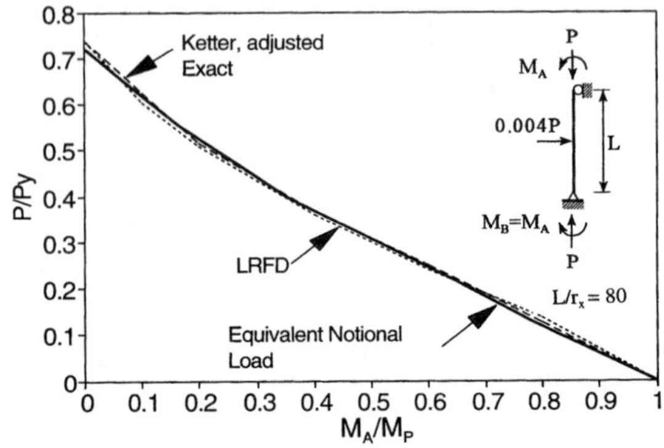

Figure 3.6 Comparison of strength curves for beam-column of $L/r = 80$ (equivalent notional load method).

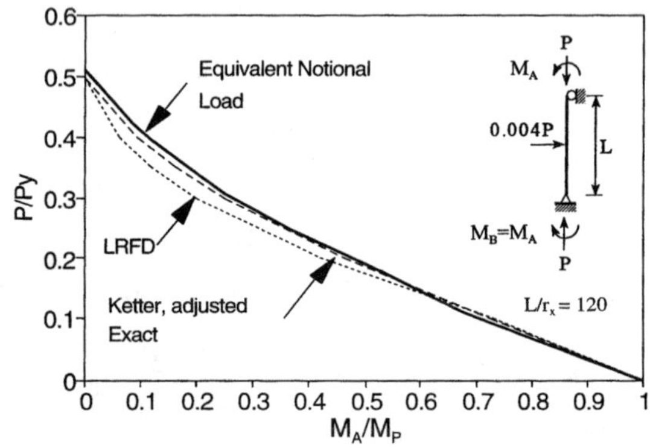

Figure 3.7 Comparison of strength curves for beam-column of $L/r = 120$ (equivalent notional load method).

ysis provides the same ultimate strengths for the four columns shown in Figure 3.10.

In the equivalent notional load method, the fixed-ended column and the hinged-ended column have the notional loads equal to 0.004 times the gravity loads at the mid-height of the columns, and the guided-ended column and the cantilever column have the notional loads of 0.002 times the gravity loads at the top of the columns. The identical strength is predicted for the four columns by advanced analysis as shown in Figure 3.11.

In the further reduced tangent modulus method, the same reduction factor of 0.85 for the four columns is used and results in identical strength estimation (Figure 3.12).

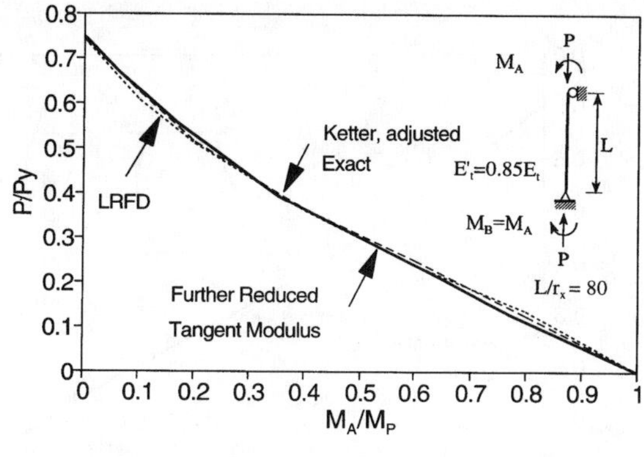

Figure 3.8 Comparison of strength curves for beam-column of $L/r = 80$ (further reduced tangent modulus method).

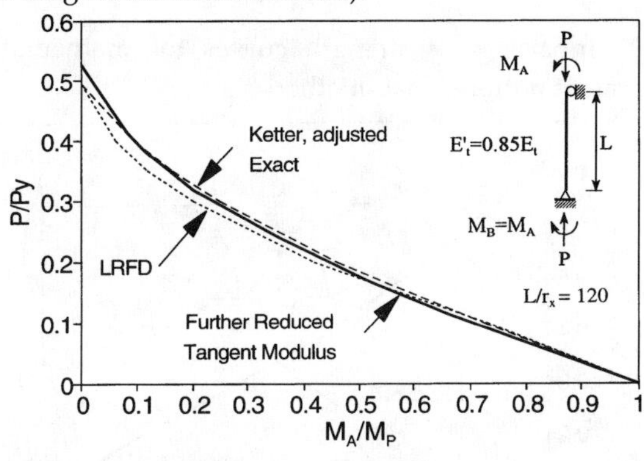

Figure 3.9 Comparison of strength curves for beam-column of $L/r = 120$ (further reduced tangent modulus method).

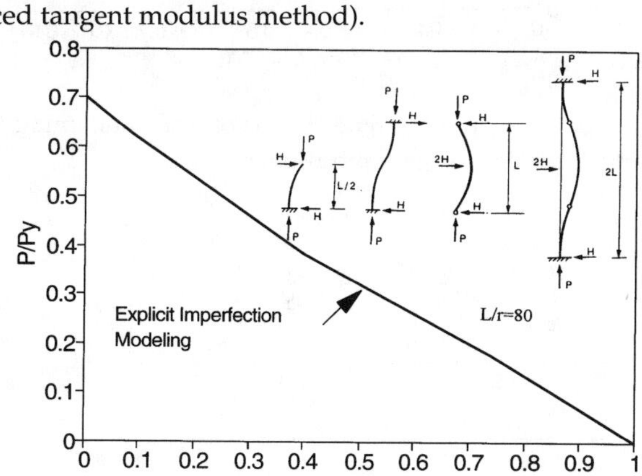

Figure 3.10 Comparison of strength curves for mathematically identical columns (explicit imperfection modeling method).

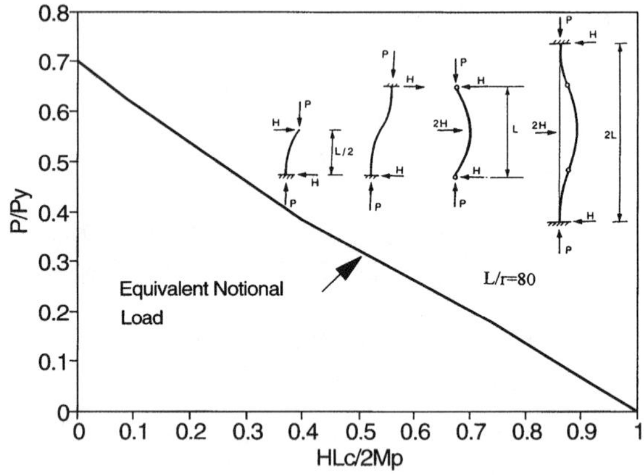

Figure 3.11 Comparison of strength curves for mathematically identical columns (equivalent notional load method).

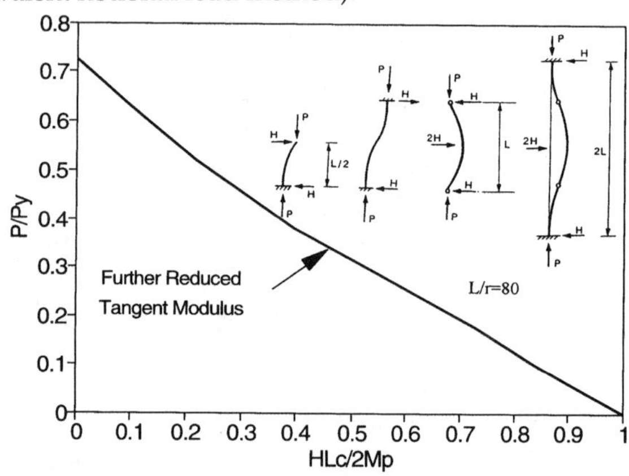

Figure 3.12 Comparison of strength curves for mathematically identical columns (further reduced tangent modulus method).

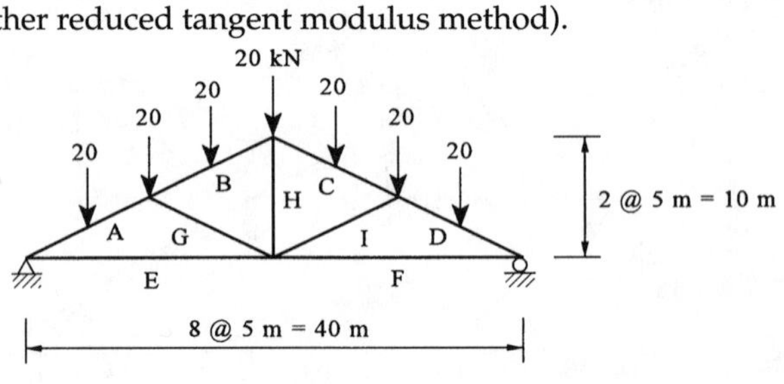

Figure 3.13 Configuration of rigidly jointed truss for verification study [5] [6].

3.5 Rigidly Jointed Truss

Clarke et al. [5] [6] studied a rigidly jointed roof truss subject to gravity loads between and at the joints of the top chord members as shown in Figure 3.13. The chord members of the truss have both axial and flexural forces since it is rigidly jointed. Therefore, this frame is an atypical frame compared to the usual rectangular portal or multistory frame subject to gravity loads. Another feature of this truss is that the W-sections are oriented such that the weak-axis is subject to the primary bending actions. The section properties are summarized in Table 3.2. Residual stresses have greater effect on W-section members bent about their weak- rather than strong-axis. The residual stress pattern is modeled as shown in Figure 2.6.

Table 3.2 Section Properties of Rigid-Jointed Truss Members

Member	Location	Section	$A(mm^2)$	$I_y(mm^4)$	$S_y(mm^3)$	$F_y(MPa)$
A, D	Top chord	250UC72.9	9290	$38.7x10^6$	$462x10^3$	250
B, C	Top chord	200UC52.2	6640	$17.7x10^6$	$264x10^3$	250
E, F	Bot. chord	100UC14.8	1890	$1.14x10^6$	$35.3x10^3$	260
G, I	Web chord	150UC23.4	2980	$4.03x10^6$	$80.9x10^3$	260
H	Web chord	100UC14.8	1890	$1.14x10^6$	$35.3x10^3$	260

Clarke et al. [5] [6] performed plastic-zone analysis of the truss. The chords were assumed to have a sinusoidal out-of-straightness with a maximum in-plane imperfection of $\delta_0 = L/1000$ at the center. Figure 3.14 shows the assumed orientation of the imperfections. Strain hardening was included to the stress-strain curve as shown in Figure 3.15, although its contribution to the frame strength is negligible. Twenty elements were used along each chord length between the joints. The number of the cross-sectional subdivisions used was 17 along the flange width, 9 along the web half-depth, 3 through the flange thickness, and 3 through the web thickness, respectively. The load factor determined by the plastic-zone method was 1.93.

In explicit imperfection modeling, each chord is divided into two elements with an initial out-of-straightness of $\delta_0 = L/1,000$ at mid-length as shown in Figure 3.16. In the equivalent notional load approach, the notional loads are applied at mid-length of the chords normal to the axis of the chords. In the further reduced tangent modulus approach, the reduced $E_t' = 0.85E_t$ is used for all members except the bottom chords subject to tensile forces. The explicit modeling and the notional load methods predict the same load factor (1.88), and the further reduced tangent modulus method predicts a load factor of 1.84. The proposed methods predict the ultimate load with conservative errors of less than

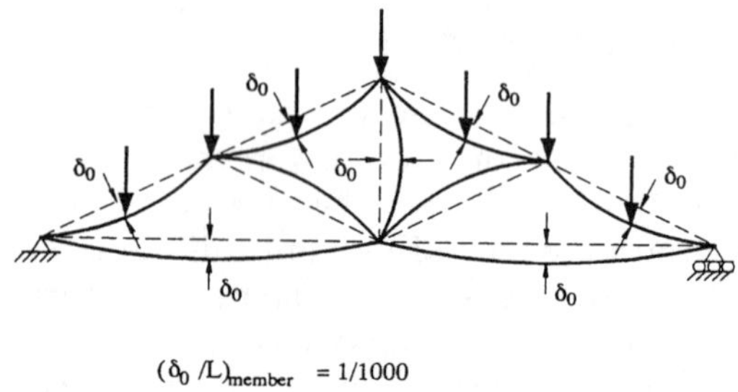

Figure 3.14 Geometric imperfections for Clarke's rigidly jointed truss [5] [6].

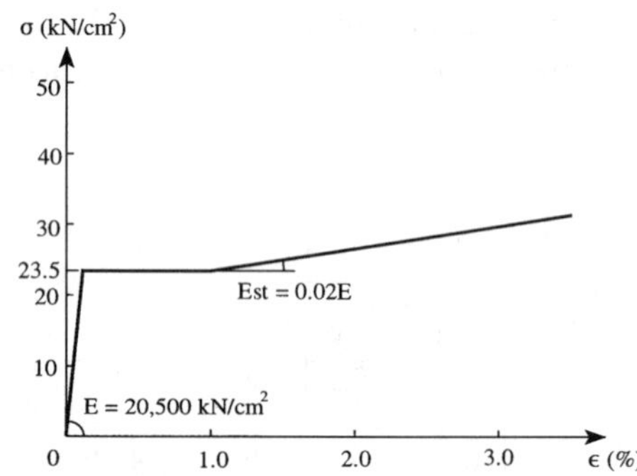

Figure 3.15 Stress-strain relationships for Clarke's truss.

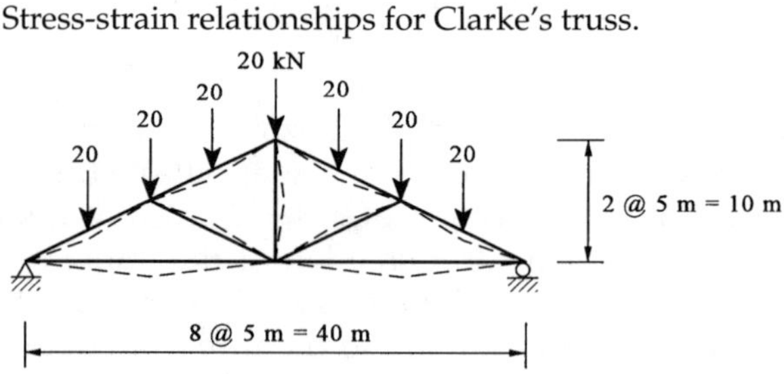

Figure 3.16 Explicit imperfection modeling of Clarke's rigidly jointed truss.

5%. The load-deflection behavior of the truss shows a good agreement between the plastic zone analysis and the practical advanced analysis as shown in Figure 3.17.

3.6 *Braced Frames*

The plastic-zone solutions reported by Chen and Zhou [4] are used as the benchmark for the braced frame. The configuration of the portal frame shown in Figure 3.18 is prevented from side-sway, and its beams and columns are rigidly connected. All members are assumed to be fully braced from out-of-plane displacement, and their cross-sections are compact. The beam is 210 in. long, and the columns are 140 in. high

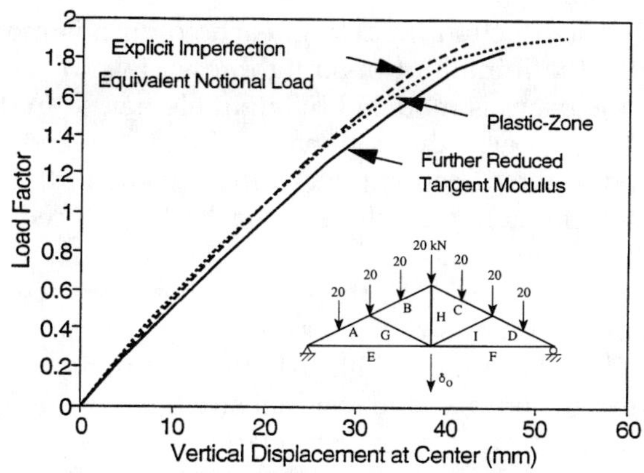

Figure 3.17 Comparison of load-displacement relationships of Clarke's truss.

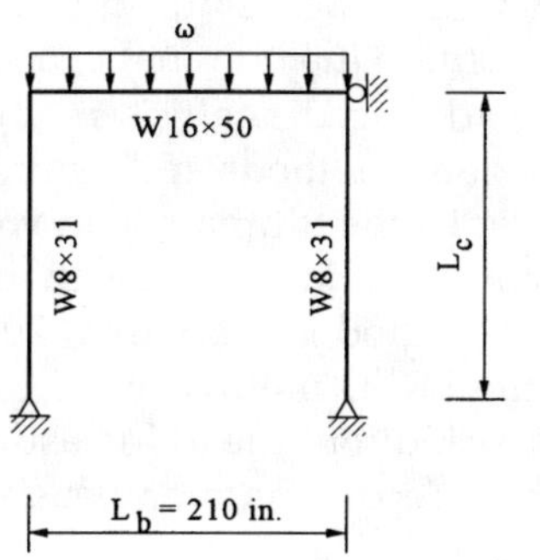

Figure 3.18 Comparison of strength for braced portal frame.

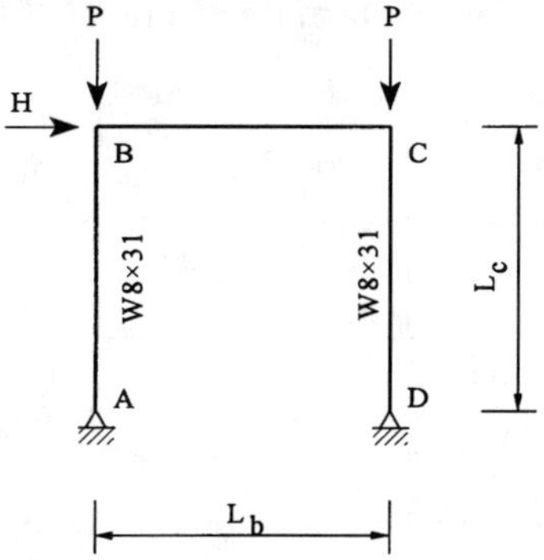

Figure 3.19 Kanchanalai's portal frame.

for case 1 and 280 in. high for case 2. The frame is subject to uniform gravity loads only.

In explicit imperfection modeling, each column is modeled by two elements with the initial out-of-straightness of $\delta_0 = L/1,000$ at mid-height, and the beam is modeled by eight elements. In the equivalent notional load approach, the notional loads are applied at mid-height of the column in the direction toward the outside of the frame. In the further reduced tangent modulus approach, the reduced $E_t' = 0.85E_t$ is used for column members.

In plastic-zone analysis [4], the column members were assumed to have a sinusoidal out-of-straightness with a maximum in-plane deflection $\delta_0 = L/1,000$ at mid-height. All cross-sections were assumed to have a maximum compressive residual stress of $0.3F_y$ (Figure 2.6). An elastic-perfectly plastic stress-strain relationship was assumed.

In the conventional LRFD method, an elastic analysis is carried out, and maximum forces in each member are then checked against the LRFD interaction equations. Ultimate loads may be obtained when the value of the interaction equations reaches unity. The resistance factors ϕ_b and ϕ_c are taken as one to obtain the nominal beam-column strength.

The three practical advanced analysis methods are compared with the LRFD and plastic-zone methods and summarized in Table 3.3. The proposed methods predict the ultimate load well, and the conservative errors in the plastic-zone solutions are less than 6%. On the other hand, the error in the LRFD method is more than 20%. The proposed methods are better than the LRFD method in ultimate strength prediction because moment redistribution due to inelasticity is considered. Thus, they should provide more economic member sizes than the conventional LRFD method.

Table 3.3 Comparison of Ultimate Strength of Braced Portal Frame

Col. length m (in.)	Proposed method		LRFD	Plastic-zone	Error		
	P_{IM}/P_y P_{NL}/P_y	P_{RE}/P_y	P_L/P_y	P_{PZ}/P_y	P_{IM}/P_{PZ} P_{NL}/P_{PZ}	P_{RE}/P_{PZ}	P_L/P_{PZ}
140 in.	0.243	0.242	0.220	0.246	0.99	0.98	0.89
280 in.	0.234	0.228	0.192	0.243	0.96	0.94	0.79

P_{IM} = ultimate strength by the explicit imperfection modeling;
P_{NL} = ultimate strength by the equivalent notional load method;
P_{RE} = ultimate strength by the further reduced tangent modulus method;
P_L = ultimate strength by the LRFD method;
P_{PZ} = ultimate strength by the plastic-zone method;
P_y = squash load.

3.7 *Sway Frames*

3.7.1 *Kanchanalai's Frames in Strong-Axis Bending*

Kanchanalai [9] performed extensive analyses of portal and leaning column frames (Figures 3.19 and 3.20), and developed exact interaction curves based on plastic-zone analyses of simple sway frames. Note that

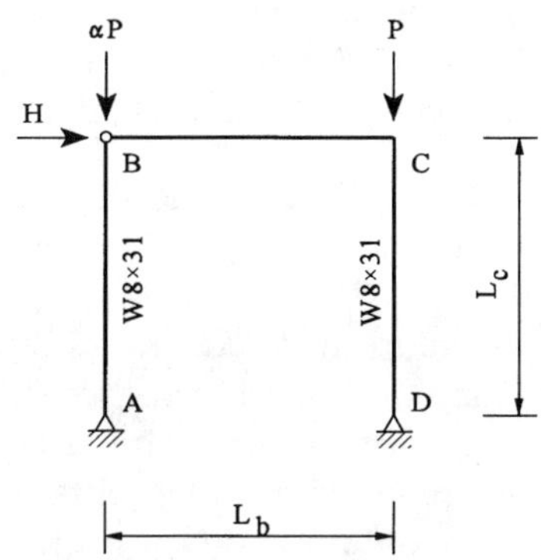

Figure 3.20 Kanchanalai's leaning column frame.

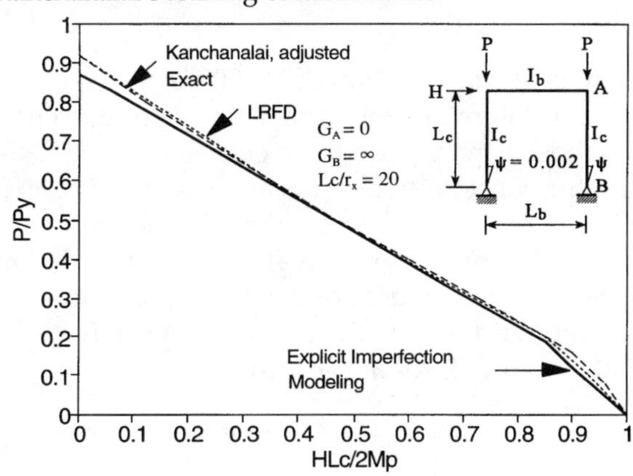

Figure 3.21 Comparison of strength curves for portal frame subject to strong-axis bending with $L_c/r_x = 20$, $G_A = 0$ (explicit imperfection modeling method).

the simple frames are more sensitive in their behavior than the highly redundant frames. His studies formed the basis of the interaction equations in AISC-LRFD design specifications [1] [2]. In his studies, the stress-strain relationship was assumed elastic-perfectly plastic with a 36 ksi yield stress and a 29,000 ksi elastic modulus. The members were assumed to have a maximum compressive residual stress of $0.3F_y$ with a pattern as shown in Figure 2.6. Initial geometric imperfections were not considered, and thus an adjustment of his interaction curves is made to account for this as described in Section 1.3.3. He further performed experimental work to verify his analysis [9] which covered a wide range of portal and leaning column frames with the slenderness ratios of 20, 30, 40, 50, 60, 70, and 80, and relative stiffness ratios (G) of 0, 3, and 4. The ultimate strength of each frame was presented in the form of interaction curves consisting of the nondimensional first-order moment ($HL_c/2M_p$ in portal frames or HL_c/M_p in leaning column frames in the x-axis) and the nondimensional axial load (P/P_y in the y-axis).

In this study, the AISC-LRFD interaction curves are used for strength comparisons. The strength calculations are based on the LeMessurier K-factor method presented in Section 1.3.4 [13] because it accounts for story buckling and results in more accurate predictions. The inelastic stiffness reduction factor τ described in Section 1.3.1 [2] is used to calculate K in the LeMessurier's procedure. The resistance factors ϕ_b and ϕ_c in the LRFD equations are taken as 1.0 to obtain the nominal strength. The interaction curves are obtained by the accumulation of a set of moments and axial forces that result in unity on the value of the interaction equation.

When a geometric imperfection of $L_c/500$ is used for unbraced frames including leaning column frames, most of the strength curves fall within an area bounded by the plastic-zone curves and the LRFD curves. In portal frames, the conservative errors are less than 5%, an improvement on the LRFD error of 11%, and the maximum unconservative error is not more than 1% as shown in Figures 3.21 through 3.25. In leaning column frames, the conservative errors are less than 12% as opposed to the 17% error of the LRFD, and the maximum unconservative error is not more than 5% as shown in Figures 3.26 through 3.29.

When a notional load factor of 0.002 is used, the strengths predicted by this method are close to those given by explicit imperfection modeling method (Figures 3.30 through 3.38).

When the reduced tangent modulus factor of 0.85 is used for portal and leaning column frames, the interaction curves generally fall between the plastic-zone and LRFD curves. In portal frames, the conservative error is less than 8% (better than 11% error of the LRFD), and the maximum unconservative error is not more than 5% (Figures 3.39 through 3.43). In leaning column frames, the conservative error is less than 7% (better than 17% error of the LRFD), and the maximum unconservative error is not more than 5% (Figures 3.44 through 3.47).

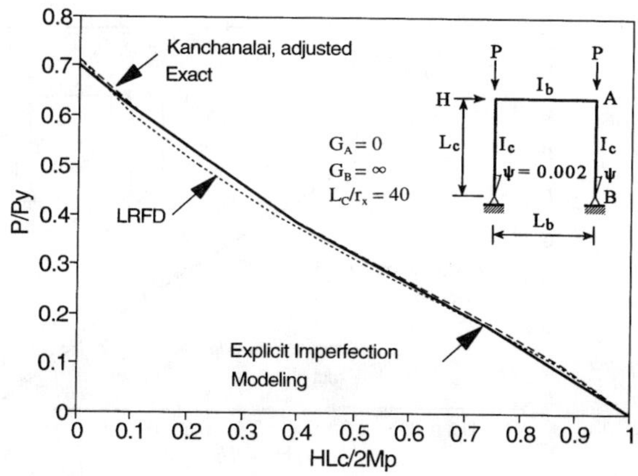

Figure 3.22 Comparison of strength curves for portal frame subject to strong-axis bending with $L_c/r_x = 40, G_A = 0$ (explicit imperfection modeling method).

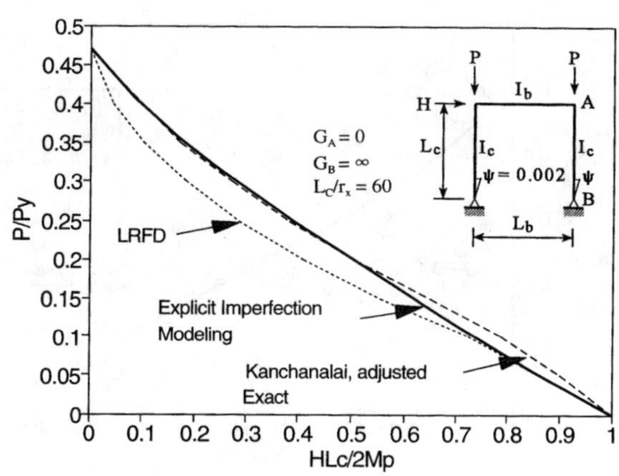

Figure 3.23 Comparison of strength curves for portal frame subject to strong-axis bending with $L_c/r_x = 60, G_A = 0$ (explicit imperfection modeling method).

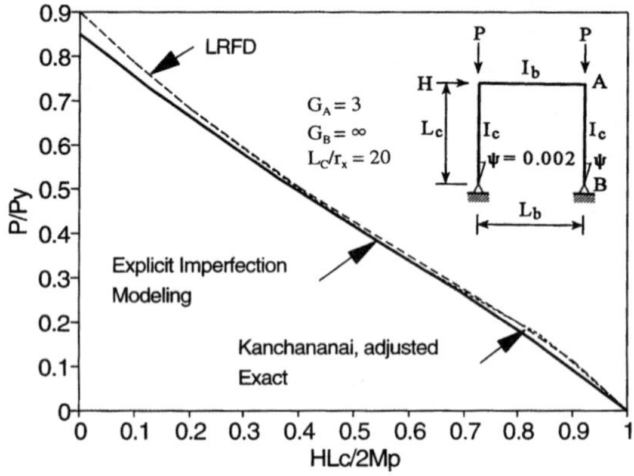

Figure 3.24 Comparison of strength curves for portal frame subject to strong-axis bending with $L_c/r_x = 20, G_A = 3$ (explicit imperfection modeling method).

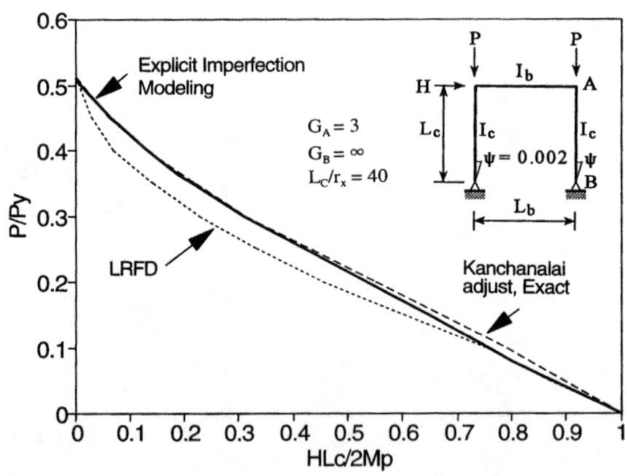

Figure 3.25 Comparison of strength curves for portal frame subject to strong-axis bending with $L_c/r_x = 40, G_A = 3$ (explicit imperfection modeling method).

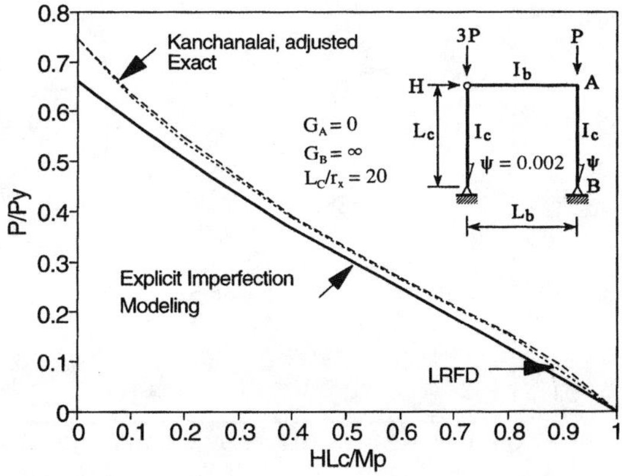

Figure 3.26 Comparison of strength curves for leaning column frame subject to strong-axis bending with $L_c/r_x = 20$, $G_A = 0$ (explicit imperfection modeling method).

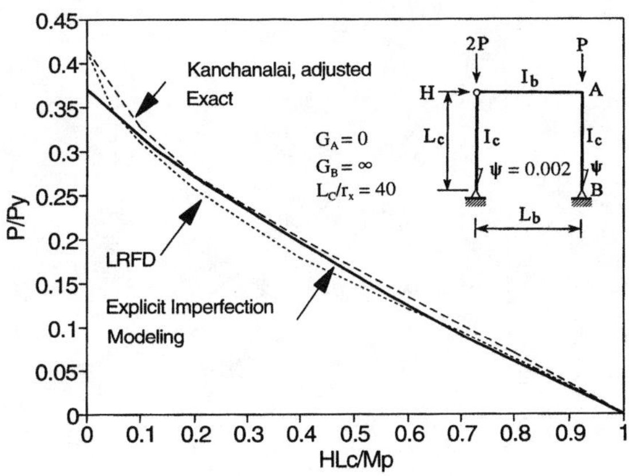

Figure 3.27 Comparison of strength curves for leaning column frame subject to strong-axis bending with $L_c/r_x = 40$, $G_A = 0$ (explicit imperfection modeling method).

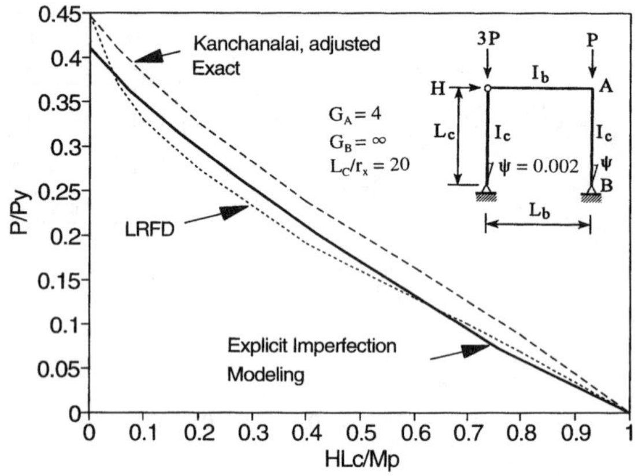

Figure 3.28 Comparison of strength curves for leaning column frame subject to strong-axis bending with $L_c/r_x = 20$, $G_A = 4$ (explicit imperfection modeling method).

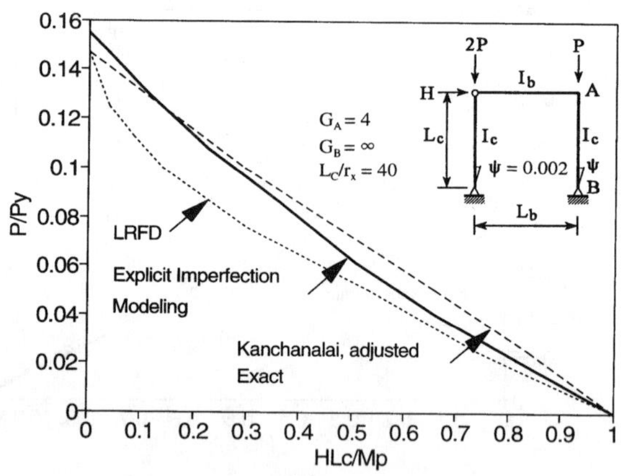

Figure 3.29 Comparison of strength curves for leaning column frame subject to strong-axis bending with $L_c/r_x = 40$, $G_A = 4$ (explicit imperfection modeling method).

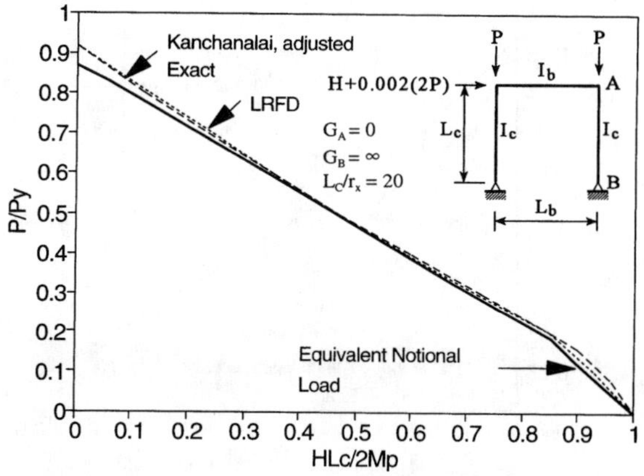

Figure 3.30 Comparison of strength curves for portal frame subject to strong-axis bending with $L_c/r_x = 20$, $G_A = 0$ (equivalent notional load method).

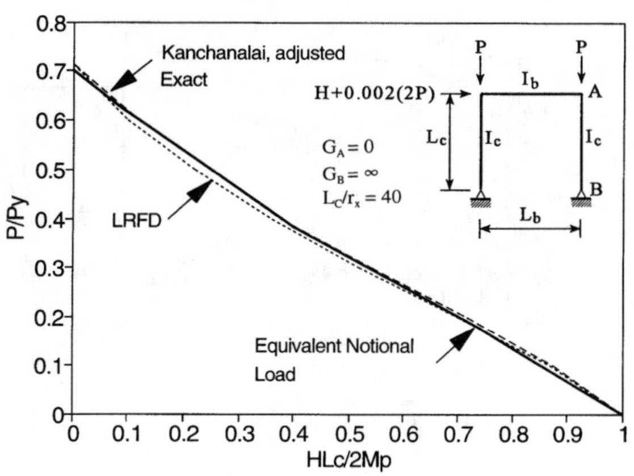

Figure 3.31 Comparison of strength curves for portal frame subject to strong-axis bending with $L_c/r_x = 40$, $G_A = 0$ (equivalent notional load method).

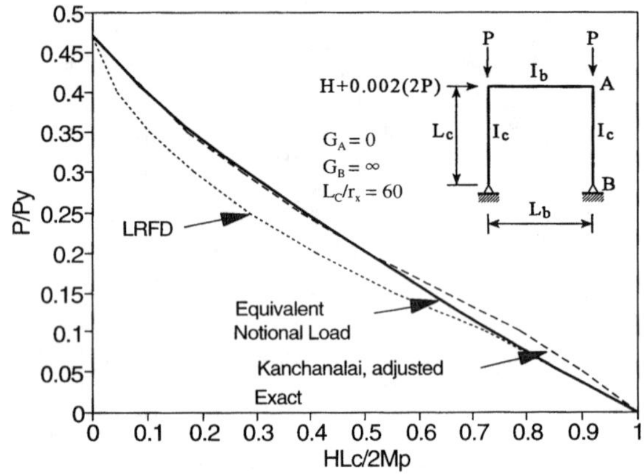

Figure 3.32 Comparison of strength curves for portal frame subject to strong-axis bending with $L_c/r_x = 60, G_A = 0$ (equivalent notional load method).

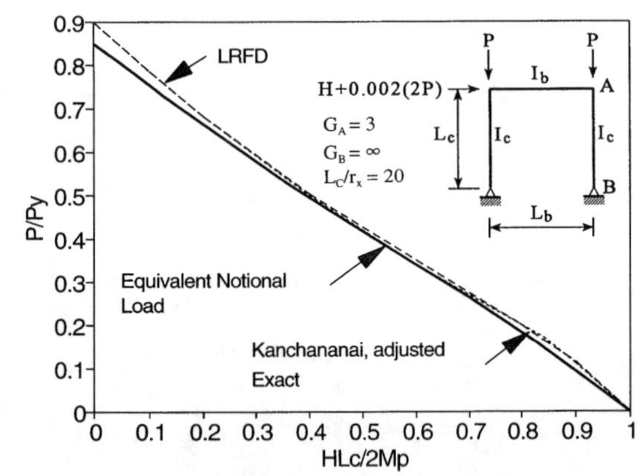

Figure 3.33 Comparison of strength curves for portal frame subject to strong-axis bending with $L_c/r_x = 20, G_A = 3$ (equivalent notional load method).

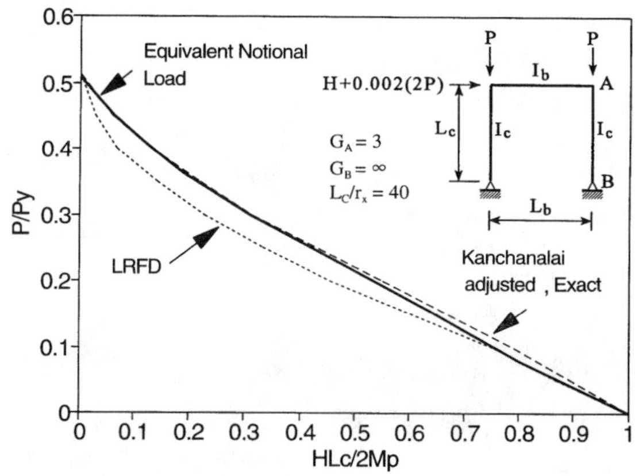

Figure 3.34 Comparison of strength curves for portal frame subject to strong-axis bending with $L_c/r_x = 40$, $G_A = 3$ (equivalent notional load method).

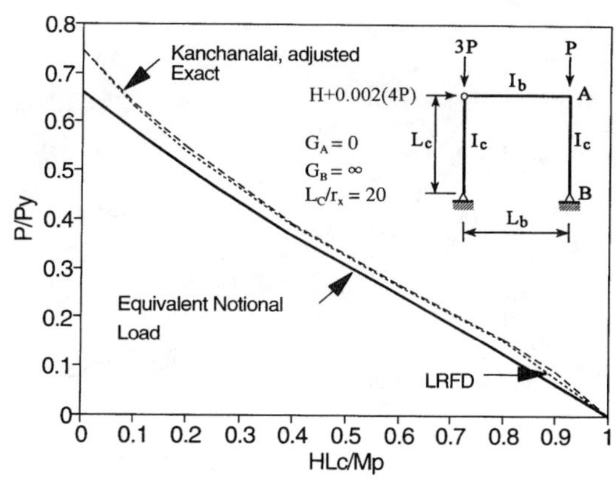

Figure 3.35 Comparison of strength curves for leaning column frame subject to strong-axis bending with $L_c/r_x = 20$, $G_A = 0$ (equivalent notional load method).

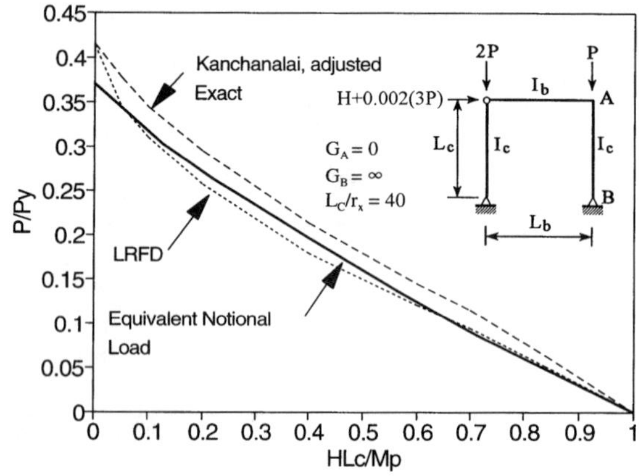

Figure 3.36 Comparison of strength curves for leaning column frame subject to strong-axis bending with $L_c/r_x = 40$, $G_A = 0$ (equivalent notional load method).

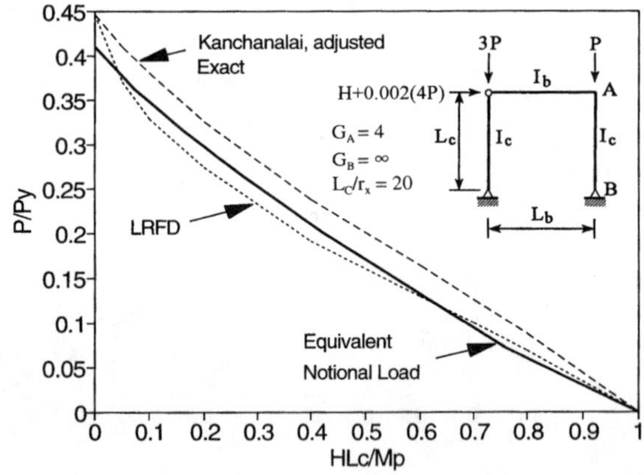

Figure 3.37 Comparison of strength curves for leaning column frame subject to strong-axis bending with $L_c/r_x = 20$, $G_A = 4$ (equivalent notional load method).

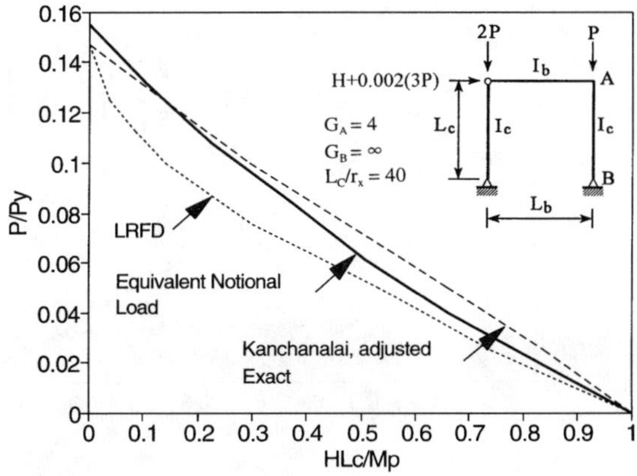

Figure 3.38 Comparison of strength curves for leaning column frame subject to strong-axis bending with $L_c/r_x = 40$, $G_A = 4$ (equivalent notional load method).

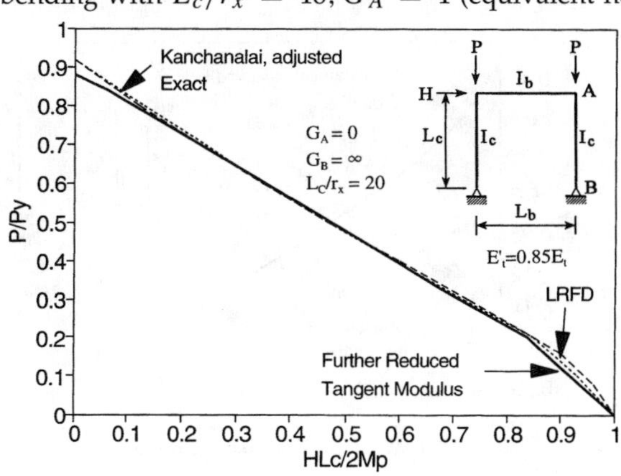

Figure 3.39 Comparison of strength curves for portal frame subject to strong-axis bending with $L_c/r_x = 20$, $G_A = 0$ (further reduced tangent modulus method).

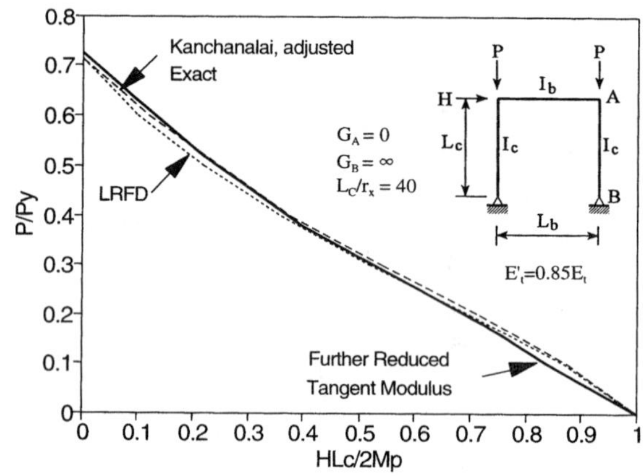

Figure 3.40 Comparison of strength curves for portal frame subject to strong-axis bending with $L_c/r_x = 40, G_A = 0$ (further reduced tangent modulus method).

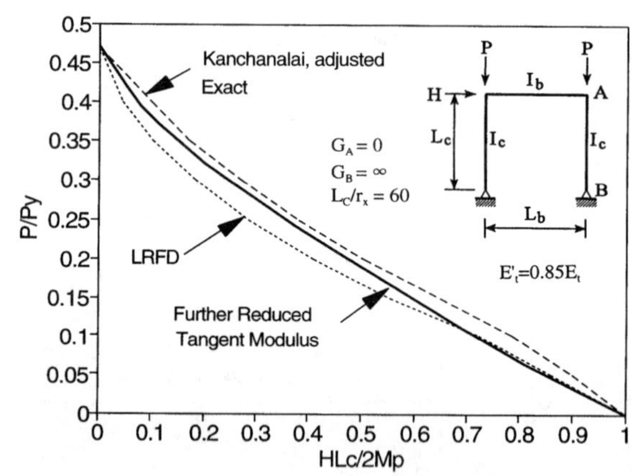

Figure 3.41 Comparison of strength curves for portal frame subject to strong-axis bending with $L_c/r_x = 60, G_A = 0$ (further reduced tangent modulus method).

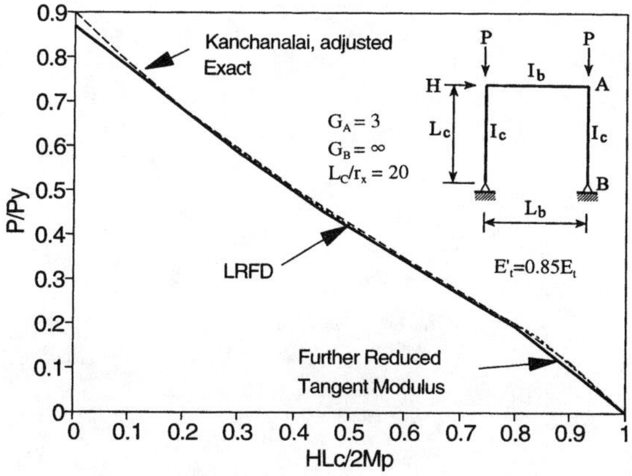

Figure 3.42 Comparison of strength curves for portal frame subject to strong-axis bending with $L_C/r_x = 20$, $G_A = 3$ (further reduced tangent modulus method).

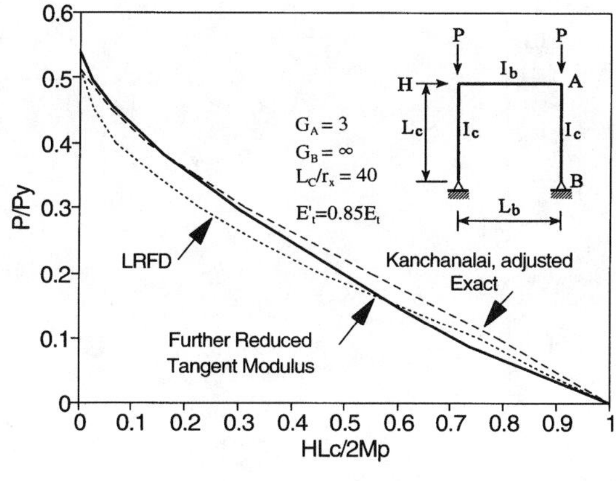

Figure 3.43 Comparison of strength curves for portal frame subject to strong-axis bending with $L_C/r_x = 40$, $G_A = 3$ (further reduced tangent modulus method).

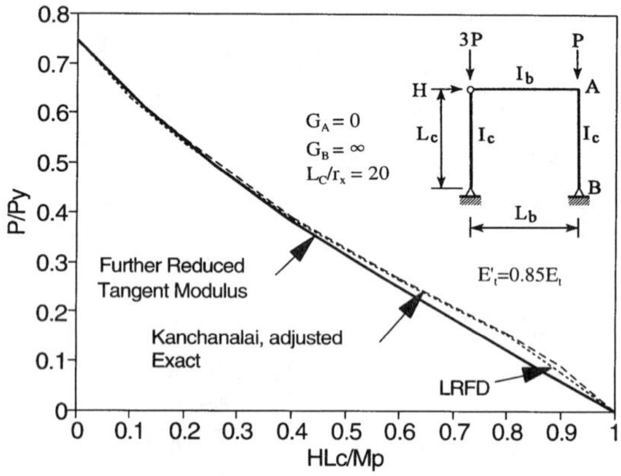

Figure 3.44 Comparison of strength curves for leaning column frame subject to strong-axis bending with $L_c/r_x = 20$, $G_A = 0$ (further reduced tangent modulus method).

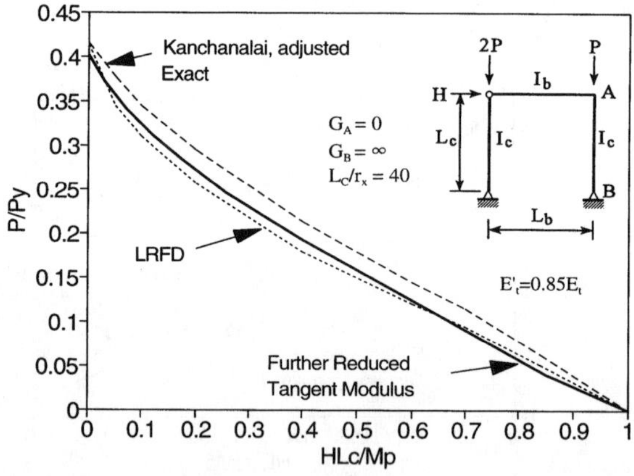

Figure 3.45 Comparison of strength curves for leaning column frame subject to strong-axis bending with $L_c/r_x = 40$, $G_A = 0$ (further reduced tangent modulus method).

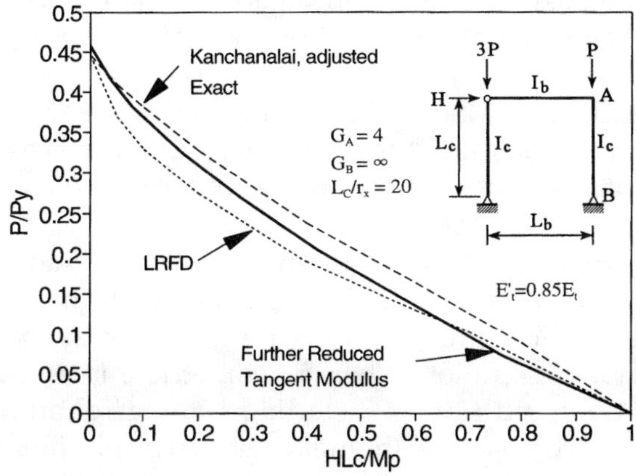

Figure 3.46 Comparison of strength curves for leaning column frame subject to strong-axis bending with $L_c/r_x = 20, G_A = 4$ (further reduced tangent modulus method).

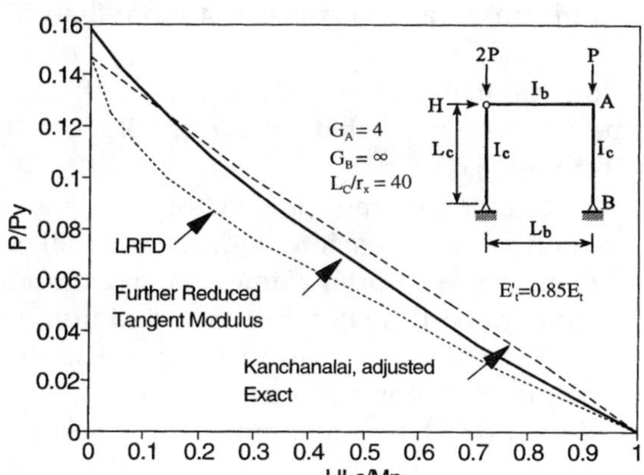

Figure 3.47 Comparison of strength curves for leaning column frame subject to strong-axis bending with $L_c/r_x = 40, G_A = 4$ (further reduced tangent modulus method).

3.7.2 Kanchanalai's Frames in Weak-Axis Bending

Kanchanalai [9] developed exact interaction curves for weak-axis bending as well. In his plastic-zone analyses at sway frames, the stress-strain relationship and the residual stresses were assumed to be identical to those of strong-axis bending. Since geometric imperfections were not modeled, an adjustment of interaction curves is necessary to account for this [16] [17].

In this study, the AISC-LRFD interaction curves are also used for strength comparisons about the weak-axis. These curves for the weak-axis are identical to those of the strong-axis for equivalent slenderness ratios.

Since the same plastic strength and stiffness degradation functions are used for both strong- and weak-axis bending, the proposed methods predict identical strengths for strong- and weak-axis bending for a given slenderness ratio. When a geometric imperfection of $L_c/500$ is used for unbraced frames including leaning column frames, most strength curves come within the area bounded by plastic-zone and LRFD curves. In portal frames, the maximum conservative error is 7% (equal to the LRFD), and the maximum unconservative error is not more than 5% as shown in Figures 3.48 through 3.52. In leaning column frames, the maximum conservative error is 12% (less than 17% by the LRFD), and maximum unconservative error is not more than 5% as shown in Figures 3.53 through 3.56.

When the notional load factor of 0.002 is used, the strengths are almost the same as those predicted by explicit imperfection modeling (Figures 3.57 through 3.65).

When the reduced tangent modulus factor of 0.85 is used for portal and leaning column frames, most interaction curves fall between plastic-zone and LRFD curves. In portal frames, the maximum conservative error is 17% (equal to the LRFD), while the maximum unconservative error is not more than 5% as shown in Figures 3.66 through 3.70. In leaning column frames, the maximum conservative error is 10% (less than the 17% of the LRFD), and maximum unconservative error is not more than 5% as shown in Figures 3.71 through 3.74.

3.7.3 Vogel's Frames

Portal Frame

Vogel [15] presented the load-displacement relationship for a portal frame using plastic-zone analysis. The frame is shown in Figure 3.75. All members are hot-rolled and their properties are in Table 3.4. Based on ECCS recommendations, the maximum compressive residual stress is $0.3F_y$ when the ratio of depth to width (d/b) is greater than 1.2, and

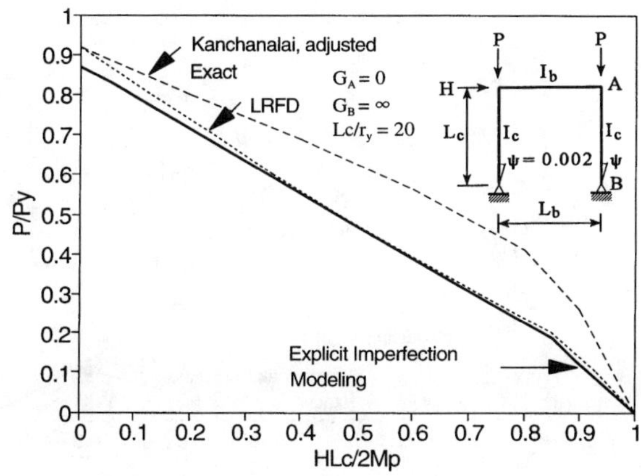

Figure 3.48 Comparison of strength curves for portal frame subject to weak-axis bending with $L_c/r_y = 20, G_A = 0$ (explicit imperfection modeling method).

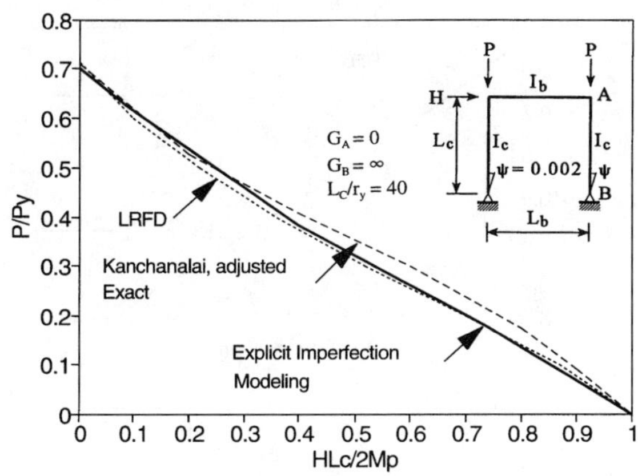

Figure 3.49 Comparison of strength curves for portal frame subject to weak-axis bending with $L_c/r_y = 40, G_A = 0$ (explicit imperfection modeling method).

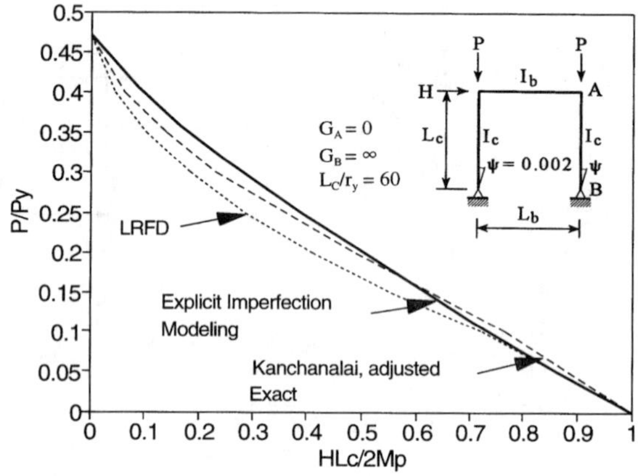

Figure 3.50 Comparison of strength curves for portal frame subject to weak-axis bending with $L_c/r_y = 60, G_A = 0$ (explicit imperfection modeling method).

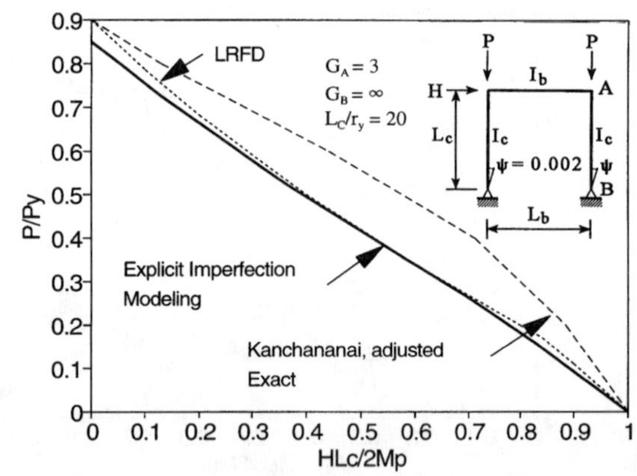

Figure 3.51 Comparison of strength curves for portal frame subject to weak-axis bending with $L_c/r_y = 20, G_A = 3$ (explicit imperfection modeling method).

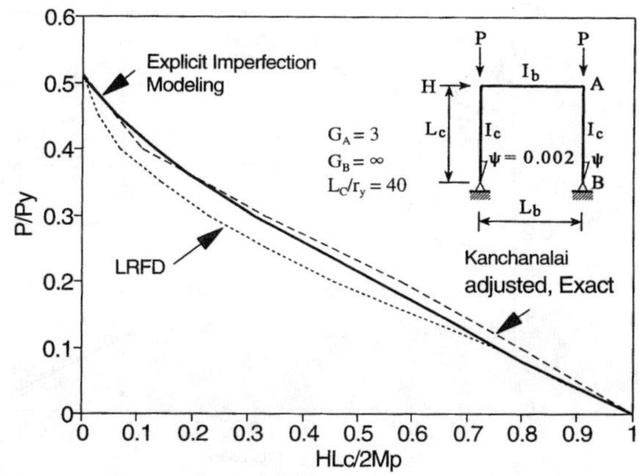

Figure 3.52 Comparison of strength curves for portal frame subject to weak-axis bending with $L_c/r_y = 40, G_A = 3$ (explicit imperfection modeling method).

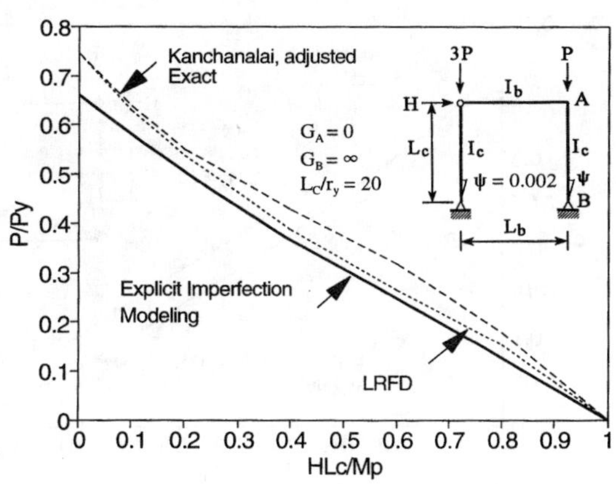

Figure 3.53 Comparison of strength curves for leaning column frame subject to weak-axis bending with $L_c/r_y = 20, G_A = 0$ (explicit imperfection modeling method).

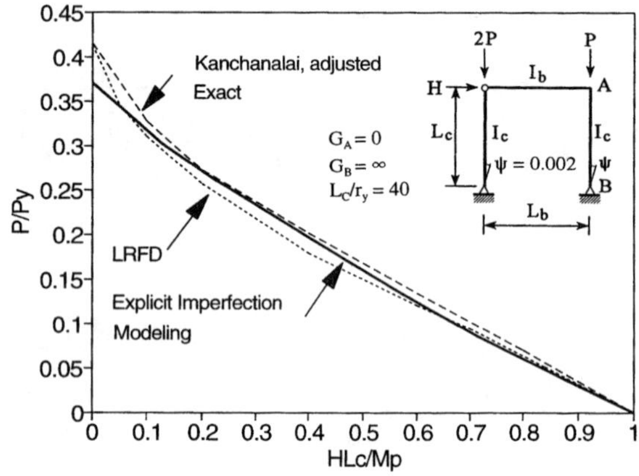

Figure 3.54 Comparison of strength curves for leaning column frame subject to weak-axis bending with $L_c/r_y = 40$, $G_A = 0$ (explicit imperfection modeling method).

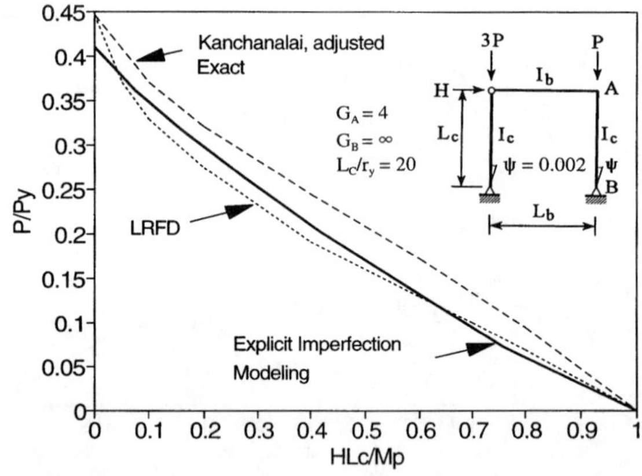

Figure 3.55 Comparison of strength curves for leaning column frame subject to weak-axis bending with $L_c/r_y = 20$, $G_A = 4$ (explicit imperfection modeling method).

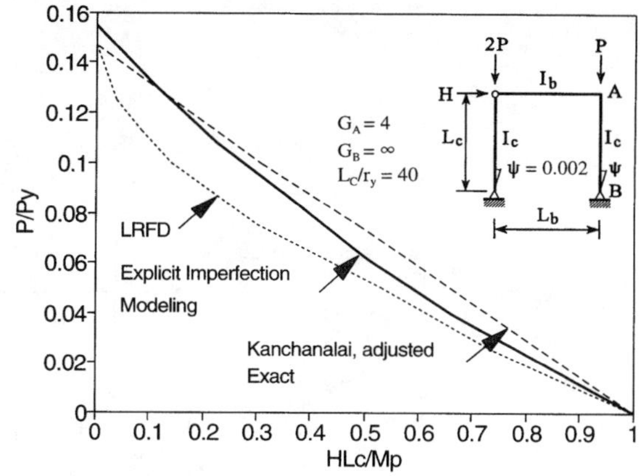

Figure 3.56 Comparison of strength curves for leaning column frame subject to weak-axis bending with $L_c/r_y = 40$, $G_A = 4$ (explicit imperfection modeling method).

Table 3.4 Cross-Section Dimensions and Properties for Vogel's Frames

Member	d (mm)	b (mm)	t_w (mm)	t_f (mm)	k (mm)	A (cm^2)	I_x (cm^4)	I_y (cm^4)	Z_x (cm^3)	$\frac{F_{rc}}{F_y}$
IPE240	240	120	6.2	9.8	15	39.1	3,892	284	367	0.3
IPE300	300	150	7.1	9.8	15	53.8	8,356	604	628	0.3
IPE330	330	160	7.5	10.7	18	62.6	11,770	788	804	0.3
IPE360	360	170	8.0	11.5	18	72.7	16,270	1,043	1,019	0.3
IPE400	400	180	8.6	12.7	21	84.5	23,130	1,318	1,307	0.3
HEB160	160	160	8.0	13.5	15	54.3	2,492	889	354	0.5
HEB200	200	200	9.0	15.0	18	78.1	5,692	2,003	643	0.5
HEB220	220	220	9.5	16.0	18	91.0	8,091	2,843	827	0.5
HEB240	240	240	10.0	17.0	21	106.0	11,260	3,923	1053	0.5
HEB260	260	260	10.0	17.5	24	118.0	14,920	5,135	1283	0.5
HEB300	300	300	11.0	19.0	27	149.0	25,170	8,563	1869	0.5
HEB340	340	300	9.5	16.5	27	133.0	27,690	7,436	1850	0.5

is $0.5F_y$ when the d/b ratio is less than 1.2 (Figure 3.76). The stress-strain relationship is elastic-plastic with strain hardening as shown in Figure 3.77. The geometric imperfection used is $L_c/400$ as calculated by the equation $\Psi_o = \frac{1}{300} r_1 r_2$ in Table 3.5. The load-displacement curves are as shown in Figure 3.78. For comparison, the geometric imperfection for this particular case is adopted as $L_c/400$ identical to Vogel's portal frame, even though the proposed imperfection for the present study is $L_c/500$ for sway frames. For the same reason, the notional load factor is used as $1/400$. The reduced tangent modulus factor of 0.85 is equivalent to imposing a geometric imperfection of $L_c/500$. The difference between the geometric imperfection in Vogel's frame ($L_c/400$) and the proposed imperfection ($L_c/500$) value is to be $L_c/2,000$. To get the

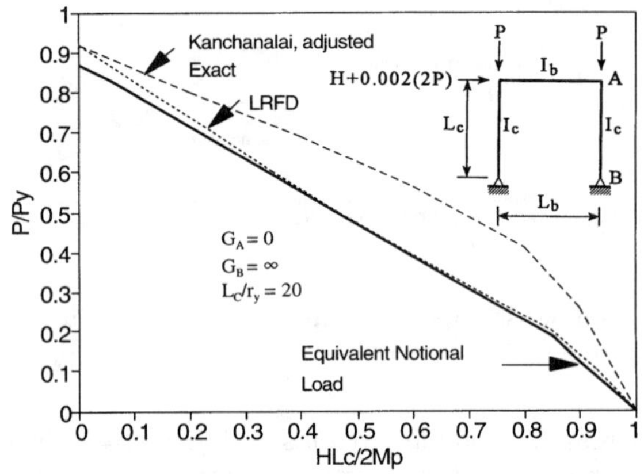

Figure 3.57 Comparison of strength curves for portal frame subject to weak-axis bending with $L_c/r_y = 20$, $G_A = 0$ (equivalent notional load method).

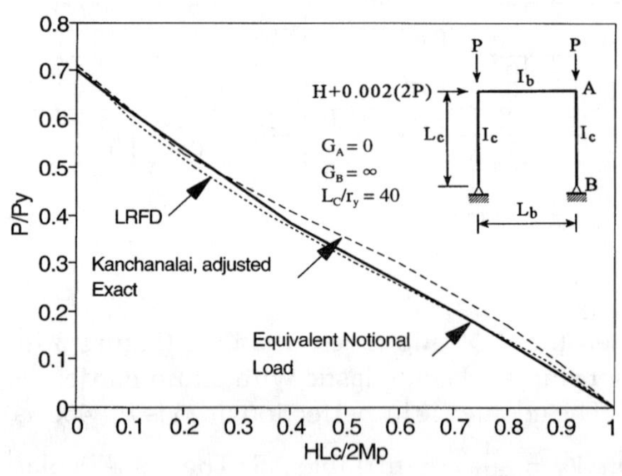

Figure 3.58 Comparison of strength curves for portal frame subject to weak-axis bending with $L_c/r_y = 40$, $G_A = 0$ (equivalent notional load method).

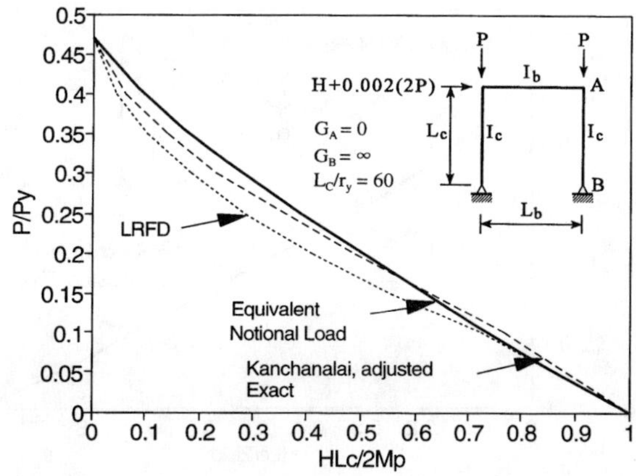

Figure 3.59 Comparison of strength curves for portal frame subject to weak-axis bending with $L_c/r_y = 60$, $G_A = 0$ (equivalent notional load method).

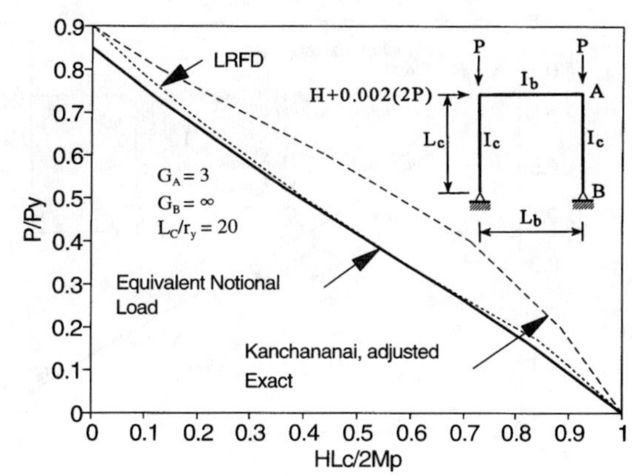

Figure 3.60 Comparison of strength curves for portal frame subject to weak-axis bending with $L_c/r_y = 20$, $G_A = 3$ (equivalent notional load method).

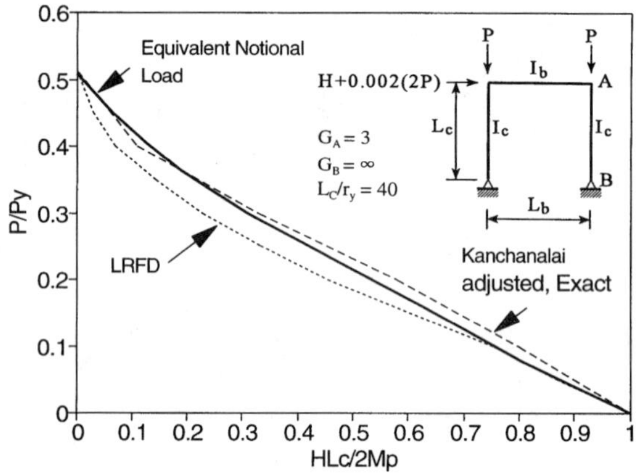

Figure 3.61 Comparison of strength curves for portal frame subject to weak-axis bending with $L_c/r_y = 40, G_A = 3$ (equivalent notional load method).

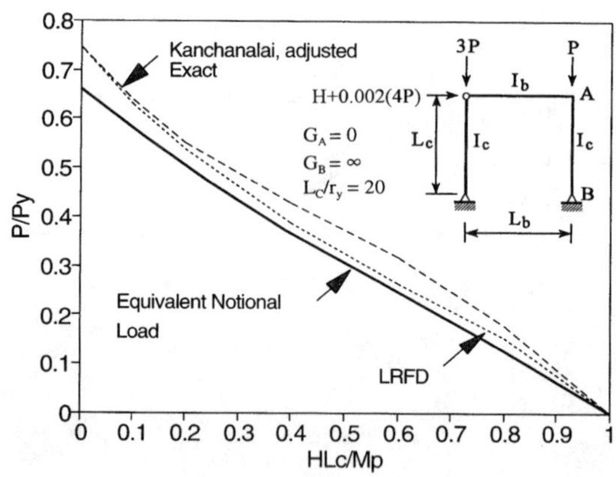

Figure 3.62 Comparison of strength curves for leaning column frame subject to weak-axis bending with $L_c/r_y = 20, G_A = 0$ (equivalent notional load method).

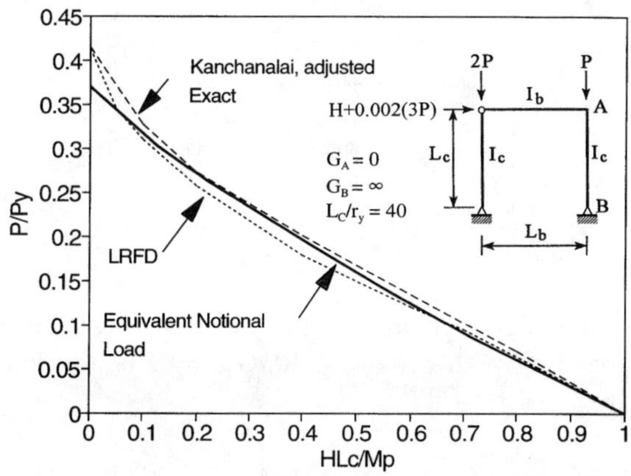

Figure 3.63 Comparison of strength curves for leaning column frame subject to weak-axis bending with $L_c/r_y = 40, G_A = 0$ (equivalent notional load method).

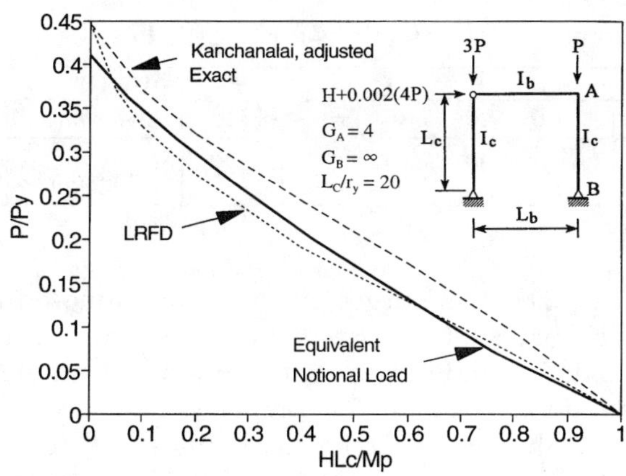

Figure 3.64 Comparison of strength curves for leaning column frame subject to weak-axis bending with $L_c/r_y = 20, G_A = 4$ (equivalent notional load method).

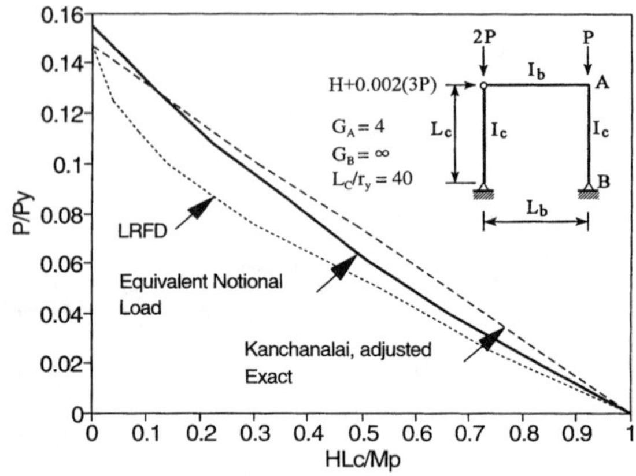

Figure 3.65 Comparison of strength curves for leaning column frame subject to weak-axis bending with $L_c/r_y = 40$, $G_A = 4$ (equivalent notional load method).

same imperfection as Vogel's, the difference $L_c/2,000$ is added to the current model through the explicit geometric imperfection in the further reduced tangent modulus method.

Table 3.5 Geometric Imperfections for Vogel's Frames

Frames	r_1	r_2	Ψ_o	Sway in top	Remark
Portal frame	1.0	$\frac{1}{2}\left(1+\frac{1}{2}\right)=\frac{3}{4}$	1/400	12.5mm	$\Psi_o = \frac{1}{300}r_1r_2$
6-Story frame	Fig. 2.47 1.0	$\frac{1}{2}\left(1+\frac{1}{3}\right)=\frac{2}{3}$	1/450	50.0mm	

The load-displacement curves given by the proposed methods and Vogel's plastic-zone analysis are compared in Figure 3.78. Explicit imperfection modeling and the equivalent notional load methods underestimate the ultimate loads by 6% compared to the Vogel's plastic-zone solution. The further reduced tangent modulus method underestimates the ultimate load by 5% shown in Table 3.6. These conservative errors are small enough for practical design. The proposed methods under-predict the lateral displacements. When the load ratio of service to ultimate load is 0.6, explicit imperfection modeling and the equivalent notional load method under-predicts the displacement by 5% at service conditions, and the further reduced tangent modulus method under-predicts it by 15% (Table 3.7).

The reduced tangent modulus method underestimates the displacement more than explicit imperfection modeling and the equivalent

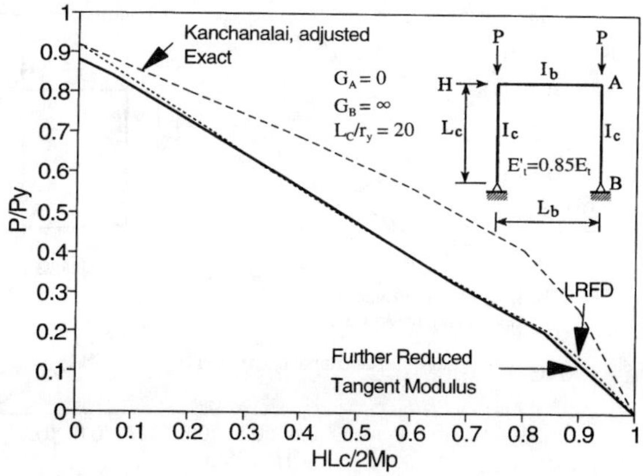

Figure 3.66 Comparison of strength curves for portal frame subject to weak-axis bending with $L_c/r_y = 20, G_A = 0$ (further reduced tangent modulus method).

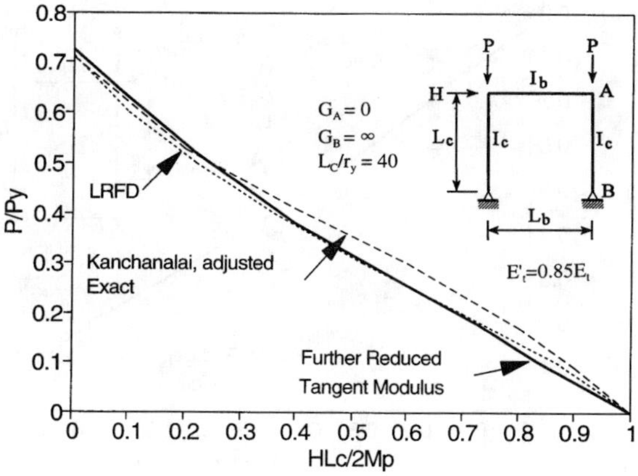

Figure 3.67 Comparison of strength curves for portal frame subject to weak-axis bending with $L_c/r_y = 40, G_A = 0$ (further reduced tangent modulus method).

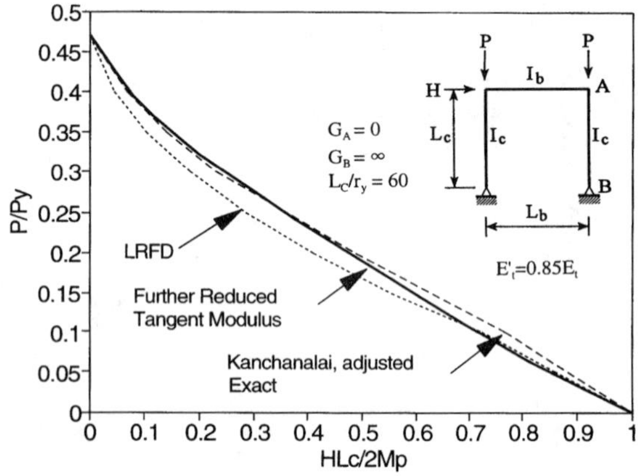

Figure 3.68 Comparison of strength curves for portal frame subject to weak-axis bending with $L_c/r_y = 60, G_A = 0$ (further reduced tangent modulus method).

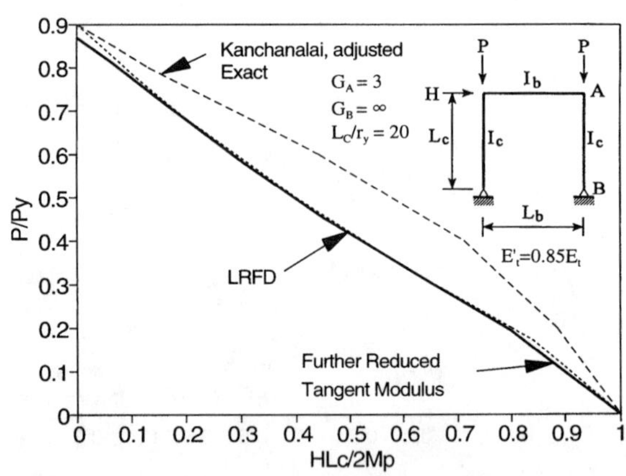

Figure 3.69 Comparison of strength curves for portal frame subject to weak-axis bending with $L_c/r_y = 20, G_A = 3$ (further reduced tangent modulus method).

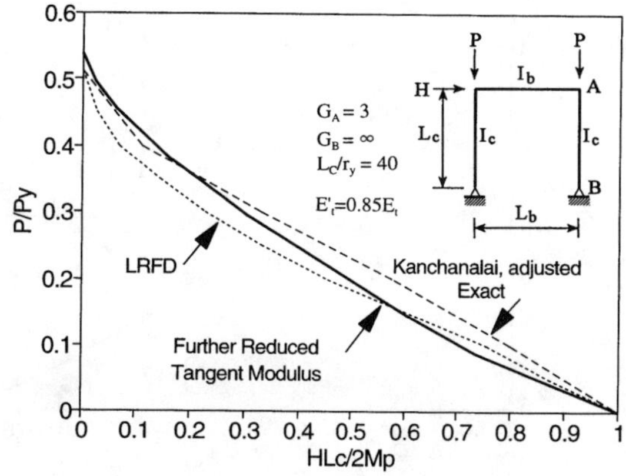

Figure 3.70 Comparison of strength curves for portal frame subject to weak-axis bending with $L_c/r_y = 40$, $G_A = 3$ (further reduced tangent modulus method).

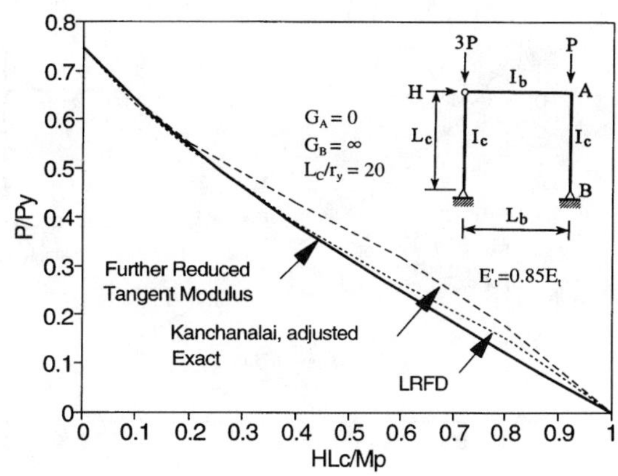

Figure 3.71 Comparison of strength curves for leaning column frame subject to weak-axis bending with $L_c/r_y = 20$, $G_A = 0$ (further reduced tangent modulus method).

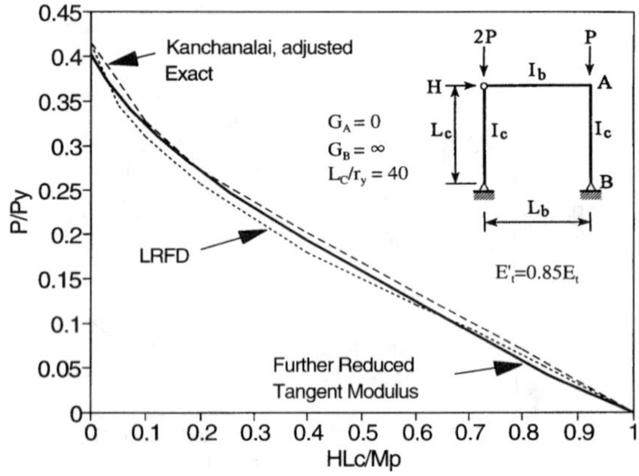

Figure 3.72 Comparison of strength curves for leaning column frame subject to weak-axis bending with $L_c/r_y = 40, G_A = 0$ (further reduced tangent modulus method).

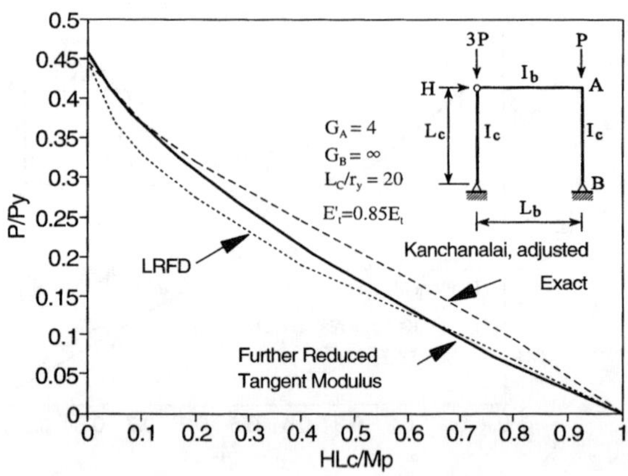

Figure 3.73 Comparison of strength curves for leaning column frame subject to weak-axis bending with $L_c/r_y = 20, G_A = 4$ (further reduced tangent modulus method).

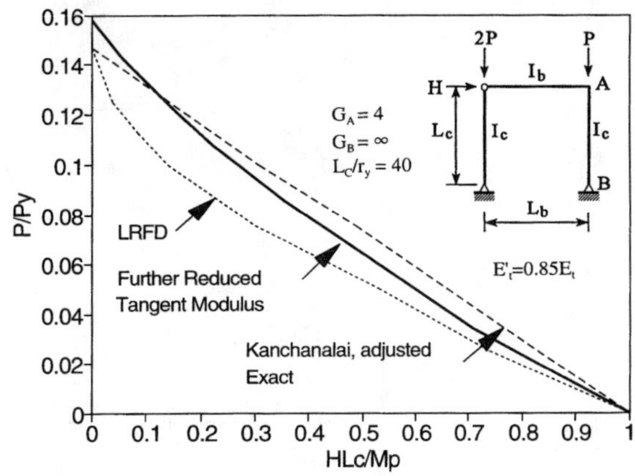

Figure 3.74 Comparison of strength curves for leaning column frame subject to weak-axis bending with $L_c/r_y = 40$, $G_A = 4$ (further reduced tangent modulus method).

Table 3.6 Comparison of Strength for Vogel's Portal Frame

Item	Exact Vogel	I.M. and N.L. method	R.E. method	$\frac{I.M.,N.L.}{Exact}$	$\frac{R.E.}{Exact}$
Load ratio	1.022	0.960	0.973	0.939	0.952

notional load method. This is because the Vogel's portal frame is subject to large gravity loads compared to small horizontal loads. The ratio of gravity to horizontal load is 160, a member rarely seen in real load cases. Under gravity loads, the explicit imperfection modeling and equivalent notional load methods give the lateral displacement resulting from the $P - \Delta$ effect by the imperfection modeling or the notional loads. On the other hand, the further reduced tangent modulus method does not produce the lateral displacement under the gravity loads because the geometric imperfections are replaced by the deterioration ($0.85E_t$) of the internal strength. As a result, the reduced tangent modulus method under-predicts the lateral displacement when gravity loads are domi-

Table 3.7 Comparison of Displacement for Vogel's Portal Frame

Load condition	Exact Vogel (cm)	I.M. and N.L. method (cm)	R.E. method (cm)	$\frac{I.M.,N.L.}{Exact}$	$\frac{R.E.}{Exact}$
Service load	0.460	0.438	0.390	0.952	0.848
Ultimate load	1.670	1.430	1.384	0.856	0.829

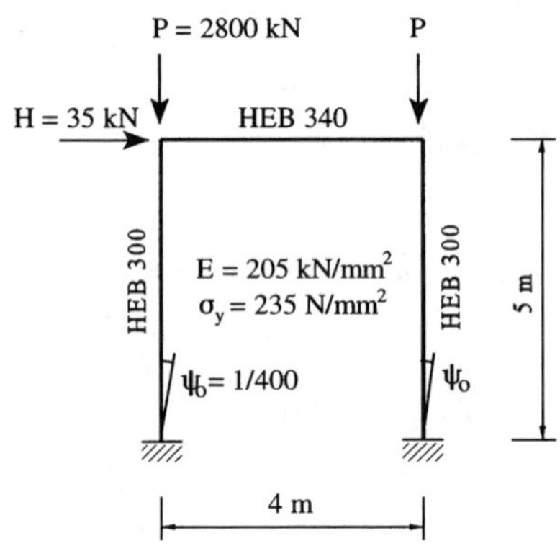

Figure 3.75 Configuration and load condition of Vogel's portal frame for verification study.

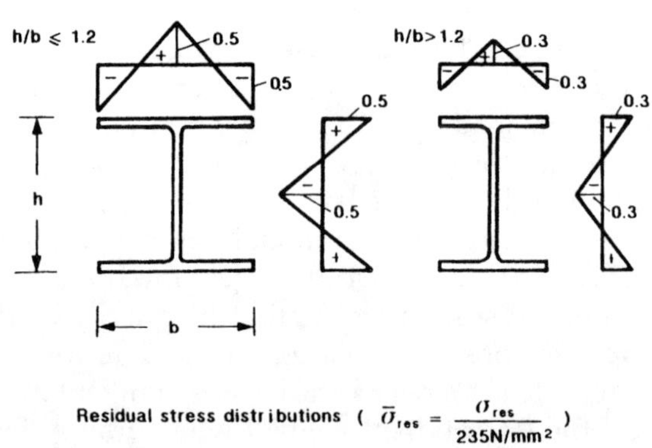

Figure 3.76 Residual stresses of cross-section for Vogel's frame.

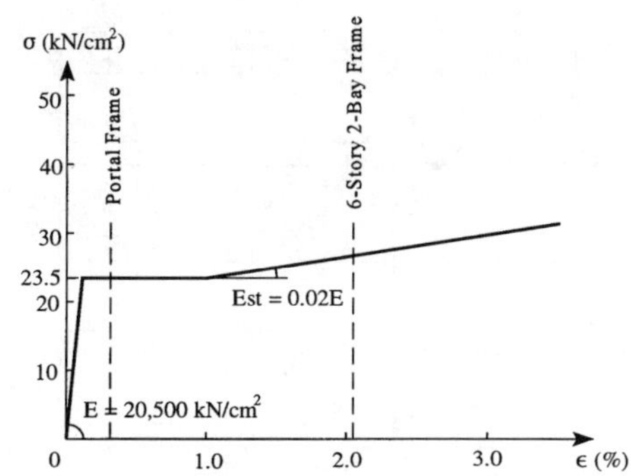

Figure 3.77 Stress-strain relationships for Vogel's frame.

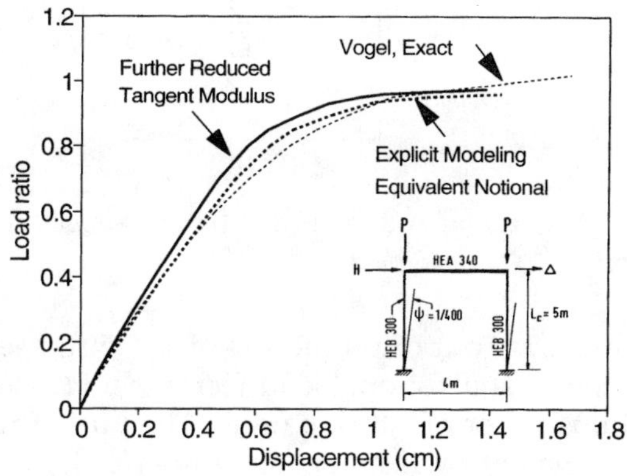

Figure 3.78 Comparison of displacements for Vogel's portal frame.

nant. Conversely, the reduced tangent modulus method over-predicts the lateral displacement up to 15% when the horizontal loads are dominant. In practical design, these errors in deflection prediction by the further reduced tangent modulus can be avoided because serviceability checks can be carried out by an elastic analysis which separates strength checks. The elastic analysis for serviceability checks can be performed easily with the proposed analysis program without using any other programs.

Six-Story Frame

Vogel [15] presented the load-displacement relationships of a 6-story frame using plastic-zone analysis. The configuration and imperfections are shown in Figure 3.79. All members are hot-rolled with properties as listed in Table 3.4. The residual stresses and the stress-strain relationship are the same as those of Vogel's portal frame. The geometric imperfections are $L_c/450$.

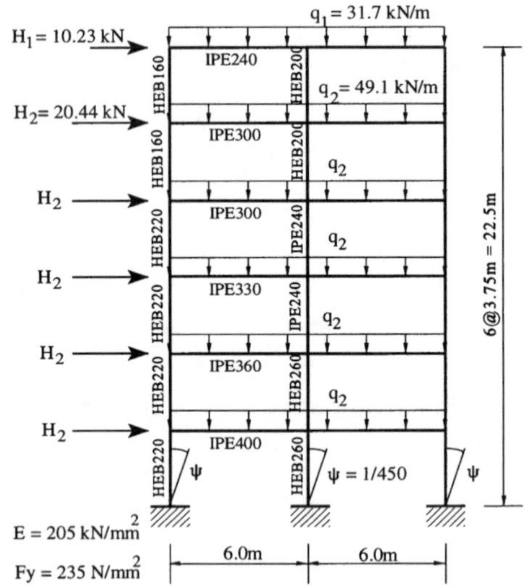

Figure 3.79 Configuration and load condition of Vogel's six-story frame for verification study.

For comparison, the out-of-plumbness of $L_c/450$ is used in the explicit modeling method. The notional load factor of 1/450 and the reduced tangent modulus factor of 0.85 are used. The further reduced tangent modulus is equivalent to the geometric imperfection of $L_c/500$. Thus, the geometric imperfection of $L_c/4500$ is additionally modeled in the further reduced tangent modulus method, where $L_c/4500$ is the difference between the Vogel's geometric imperfection of $L_c/450$ and the proposed geometric imperfection of $L_c/500$.

The load-displacement curves and the proposed methods, together with the Vogel's plastic-zone analysis, are compared in Figure 3.80. The errors in strength prediction by the proposed methods are less than 1% as shown in Table 3.8. Explicit imperfection modeling and the equivalent notional load method under-predict lateral displacements by 3%, and the further reduced tangent modulus method shows a good agreement in displacement with the Vogel's exact solution. Vogel's frame is a good example of how the reduced tangent modulus method predicts lateral displacement well under reasonable load combinations (Table 3.9).

Table 3.8 Comparison of Strength for Vogel's 6-Story Frame

Item	Exact Vogel	I.M. and N.L. method	R.E. method	$\frac{I.M.,N.L.}{Exact}$	$\frac{R.E.}{Exact}$
Load ratio	1.111	1.117	1.120	1.005	1.008

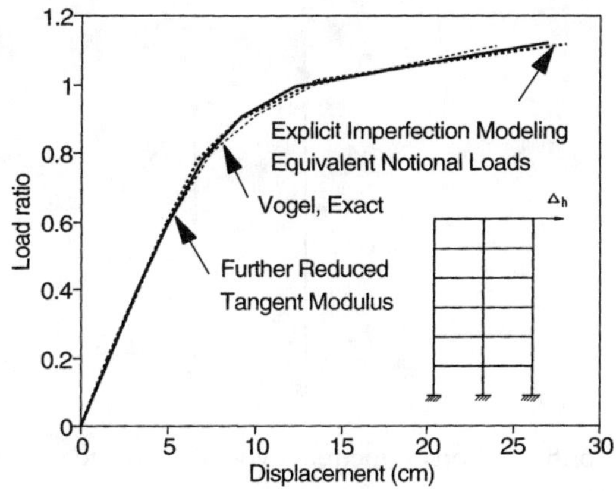

Figure 3.80 Comparison of displacements for Vogel's six-story frame.

Table 3.9 Comparison of Displacement for Vogel's 6-Story Frame

Load condition	Exact Vogel (cm)	I.M. and N.L. method (cm)	R.E. method (cm)	$\frac{I.M.,N.L.}{Exact}$	$\frac{R.E.}{Exact}$
Service load	5.0	4.88	5.01	0.976	1.002
Ultimate load	24.0	28.1	27.0	1.171	1.125

3.8 Special Frames

In general, braced frames have an effective length factor K less than 1.0, and unbraced frames have one greater than 1.0. However, this general rule may be violated when the frames are anti-symmetric geometry. In the following, the strengths are calculated for the special frames including a two-story braced column with a K greater than 1.0 and an unbraced frame with a K-factor less than 1.0.

3.8.1 Braced Column with K-factor Greater Than 1.0

Figure 3.81 shows a two-story column having a lower of 15 ft and an upper of 10 ft subject to an axial force P. All members are laterally braced, and W8x31 is used. Since the shorter upper column helps the strength of the longer lower column, the upper column has the K-factor greater than 1.0 although it is braced.

When the strength of this column is calculated by AISC-LRFD procedures, the alignment chart method does not accurately predict the strength of a column with unequal braced lengths, and thus differential

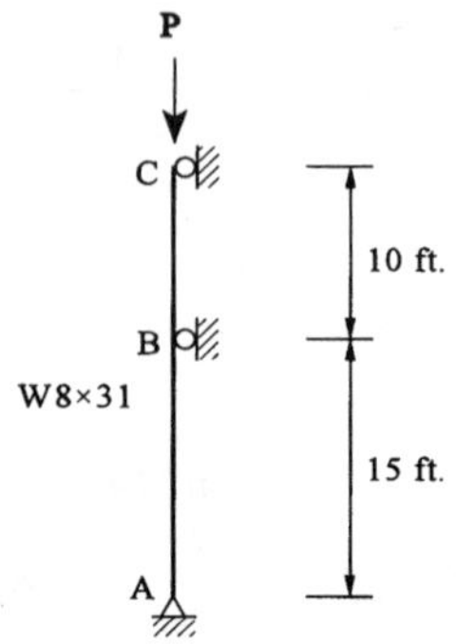

Figure 3.81 Configuration of unequal two-story braced column with K-factor greater than 1.0.

equation or slope-deflection equations should be used to calculate K. In the first case, a fourth-order differential equation can be solved for K. The column is divided into two segments, (1 and 2 shown in Figure 3.82), and equations are written for each. For convenience, two sets

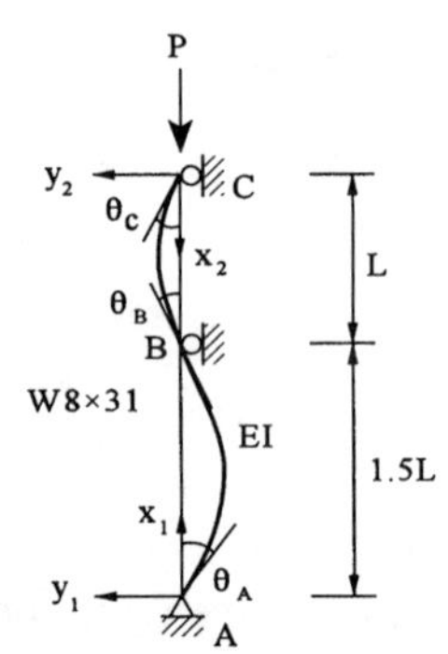

Figure 3.82 Coordinates of unequal two-story braced column.

of coordinate axes are established: $x_1 - y_1$ for segment 1 and $x_2 - y_2$ for segment 2. The fourth-order differential equation for segment 1 is:

$$y_1^{iv} + k^2 y_1'' = 0 \tag{3.2}$$

where $k^2 = P/EI$ and the general solution is

$$y_1 = A \sin kx_1 + B \cos kx_1 + Cx_1 + D \tag{3.3}$$

Similarly, the general solution for segment 2 is

$$y_2 = E \sin kx_2 + F \cos kx_2 + Gx_2 + H \tag{3.4}$$

By using boundary conditions

$$y_1(0) = 0 \tag{3.5a}$$

$$y_1''(0) = 0 \tag{3.5b}$$

$$y_1\left(\frac{3}{2}L\right) = 0 \tag{3.5c}$$

$$y_2(0) = 0 \tag{3.5d}$$

$$y_2''(0) = 0 \tag{3.5e}$$

$$y_2(L) = 0 \tag{3.5f}$$

and the continuity conditions

$$y_1'\left(\frac{3}{2}L\right) = -y_2'(L) \tag{3.6a}$$

$$y_1''\left(\frac{3}{2}L\right) = y_2''(L) \tag{3.6b}$$

It can be concluded that

$$B = D = F = H = 0 \tag{3.7}$$

and

$$\begin{bmatrix} \sin \frac{3}{2}L & \frac{3}{2}L & 0 & 0 \\ 0 & 0 & \sin kL & L \\ k \cos \frac{3}{2}kL & 1 & k \cos kL & 1 \\ -\sin \frac{3}{2}kL & 0 & \sin kL & 0 \end{bmatrix} \begin{bmatrix} A \\ C \\ E \\ G \end{bmatrix} = \begin{bmatrix} 0 \\ 0 \\ 0 \\ 0 \end{bmatrix} \tag{3.8}$$

The minus sign in Equation (3.6a) indicates a positive slope at the intermediate support with respect to the $x_1 - y_1$ axes and corresponds to a negative slope in the $x_2 - y_2$ axes. For a nontrivial solution, the determinant of the coefficient matrix of Equation (3.8) must vanish

$$\det\begin{vmatrix} \sin\frac{3}{2}kL & \frac{3}{2}L & 0 & 0 \\ 0 & 0 & \sin kL & L \\ k\cos\frac{3}{2}kL & 1 & k\cos kL & 1 \\ -\sin\frac{3}{2}kL & 0 & \sin kL & 0 \end{vmatrix} = 0 \tag{3.9}$$

Simplifying, the characteristic equation is

$$5\sin\frac{3}{2}kL - 3kL\sin\frac{5}{2}kL = 0 \tag{3.10}$$

By trial and error, the smallest value of k that satisfies the characteristic equation is

$$k = \frac{2.427}{L} \tag{3.11}$$

Using $k^2 = P/EI$, we have

$$P_{cr} = 5.89\frac{EI}{L^2} \tag{3.12}$$

The Euler buckling load of the lower column (segment)

$$P_{e1} = \frac{\pi^2 EI}{(1.5L)^2} = 4.386\frac{EI}{L^2} \tag{3.13}$$

Therefore, the effective length factor K is

$$K_1 = \sqrt{\frac{P_{e1}}{P_{cr}}} = 0.863 \tag{3.14}$$

Similarly, the buckling load of the upper column (Segment 2) is

$$P_{e_2} = \frac{\pi^2 EI}{L^2} \tag{3.15}$$

and the K-factor is

$$K_2 = \sqrt{\frac{P_{e2}}{P_{cr}}} = 1.294 \tag{3.16}$$

The K-factor can also be determined by the slope-deflection method. Referring to Figure 3.82 and using slope-deflection equations, we have:

Column AB

$$M_{AB} = \frac{EI}{L_1}\left(s_{ii_1}\theta_A + s_{ij_1}\theta_B\right) \tag{3.17a}$$

$$M_{BA} = \frac{EI}{L_1}\left(s_{ii_1}\theta_B + s_{ij_1}\theta_A\right) \tag{3.17b}$$

Column BC

$$M_{BC} = \frac{EI}{L_2}\left(s_{ii_2}\theta_B + s_{ij_2}\theta_C\right) \tag{3.18a}$$

$$M_{CB} = \frac{EI}{L_2}\left(s_{ii_2}\theta_C + s_{ij_2}\theta_B\right) \tag{3.18b}$$

The conditions $M_{AB} = 0$ and $M_{CB} = 0$ lead to

$$\theta_A = -\frac{s_{ij_1}}{s_{ii_1}}\theta_B \tag{3.19a}$$

$$\theta_C = -\frac{s_{ij_2}}{s_{ii_2}}\theta_B \tag{3.19b}$$

Therefore, the end moments M_{BA} and M_{BC} are reduced to

$$M_{BA} = \frac{EI}{L_1}\left(s_{ii_1} - \frac{s_{ij_1}^2}{s_{ii_1}}\right)\theta_B \tag{3.20a}$$

$$M_{BC} = \frac{EI}{L_2}\left(s_{ii_2} - \frac{s_{ij_2}^2}{s_{ii_2}}\right)\theta_B \tag{3.20b}$$

From joint equilibrium

$$M_{BA} + M_{BC} = 0 \tag{3.21}$$

$$\frac{EI}{L_1}\left(s_{ii_1} - \frac{s_{ij_1}^2}{s_{ii_1}}\right)\theta_B + \frac{EI}{L_2}\left(s_{ii_2} - \frac{s_{ij_2}^2}{s_{ii_2}}\right)\theta_B = 0 \tag{3.22}$$

Simplified,

$$\frac{1}{1.5}\left(s_{ii_1} - \frac{s_{ij_1}^2}{s_{ii_1}}\right) + \left(s_{ii_2} - \frac{s_{ij_2}^2}{s_{ii_2}}\right) = 0 \tag{3.23}$$

Using Table 2.1, we obtain by trial-and-error

$$kL_1 = 3.637 \qquad kL_2 = 2.425 \tag{3.24}$$

We have

$$P_{cr} = 13.23\frac{EI}{L_1^2} = 5.88\frac{EI}{L_2^2} \tag{3.25}$$

The K-factors are

$$K_1 = \sqrt{\pi^2/13.23} = 0.864 \tag{3.26a}$$

$$K_2 = \sqrt{\pi^2/5.88} = 1.296 \tag{3.26b}$$

The K-factor is greater than 1.0 for Column 2 in the braced frame, and this constitutes a "special" frame. The same K is obtained by the use of both the differential and the slope-deflection equations. The slenderness parameter λ_c is

$$\lambda_{c1} = \frac{K_1 L_1}{\pi r_x}\sqrt{\frac{F_y}{E}} = \frac{(0.864)(15)(12)}{(\pi)(3.47)}\sqrt{\frac{36}{29000}} = 0.503 \tag{3.27a}$$

$$\lambda_{c2} = \frac{K_2 L_2}{\pi r_x}\sqrt{\frac{F_y}{E}} = \frac{(1.296)(10)(12)}{(\pi)(3.47)}\sqrt{\frac{36}{29000}} = 0.503 \tag{3.27b}$$

Since the slenderness parameters are the same for the upper and lower column, the nominal strength is identical as:

$$\begin{aligned} P_n &= A_g\left(0.658^{\lambda_c^2}\right)F_y \\ &= (9.13)(0.658)^{(0.503)^2}(36) \qquad (3.28) \\ &= 296 \text{ kips} \end{aligned}$$

In explicit imperfection modeling, the two-element columns are assumed to have an initial geometric imperfection of 0.01 and 0.015 feet at mid-height of the upper and lower column. The predicted column strength is 280 kips, and the error is about 5% compared to the LRFD method. In the equivalent notional load method, the notional load of 0.004 times the gravity load is applied at the mid-height of the upper and lower column. The strength is identical to that of explicit imperfection modeling. In the further reduced tangent modulus method, the reduced tangent modulus factor of 0.85 leads to a strength of 284 kips corresponding well with the LRFD column strength with the error of less than 4%.

3.8.2 Unbraced Frame with K-factor Less Than 1.0

Figure 3.83 shows an unbraced portal frame consisting of a rigid beam and W8x31 columns. The left column is weaker than the right column because the weak-axis of the left column is in the major bending plane. The frame is fully braced from out-of-plane movement. The frame is subject to gravity and one lateral load. The right column helps the left resist story buckling because it is stronger. The effect is the left column has K less than 1.0.

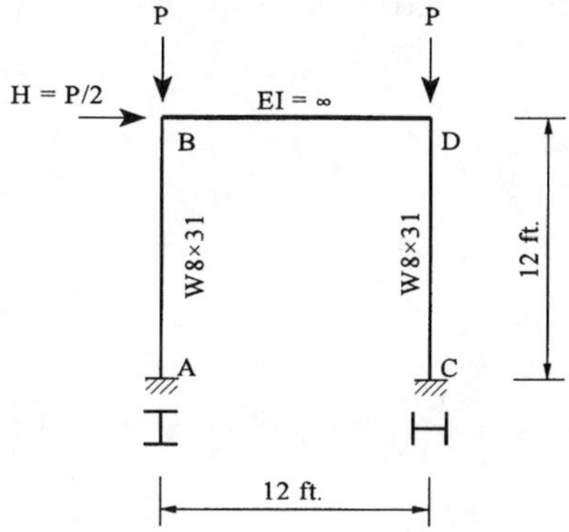

Figure 3.83 Configuration of unbraced frame with K-factor less than 1.0.

In this study, the AISC-LRFD interaction curves are used for strength comparisons. The strength calculations are based on LeMessurier's K method [13] because it accounts for story buckling and gives a more accurate beam-column strength prediction.

Using the nomograph shown in Figure 1.12, we obtain

$$K_{AB} = K_{CD} = 1 \tag{3.29}$$

The axial forces in the column, from simple statics (Figure 3.84), are

$$P_{AB} = \frac{3}{4}P, \qquad P_{CD} = \frac{5}{4}P \tag{3.30}$$

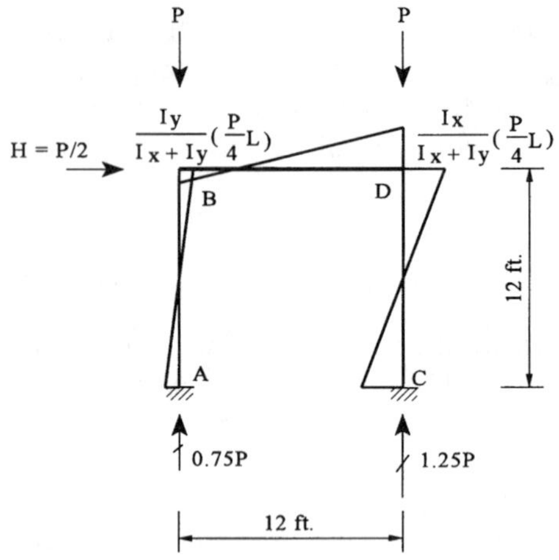

Figure 3.84 Member forces of the unbraced frame.

The modified effective length factors for the columns are

$$K'_{AB} = \sqrt{\frac{\sum P_u}{P_{ui}} \times \frac{I_i}{\sum (I_i/K_i^2)}} = \sqrt{\frac{2P}{(3/4)P} \times \frac{37.1}{110 + 37.1} \times (1.0)^2} = 0.820 \tag{3.31a}$$

$$K'_{CD} = \sqrt{\frac{2P}{(5/4)P} \times \frac{110}{110 + 37.1} \times (1.0)^2} = 1.094 \tag{3.31b}$$

The K-factor is less than 1.0 for Column AB in an unbraced frame, and therefore it is considered a "special" frame.

The nominal load P can now be determined by using interaction equations. For Column CD, the slenderness parameter is

$$\lambda_{cx} = \frac{K'_{CD} L}{\pi r_x}\sqrt{\frac{F_y}{E}} = \frac{(1.094)(12 \times 12)}{(\pi)(3.47)}\sqrt{\frac{36}{29,000}} = 0.509 \tag{3.32}$$

The nominal axial strength is

$$P_n = A_g \left(0.658^{\lambda_c^2}\right) F_y = (9.13)(0.658)^{(0.509)^2}(36) = 294.9 \text{ kips} \tag{3.33}$$

The second-order moment is

$$M_u = B_1 M_{nt} + B_2 M_{lt} = 0 + B_2 M_{lt} = B_2 M_{lt} \tag{3.34}$$

in which, the B_2-factor is

$$B_2 = \frac{1}{1 - \frac{\sum P}{\sum P_{e2}}} = \frac{1}{1 - \frac{2P}{2{,}030}} = \frac{1}{1 - 0.001P} \tag{3.35}$$

where

$$\sum P_{e_2} = \frac{\pi^2 E\left(I_x + I_y\right)}{L^2} = \frac{\pi^2(29,000)(110 + 37.1)}{(12)(12)} = 2,030 \tag{3.36}$$

$$M_{lt} = \tfrac{I_x}{I_x+I_y}\left(\tfrac{P}{4}L\right) = \tfrac{110}{110+37.1} \times \tfrac{P}{4} \times (12)(12) = 26.9P \tag{3.37}$$

Therefore, we have

$$M_u = \frac{26.9P}{1 - 0.001P} \tag{3.38}$$

Since the column is laterally braced, the nominal flexural strength is equal to the full plastic strength as

$$M_p = (Z)\left(F_y\right) = (30.4)(36) = 1,094.4 \tag{3.39}$$

Use the interaction equation assuming $P/P_n < 0.2$

$$\frac{P_u}{2P_n} + \frac{M_u}{M_n} = \frac{P}{(2)(294.9)} + \frac{\frac{26.9P}{1-0.001P}}{1094.4} = 1 \tag{3.40}$$

Note that the resistance factors are taken as unity for nominal strengths.

Solving Equation (3.40), we obtain a nominal load P of 36.7 kips satisfying the assumption $P/P_n < 0.2$.

We have to check the adequacy of Column AB with $P = 36.7$ kips in the following.

$$\lambda_c = \frac{\left(K'_{AB}\right)(L)}{\pi r_y}\sqrt{\frac{F_y}{E}} = \frac{(0.820)(12)(12)}{(\pi)(2.02)}\sqrt{\frac{36}{29,000}} = 0.656$$

$$P_n = (9.13)(0.658)^{(0.656)^2}(36) = 274.5 \text{ kips}$$

$$M_{lt} = \frac{I_y}{I_x + I_y}\left(\frac{P}{4}L\right) = \frac{37.1}{110 + 37.1}\left(\frac{36.7}{4}\right)(12)(12) = 333.2 \text{ k-in}$$

$$B_2 = \frac{1}{1 - 0.001P} = \frac{1}{1 - 0.001 \times 36.7} = 1.038$$

$$M_u = B_2 M_{lt} = (1.038)(333.2) = 345.9 \text{ k-in}$$

$$M_n = M_p = (14.1)(36) = 507.6 \text{ k-in}$$

$$\frac{P}{P_n} = \frac{36.7}{274.5} = 0.134 < 0.2$$

$$\frac{P}{2P_n} + \frac{M_u}{M_n} = \frac{0.134}{2} + \frac{345.9}{507.6} = 0.748 < 1.0$$

Therefore, Column AB is adequate. The nominal load is determined as 36.7 kips.

When a geometric imperfection of $L_c/500$ is used, the predicted strength is 38.2 kips (4% greater than the AISC-LRFD method estimate). This is because advanced methods allow for inelastic moment redistribution. This implies that the advanced analysis uses the full strength of Column AB as well as Column CD but the LRFD method use the full strength of Column CD but not AB (remaining strength equal to 25.2% = 1 − 0.748). When a notional load factor of 0.002 is used, the strength predicted by this method is the same as that predicted by the explicit imperfection modeling method. When the reduced tangent modulus factor of 0.85 is used, the strength is 38.1 kips (4% greater than that of the AISC-LRFD method).

3.9 Semi-Rigid Frames

In the open literature, no benchmark problems solving semi-rigid frames with geometric imperfections are available for a verification study. An alternative is to separate the effects of semi-rigid connections and geometric imperfections. In previous sections, the geometric imperfections were studied and comparisons between proposed methods, plastic-zone analyses and conventional LRFD methods were made. Herein, the effect of semi-rigid connections will be verified by comparing analytical and experimental results.

3.9.1 Displacement Characteristics

Figure 3.85 shows the displacement of the frames with hinged, rigid, and semi-rigid connections. The frames with hinged and semi-rigid connections are modifications of the Vogel's portal frame with rigid connections (Figure 3.75). The three parameters relating to semi-rigid connections are: (1) an initial connection stiffness R_{ki} of 280,000 kip-in/rad, (2) an ultimate moment capacity M_u of 1,250 kip-in, and (3) a shape parameter

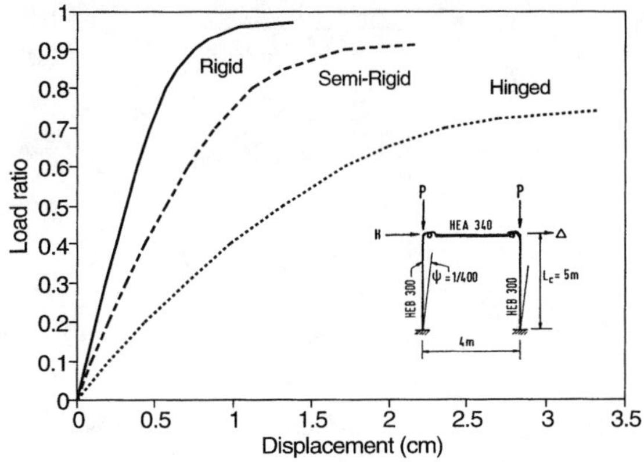

Figure 3.85 Comparison of displacement characteristics for hinged, rigid, and semi-rigid frame.

n of 0.98. As expected, the displacement and the strength of the semi-rigid frame are between those of the rigid and the hinged frames. If the initial connection stiffness R_{ki} and the ultimate moment capacity M_u of the semi-rigid frame approach infinity, the displacement and strength approach those of the rigid frame. Conversely, if they approach zero, the displacement and strength of the frame match those of the hinged frame. Through this study, the envelopes bounding the upper and lower strengths of a semi-rigid frame with various connection stiffness are checked and compared to the rigid and hinged frames.

3.9.2 Comparison with Analytical Result

Barakat [3] studied several frames with semi-rigid connections. The two-story, one-bay frame among those frames is selected as a benchmark for the present study. The configuration of the frame is shown in Figure 3.86. The connection is a top and seat angle with double web angles. The experimental moment-rotation relationship is shown in Figure 3.87. In Barakat's studies, after gravity loads are applied as the first loading sequence, the horizontal loads are added. Exact second-order elastic analyses were carried out using the computer program FLFRM developed by Goto and Chen [8]. It should be mentioned that the program FLFRM does not consider geometric imperfections or residual stresses.

To compare the results of the program FLFRM, geometric imperfections are not now considered. The three-parameters power model is used to represent the experimental connection curve. Three parameters, the same as those presented in Section 3.9.1, are used. The com-

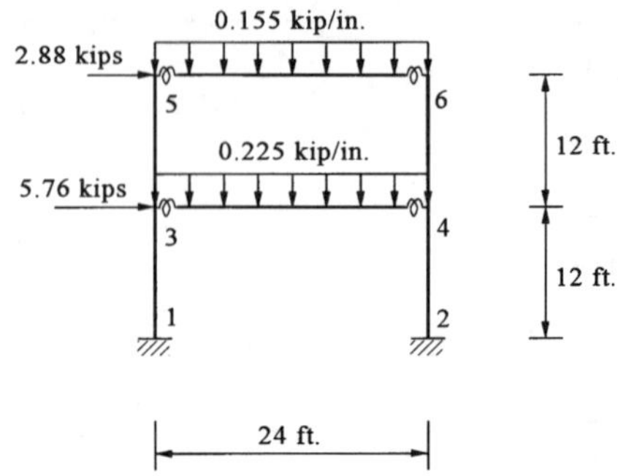

Figure 3.86 Configuration and load condition of two-story semi-rigid frame for verification study.

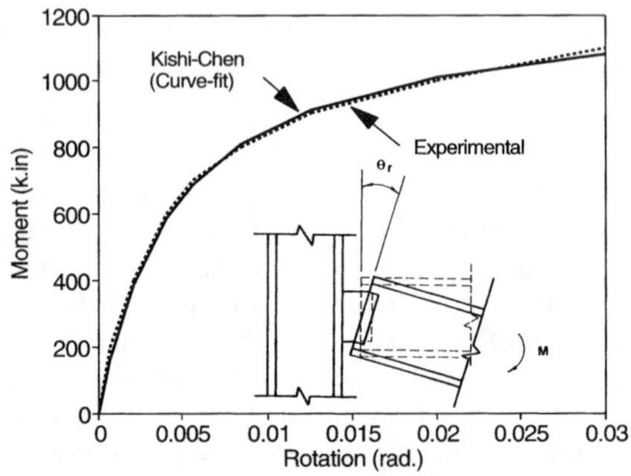

Figure 3.87 Comparison of moment-rotation relationships of semi-rigid connection by experiment and curve-fitting.

parison between the experimental curve and the curve using the three-parameter power model (by a curve-fit) is shown in Figure 3.87. The stress state under the given loading condition is confirmed as elastic, and thus corresponds to Barakat's results. The column moments are compared in Table 3.10. The proposed method predicts the column moments, with less than 1% error when compared to exact solutions.

3.9.3 *Comparison with Experimental Result*

Stelmack [14] studied the experimental response of two flexibly connected steel frames. A two-story, one-bay frame in his study is selected as a benchmark for the present study. The frame was fabricated from A36 W5x16 sections, with pinned base supports (Figure 3.88). The connections were bolted top and seat angles (L4x4x1/2) made of A36 steel

Table 3.10 Comparison of Column Moment of Semi-Rigid Frame

Column moment	Exact [8] (kip-in)	Proposed method (kip-in)	Proposed/exact
M_2	543	536	0.99
M_6	691	682	0.99

and A325 3/4"D bolts (Figure 3.89). The experimental moment-rotation relationship is shown in Figure 3.90. A gravity load of 2.4 kips was applied at third points along the beam at the first level, followed by a lateral load application. The lateral load-displacement relationship was provided in Stelmack's.

Herein, the three parameters of the power model are determined by curve-fitting and the program 3PARA.f is presented in Section 2.5.5. The three parameters obtained by the curve-fit are R_{ki} = 40,000 k-in/rad., M_u = 220 k-in, and n = 0.91. To determine the parameters by the program, the input data for connection detail shown in Figure 3.89 is as shown in Table 3.11. We obtain three parameters of R_{ki} = 29,855 kips/rad, M_u = 185 k-in, and n = 1.646 with 3PARA.f. The output CONN.OUT is listed in Table 3.12.

Table 3.11 Input Data for the Top- and Seat-Angle Connection

1	36	29000			
5	0.5	0.875	2.5	1.25	5.01

The moment-rotation curves given by experiment and curve-fitting show good agreement (Figure 3.90). The parameters given by the Kishi-Chen equations and by experiment show some diviation (Figure 3.90). In spite of this difference, the Kishi-Chen equations, using the computer program 3PARA.f, are a more practical alternative in design since experimental moment-rotation curves are not usually available [12]. In the analysis, the gravity load is first applied, then the lateral load. The lateral displacements given by the proposed methods and by the experimental compare well (Figure 3.91). The proposed method adequately predicts the behavior and strength of semi-rigid connections.

3.10 Summary

The proposed practical advanced analysis methods have been verified for a wide range of frames including columns with various slenderness ratios; braced, unbraced, leaning column, and semi-rigid frames;

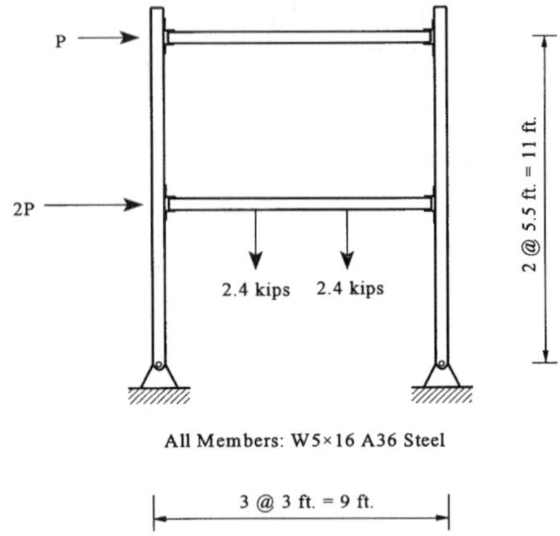

Figure 3.88 Configuration and load condition of Stelmack's two-story semi-rigid frame.

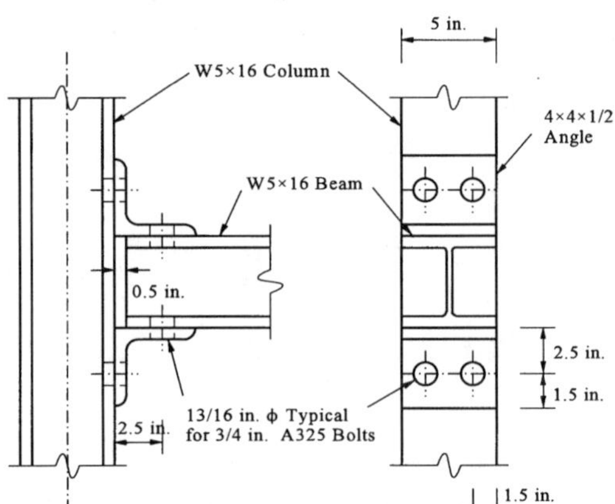

Figure 3.89 Top- and seat-angle connection details.

and strong- and weak-axis bending. Verification studies are performed comparing with plastic-zone solutions available in the literature as well as the conventional LRFD solutions. The methods adequately assess the strengths when compared with exact plastic-zone solutions. The maximum unconservative errors are no more than 5% for a wide range of columns, beam-columns, and frames. The strength predictions by the proposed methods are equivalent to or better than those predicted by the LRFD method. Each of the three imperfection models (explicit imperfection, equivalent notional loads, and further reduced tangent modulus) predicts exactly the same ultimate strength for mathematically identical column members with various boundary conditions. The proposed methods predict well the displacements of frames and

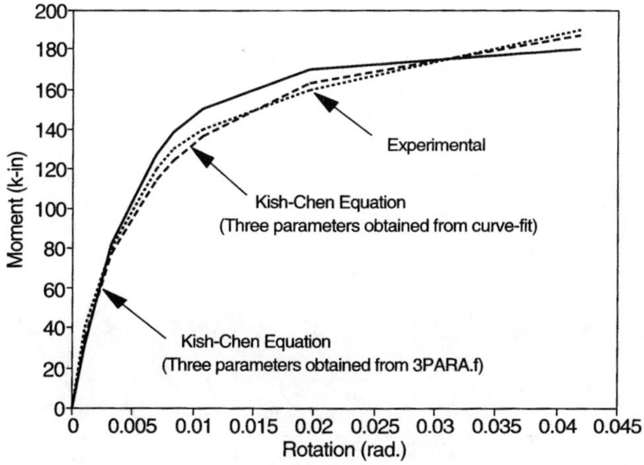

Figure 3.90 Comparison of moment-rotation relationships of semi-rigid connection by experiment and Kishi-Chen equation.

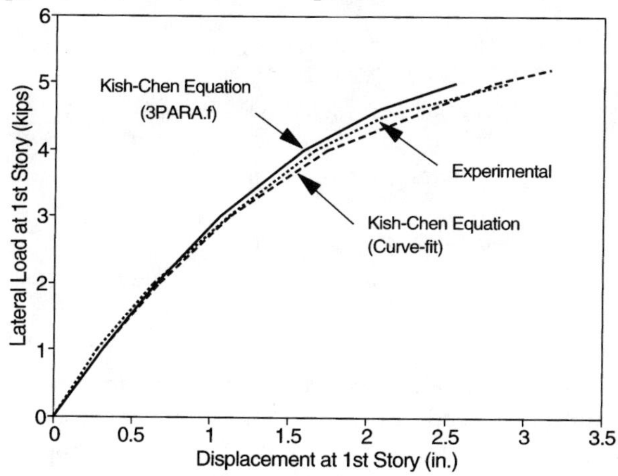

Figure 3.91 Comparison of displacements of Stelmack's two-story semi-rigid frame.

Table 3.12 Output of Three Parameters for the Top and Seat-Angle Connection

```
INPUT DATA
==========

    CONNECTION TYPE:
    ----------------

       TOP- AND SEAT- ANGLE CONNECTION

    MATERIAL PROPERTY:
    ------------------

       Fy = 36.0000     E  = 29000.0

    DIMENSIONS RELATED TO TOP ANGLE:
    --------------------------------

       lt =  5.0000     tt =  0.5000    kt =  0.8750

       gt =  2.5000     W  =  1.2500    d  =  5.0100


    DIMENSIONS RELATED TO BOTTOM ANGLE:
    -----------------------------------

       ls =  5.0000     ts =  0.5000    ks =  0.8750

       gs =  2.5000     W  =  1.2500


  OUTPUT OF THREE PARAMETERS
  ==========================

       INITIAL STIFFNESS, Rki =   29855.14


       ULTIMATE MOMENT,    Mu  =     184.65


       SHAPE PARAMETER,    n   =      1.646
```

rigidly jointed trusses when compared with plastic-zone solutions. The proposed methods predict the strength of special frames, including a two-story braced column with the K-factor greater than 1.0 and an unbraced frame with the K-factor less than 1.0, with reliable accuracy. The connection parameters (such as the initial stiffness R_{ki}, the ultimate moment M_u, and the shape parameter n) may be easily determined by the proposed program (3PARA.f) with an acceptable error compared to experimental data. The behavior and strength of semi-rigid frames may be predicted by any of these methods. The appropriate magnitudes of imperfections associated with the three modelings are recommended, based on the foregoing verification study, as follows:

- Explicit imperfection modeling

$$: \delta_0 = L_c/1,000 \text{ for braced frames}$$

$$: \delta_0 = L_c/500 \text{ for unbraced frames}$$

- Equivalent notional load factor

$$: \alpha = 1/250 \text{ for braced frames}$$

$$: \alpha = 1/500 \text{ for unbraced frames}$$

- Further reduced tangent modulus

$$: E_t' = 0.85E_t$$

References

[1] American Institute of Steel Construction (AISC, 1986) *Load and Resistance Factor Design, Manual of Steel Construction*, 1st ed., Chicago, IL.

[2] American Institute of Steel Construction (AISC, 1994) *Load and Resistance Factor Design, Manual of Steel Construction*, Vol. 1 and 2, 2nd ed., Chicago, IL.

[3] Barakat, M. and Chen, W. F. (1990) Practical analysis of semi-rigid frames, *Eng. J., AISC*, 27(2), 2nd. Quarter, Chicago, pp. 54-68.

[4] Chen, W. F. (1987) *Joint Flexibility in Steel Frames*, Elsevier Science Publication Co., NY, 290 pp.

[5] Clarke, M. J., Bridge, R. Q., Hancock, G. J., and Trahair, N. S. (1993) Australian trends in the plastic analysis and design of steel building frames, in *Plastic Hinge-Based Methods for Advanced Analysis and Design of Steel Frames: An Assessment of the State-of-the-Art*, D. W. White and W. F. Chen, Eds., Structural Stability Research Council, Lehigh University, Bethlehem, PA, pp. 65-93.

[6] Clarke, M. J., Bridge, R. Q., Hancock, G. J., and Trahair, N. S. (1992) Advanced analysis of steel building frames, *J. Constr. Steel Res.*, 23(1-3), 1-29.

[7] Galambos, J. M. and Ketter, R. L. (1959) Columns under combined bending and thrust, *J. Eng. Mech. Div. Proc., ASCE*, 85(EM2), 1-30.

[8] Goto, Y. and Chen, W. F. (1986) Documentation of Program FL-FRM for Design Analysis of Flexibly Jointed Frames, Structural Engineering Report No. CE-STR-86-34, School of Civil Engineering, Purdue University, West Lafayette, IN.
[9] Kanchanalai, T. (1977) The Design and Behavior of Beam-Columns in Unbraced Steel Frames, AISI Project No. 189, Report No. 2, Civil Engineering/Structures Research Lab., University of Texas, Austin, TX, 300 pp.
[10] Ketter, R. (1961) Further studies of the strength of beam-columns, *J. Struc. Div. Proc.* ASCE, 87(ST6), 135-152.
[11] Kim, S. E. (1996) Practical Advanced Analysis for Steel Frame Design, Ph.D. dissertation, School of Civil Engineering, Purdue University, West Lafayette, IN, 271 pp.
[12] Kim, S. E. and Chen, W. F. (1996) Practical advanced analysis for semi-rigid frame design, *AISC Eng. J.* (Accepted for publication).
[13] LeMessurier. W. J. (1977) A practical method of second order analysis, Part 2 - Rigid Frames. *AISC Eng. J.*, 2nd quarter, 14(2), 49-67.
[14] Stelmack T. W. (1982) Analytical and Experimental Response of Flexibly Connected Steel Frames, M.S. dissertation, Department of Civil, Environmental, and Architectural Engineering, University of Colorado, 134 pp.
[15] Vogel, U. (1985) Calibrating frames. Stahlbau 10.
[16] White, D. W. (1993) Plastic hinge methods for advanced analysis of steel frames, *J. Constr. Steel Res.*, 24(2), 121-152.
[17] Yura, J. A. (1988) Elements for Teaching Load and Resistance Factor Design, University of Texas at Austin, TX, pp. 55-71.

chapter four

Analysis and Design Principles

4.1 Introduction

In the preceding chapter, the proposed advanced analysis method was verified using several benchmark problems available in the literature. Verification studies were carried out by comparing it to the plastic-zone and conventional LRFD solutions. It was shown that practical advanced analysis predicted the behavior and failure mode of a structural system with reliable accuracy.

In this chapter, analysis and design principles are summarized for the practical application of the advanced analysis method. Step-by-step analysis and design procedures for the method are presented. The procedures and their key considerations are briefly summarized as follows.

1. Design format: LRFD
2. Loads: Dead, live, impact, wind, earthquake, snow, and rain load
3. Load combinations: LRFD
4. Resistance factors: $\phi_b = 0.9$, $\phi_c = 0.85$, built into the program
5. Establishment of structural systems: Braced and unbraced frames
6. Section types: W, tube, and pipe shape
7. Preliminary member sizing: Using inflection point and equivalent axial force
8. Modeling of structural members: One or two elements per member
9. Modeling of geometric imperfections: Choice of explicit imperfection modeling, equivalent notional load modeling, and further reduced tangent modulus modeling

10. Load application: Proportional load and incremental load
11. Analysis: Preparation of input data and execution of program
12. Load-carrying capacity: Choice of with or without significant inelastic moment redistribution
13. Serviceability limits: Beam deflections and story drift
14. Ductility requirements: Compactness and lateral torsional buckling
15. Adjustment of member sizes: Check adequacy of a structural system and revise member size

These procedures describe the steps involved in performing an advanced analysis, and are fundamental to a thorough understanding of the approximations and assumptions, as well as the power and potential of this method.

4.2 Design Format

Advanced analysis follows the format of LRFD. In LRFD, the factored load effect does not exceed the factored nominal resistance of structure. Two safety factors are used: one is applied to loads, the other to resistances. This approach is an improvement on other models (ASD and PD) because both the loads and the resistances have unique factors for unique uncertainties. The load and resistance factor design has the format

$$\phi R_n \geq \sum_{i=1}^{m} \gamma_i Q_{ni} \tag{4.1}$$

where

R_n = nominal resistance of the structural member
Q_n = nominal load effect (e.g., axial force, shear force, bending moment)
ϕ = resistance factor ($\leq$ 1.0) (e.g., 0.9 for beams, 0.85 for columns)
γ_i = load factor (usually > 1.0) corresponding to Q_{ni} (e.g., $1.4D$ and $1.2D + 1.6L + 0.5S$)
i = type of load (e.g., D = dead load, L = live load, S = snow load)
m = number of load type

Note that the LRFD [3] uses separate factors for each load and therefore reflects the uncertainty of different loads and combinations of loads. As a result, a relatively uniform reliability is achieved.

The LRFD Specification was developed on the basis of probability concept, calibrations to the AISC-ASD Specification [2], design office studies of representative structures, judgment, and experience. The main difference between conventional LRFD methods and advanced analysis methods is that the left side of Equation (4.1), (ϕR_n), in the LRFD method is the resistance or strength of the component of a structural system, but in the advanced analysis method it represents the resistance or the load-carrying capacity of the whole structural system. In the advanced analysis method, the load-carrying capacity is obtained from applying incremental loads until a structural system reaches its strength limit state such as yielding or buckling. The size of the "incremental load" defined by users is increased automatically until "failure" of the structural system. The right-hand side of Equation (4.1), ($\gamma_i Q_{ni}$), represents the load effect in both LRFD and advanced analysis methods.

4.3 Loads

Structures are subjected to various loads including dead, live, impact, snow, rain, wind, and earthquake loads. Structures must be designed to prevent failure and limit excessive deformation; thus, an engineer must anticipate the loads a structure may experience over its service-life with reliability.

Loads may be classified as static or dynamic. Dead loads are typical of static loads, and wind or earthquake loads are dynamic. Dynamic loads are usually converted to equivalent static loads in conventional design procedures, and it may be adopted in advanced analysis as well. This section briefly outlines those loads most commonly considered in engineering design.

4.3.1 Dead Load

Dead loads consist of the weight of the structural frame and permanent appurtenance to the structure (pipes, electrical conduits, ducts, lighting fixtures, etc.). The dead load is not usually known with much accuracy until the design is completed and the member sizes confirmed. The first estimate is always compared to the finalized design weight. The analysis and design should be repeated with the updated weight, until the initial assumptions are close to the resulting values. Typical unit weights for construction and other materials are listed in Table 4.1.

4.3.2 Live Load

Live loads are gravity loads that vary in location and magnitude. Some examples of live loads are human occupants, vehicles, furniture, mov-

Table 4.1 Unit Weights of Construction and Other Materials, lb/ft^3

Metals		Liquids	
Steel	490	Alcohol	49
Aluminum	165	Vegetable oil	58
Iron		Water	
Cast	450	Maximum density	62
Wrought	485	Ice	56
Lead	710	100° C	60
Nickel	565	Seawater	64
Zinc	440	Snow, fresh fallen	8
Concrete		Soils	
Lightweight	70-110	Clay	
Semi-light weig	120-130	Dry	63
Normal	150	In water	80
		Damp, plastic	110
Brick		Clay and gravel	
Pressed	140	Dry	100
Common	120	In water	65
Soft	100	Sand, gravel	
		Dry, loose	90-105
Stone		Dry, packed	100-120
Granite, ashlar	165	Wet	118-120
Gypsum, alabast	159	In water	60
Limestone, marble	165		
Sandstone, bluestone	147	Coal	
Shale, slate	175	Anthracite	47-58
		Bituminous	40-54
Wood		Coke	23-32
Green	33-64		
Air-dry (12% moisture)	25-47	Bituminous substances	
		Asphalt	80
Glass		Pitch	69
Common	156	Tar	75
Plate	161		

Adopted from ANSI, 1988.

able equipment, and store goods. Since realistic magnitudes and positions of these loads are difficult to determine, they are conservatively specified by state and local codes on the basis of experience and accepted practice. To ensure adequate safety, live loads may be fully or partially applied to estimate maximum load effect on each of the structural element being considered. Minimum uniform and concentrated live loads are listed in Tables 4.2 and 4.3 [8].

Table 4.2 Typical Minimum Uniformly Distributed Live Load

Occupancy or use	Live load, psf
1. Hotel guest rooms School classrooms Private apartments Hospital private rooms	40
2. Offices	50
3. Assembly halls, fixed seat Library reading rooms	60
4. Corridors, above first floor in schools, libraries, and hospitals	80
5. Assembly areas; theater lobbies Dining rooms and restaurants Office building lobbies Main floor, retail stores Assembly hall, movable seats	100
6. Wholesale stores, all floors Manufacturing, light Storage warehouses, light	125
7. Armories and drill halls Stage floors Library stack rooms	150
8. Manufacturing, heavy Sidewalks and driveways subject to trucking Storage warehouses, heavy	250

Adopted from ANSI, 1988.

Since it is unlikely that an entire area of a structure be uniformly and simultaneously supporting live loads, live loads may be reduced to

Table 4.3 Minimum Concentrated Loads

Location	Load (lb)
Elevator machine room grating (on area of 4 in.^{2})	300
Finish light floor plate construction (on area of 1 in.^{2})	200
Garages	*
Office floors	2,000
Scuttles, skylight ribs, and accessible ceilings	200
Sidewalks	8,000
Stair treads (on area of 4 in.^{2} at center of tread)	300

*Floors in garages or portions of buildings used for the storage of motor vehicles shall be designed for the uniformly distributed live loads of Table 4.2 or the following concentrated loads: (1) for passenger cars accommodating not more than nine passengers, 2,000 lb acting on an area of 20 in.^{2}; (2) mechanical parking structures without slab or deck, passenger cars only, 1,500 lb per wheel; or (3) for trucks or buses, maximum axle load on an area of 20 in.^{2}.

Adopted from ANSI, 1988.

some degree. ASCE 7-93 [9] provides the following guideline:

$$L = L_0 \left(0.25 + \frac{15}{\sqrt{A_I}} \right) \tag{4.2}$$

where

L = reduced design live load per square foot of area supported by the member

L_0 = base design live load per square foot of area supported by the member

A_I = influence area, in square feet

The influence area A_I may be defined as floor area that influences load effects. More consistent reliability is achieved by using influence areas as opposed to tributary areas. The influence area for a column is four times its tributary area, while for a beam it is only twice as big. For example, the influence area is the total area of the four surrounding bays for an interior column, and it is the total area of the two contributing bays for an interior beam. For edge columns and girders, the influence area is one half the influence area of the interior members, i.e., two bays for columns and one for girders; for a corner column, the influence area is one bay. Typical influence areas for a structure with regular bay spacings are illustrated in Figure 4.1. For multiple floors, the influence area for a column supporting more than one floor must be the total influence area considering all the floors above it.

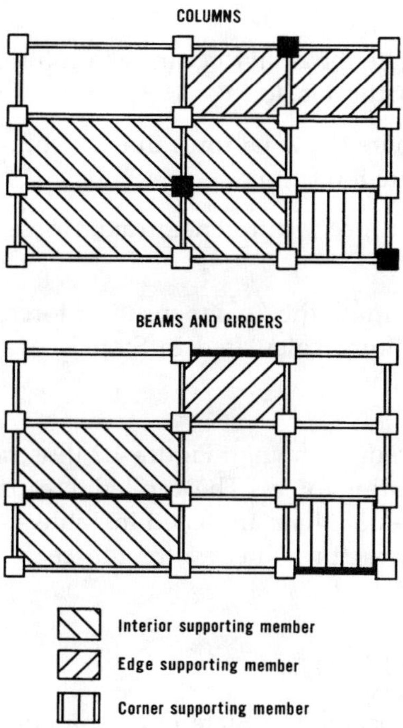

Figure 4.1 An illustration of influence area (Adopted from [8]).

The reduced design live load L [Equation (4.2)] should be greater than 40% of the live load L_0 except that for members supporting only one floor, the maximum reduction may be as large as 50%. When live loads are less than 100 lb/ft^2 on buildings such as public assembly halls, garages, and one-way slabs, the live load reduction is not permitted. When live loads are greater than 100 lb/ft^2, design live loads on members supporting more than one floor may be reduced up to 20% in garages for passenger cars only, but in other cases may not be reduced except as permitted by the governing authority.

Since the influence area for beams and columns are different, careful application of Equation (4.2) is necessary for the analysis of a structural system. Among several methods that incorporate this live load reduction [32] [33] [40] [42], a procedure proposed by Ziemian and McGuire is adopted because it is more comprehensive for incorporating live load reduction in analysis of a structural system than other methods. It uses a concept of "compensating forces" calculated on different reduction factors for columns and beams.

The compensating forces act upward on the columns that have a larger live load reduction than the beams. The procedures to determine the compensating forces may be outlined as follows:

Step 1: Estimate the axial force in each column based on a column live load reduction factor.

Step 2: Estimate the axial force in each column based on a beam live load reduction factor.

Step 3: Determine the difference in axial forces calculated in Steps 1 and 2.

Step 4: Determine the "compensating force", i.e., the difference between the force calculated in Step 3 for one column and the column above it.

The two-story frame shown in Figure 4.2 illustrates the procedure to calculate compensating forces. The applied live loads on both floors are 75 psf. Detailed calculations are listed in Table 4.4 for Section 2 of the two-dimensional structure. The resulting force distribution is shown in Figure 4.3.

4.3.3 Highway Live Load

The American Association of State Highway and Transportation Officials [1] specifies highway vehicle loads for the U.S. in terms of standard truck loads and lane loads that approximate a series of truck loads. The weight and spacings of axles and wheels are as specified by AASHTO [1] and pictured in Figure 4.4. The spacing between the two 32.0 kips axles is varied between 14.0 ft and 30.0 ft to produce extreme force effects. The transverse spacing of wheels is 6.0 ft. The design lane load is a uniformly distributed line load of 0.64 kips/ft in the longitudinal direction, acting over a 10-ft width. In addition to the lane load, a concentrated 18 kips force for moment and 26 kips force for shear are positioned to give the maximum effect as shown in Figure 4.5. An impact load is considered with truck loads but not lane loads.

In bridge design, the design loadings should be placed for the most severe effect for each structural component examined. For illustration, every other span loading on a continuous structure may produce severe positive bending. The location and distribution of forces producing

Table 4.4 Example of Compensating Forces for Column c_1 and c_2

Item	Column C_1	Column C_2
Step 1:		
(1) Tributary area of column (ft^2)	$A_1 = (12.5)(25) = 312.5$	$A_2 = (12.5)(25) = 312.5$
(2) Unreduced live load (kips)	$L_{01} = (0.075)(312.5) = 23.44$	$L_{02} = (0.075)(312.5) = 23.44$
(3) Influence area of column (ft^2)	$A_{I1} = (25)(50)(2) = 2500$	$A_{I2} = (25)(50) = 1250$
(4) Reduction factor	$RF_1 = (0.25 + 15/\sqrt{2500}) = 0.550$	$RF_1 = (0.25 + 15/\sqrt{1250}) = 0.6743$
(5) Reduced live load (kips)	$L_1 = (23.44)(0.55) = 12.89$	$L_2 = (23.44)(0.6743) = 15.80$
Step 2:		
(6) Tributary width of beam (ft)	$W_1 = 25$ ft.	$W_2 = 25$ ft.
(7) Unreduced live load on beam (k/ft)	$W_{01} = (0.075)(25) = 1.875$	$W_{02} = 1.875$
(8) Influence area of beam (ft^2)	$A_{I1} = (25)(50) = 1250$	$A_{I2} = 1250$
(9) Reduction factor	$RF_1 = (0.25 + 15/\sqrt{1250}) = 0.6743$	$RF_2 = 0.6743$
(10) Reduced live load on beam (k/ft)	$W_1 = (1.875)(0.6743) = 1.264$	$W_2 = 1.264$
(11) Reduced live load on columns (kips)	$L_1 = (1.875)(12.5)(0.6743) = 15.80$	$L_2 = 15.80$
Step 3:		
(12) Difference axial forces = (11) − (5)	$L'_1 = 15.80 - 12.89 = 2.91$	$L'_2 = 0.0$
Step 4:		
(13) Compensating force	$F_1 = L'_1 - L'_2 = 2.91$	$F_2 = L'_2 - L'_3 = 0.0$

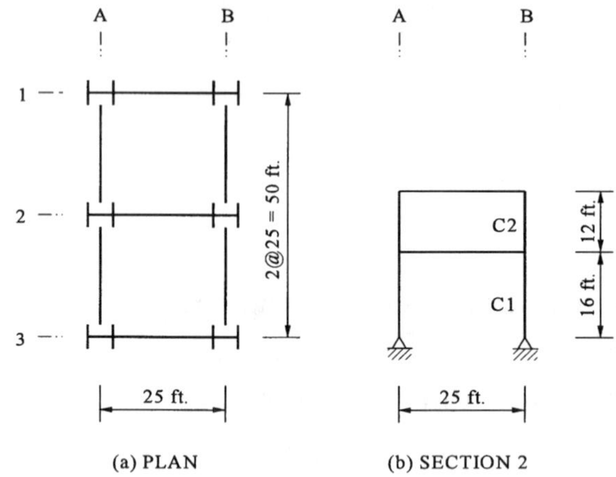

Figure 4.2 Configuration of example frame.

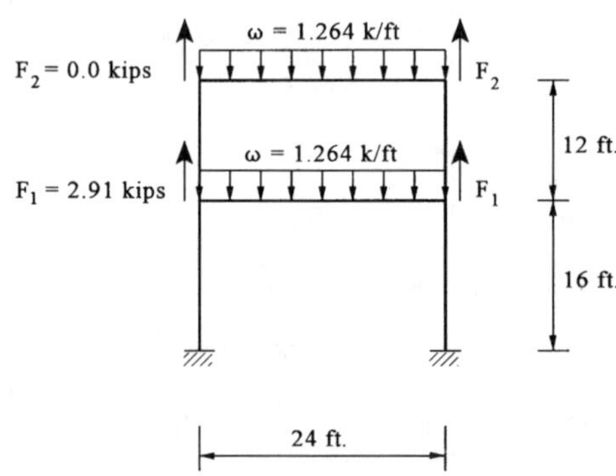

Figure 4.3 Resulting load distribution including compensating forces.

the maximum bending moment and the maximum shear force must be found. In general, the design of short spans is governed by single truck loads with a heavy concentrated load, and the design of long spans is controlled by the lane load.

4.3.4 Impact Load

Sudden loads applied by moving objects cause impact on a structure. Dead loads are static, but live loads may produce a dynamic repercussion. A pedestrian load and design lane load are not considered to cause dynamic effect, but vehicles, cranes, and machinery produce dynamic loads.

The ordinary design procedure to account for dynamic effects is to increase the live loads by an "impact factor" and thus avoid dynamic analyses. For buildings, the AISC-LRFD Specification [3] specifies the

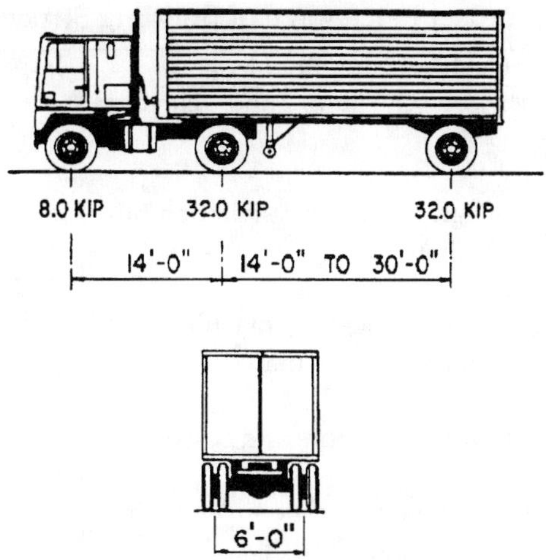

Figure 4.4 Design truck load (AASHTO, 1994).

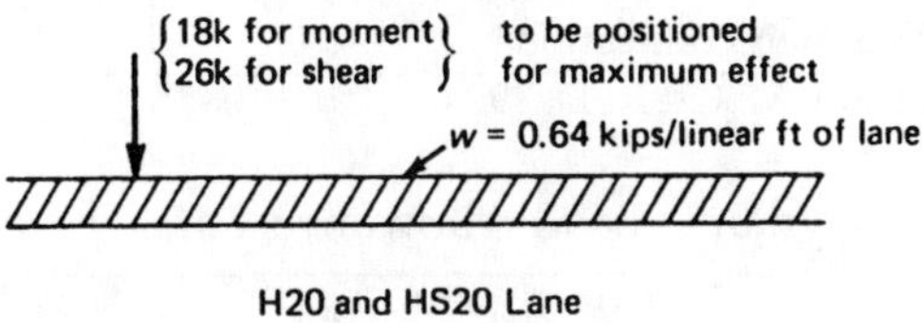

Figure 4.5 Design lane load (AASHTO, 1994).

impact factor in the design of supports for cranes and heavy machinery. This is summarized in Table 4.5.

For bridge design, impact is caused by vehicles traveling on the structure. AASHTO [1] recommends that live loads should be increased by the percentage specified in Table 4.6. The factor applied to the static load should be taken as $(1 + IM)$, where IM is an impact factor.

4.3.5 Wind Load

Wind load is a typical lateral load. Every building and long bridge should be designed to resist the wind effect. Wind is assumed to come from any horizontal direction, and no reduction in wind pressure may be taken for the shielding effect of adjacent structures [38].

The velocity pressure may be written as [8]:

$$q_s = 0.00256 K_z (IV)^2 \tag{4.3}$$

Table 4.5 Impact Percentage of Building Structure

Structural component	Impact percentage
For supports of elevators and elevator machinery	100
For supports of light machinery, shaft or motor driven, not less than	20
For supports of reciprocating machinery or power driven units, not less than	50
For hangers supporting floors and balconies	33
For cab-operated traveling crane support girders and their connections	25
For pendant-operated traveling crane support girders and their connections	10

Adopted from AISC-LRFD, 1994.

Table 4.6 Dynamic Load Allowance, IM

Component	IM(%)
Deck joints - all limit states	75
All other components	
Fatigue and fracture limit state	15
All other limit states	33

Adopted from AASHTO, 1989.

where

q_s = velocity pressure, psf
K_z = velocity pressure explosure coefficient varying 0.12 to 2.41
I = importance factor varying 0.95 to 1.11
V = wind velocity, mph

Design wind pressure for structures or elements of structures shall be determined for any height in accordance with the following formula [38]:

$$P = C_e C_q q_s I \tag{4.4}$$

where

P = design wind pressure, psf

C_e = combined height, exposure and gust factor coefficient varying 0.7 to 2.2

C_q = pressure coefficient varying 0.3 to 4 for the structure or portion of structure under consideration.

As an example, if $C_e = 1.62$, $C_q = 1.0$, $I = 1.0$, $K_z = 1$, and $V =$ 70 mph, the resulting velocity pressure is obtained as 12.5 psf from Equation (4.3) and the design wind pressure may be calculated as 20 psf using Equation (4.4), which is the commonly used wind pressure. The minimum basic wind speed at any site should be greater than 70 mph, and the wind pressure should be greater than 10 psf [8] [38].

4.3.6 Earthquake Load

An earthquake produces both horizontal and vertical ground motions. The horizontal motion is larger in magnitude and more severely effects the structure than the vertical motion. While a structure may be examined using the dynamics of earthquake action, most codes allow that an equivalent lateral force may be used to simplify procedures. Detail applications of earthquake loading are described in the UBC [38] as follows:

The design seismic forces do not act simultaneously in the direction of each principal axis of the structure. The seismic dead load, W, is the total dead load and is applicable to portions of other loads as listed below.

1. In storage and warehouse occupancies, a minimum of 25% of the floor live load shall be applied.
2. Where a partition load is used in the floor design, a load of not less than 10 lb per square foot (psf) shall be included.
3. Where the snow load is greater than 30 psf, the snow load shall be included. Where considerations of siting, configuration, and load duration warrant, the snow load may be reduced up to 75% when approved by the building official.
4. Total weight of permanent equipment shall be included.

The total design base shear in a given direction shall be determined from the following formula:

$$V = \frac{ZIC}{R_w} W \tag{4.5}$$

$$C = \frac{1.25S}{T^{2/3}} \tag{4.6}$$

where

V = total design lateral force or shear at the base
Z = seismic zone factor varying 0.075 to 0.40
I = importance factor varying 1.0 to 1.25
R_w = numerical coefficient varying 4 to 12
W = total seismic dead load
S = site coefficient varying 1.0 to 2.0
T = fundamental period of vibration in seconds

The value of C need not exceed 2.75 and may be used for any structure without regard to soil type or structure period. Except for those provisions where code-prescribed forces are scaled up by 3 ($R_w/8$) the minimum value of the ratio C/R_w shall be 0.075.

The value of T shall be determined from one of the following methods:

Method A: For all buildings, the value T may be approximated from the following formula:

$$T = C_t\,(h_n)^{3/4} \tag{4.7}$$

where

h_n = height in feet above the base to level n
C_t = 0.035 for steel moment-resisting frames
C_t = 0.030 for reinforced concrete moment-resisting frames and eccentrically braced frames
C_t = 0.020 for all other buildings.

Alternatively, the value of C_t for structures with concrete or masonry shear walls may be taken as $0.1/\sqrt{A_c}$. The value of A_c shall be determined from the following formula:

$$A_c = \sum A_e\left[0.2 + (D_e/h_n)^2\right] \tag{4.8}$$

where

A_c = combined effective area of the shear wall in the first story of the structure
A_e = minimum cross-section shear area in any horizontal plane in the first story
D_e = length of a shear wall in the first story in the direction parallel to the applied forces

The value of D_e/h_n used in Equation (4.8) shall not exceed 0.9.

Method B: The fundamental period T may be calculated using the structural properties and deformation characteristics of the resisting elements in a properly substantiated analysis. This requirement may be satisfied by using the following formula:

$$T = 2\pi \sqrt{\left(\sum_{i=1}^{n} w_i \delta_i^2\right) \div \left(g \sum_{i=1}^{n} f_i \delta_i\right)} \tag{4.9}$$

The values of f_i represent any lateral force distributed approximately in accordance with the principles of Equations (4.10 through 4.12) or any other rational distribution. The elastic deflections, δ_i shall be calculated using the applied lateral forces, f_i. The value of C in Equation (4.6) shall be not less than 80% of the value obtained using T from Method A.

The total force shall be distributed over the height of the structure in conformance with Equations (4.10 through 4.12) in the absence of a more rigorous procedure

$$V = F_t + \sum_{i=1}^{n} F_i \tag{4.10}$$

The concentrated force F_t, at the top, applied in addition to F_n, shall be determined from the formula:

$$F_t = 0.07TV \tag{4.11}$$

The value of T used to calculate F_t may be the period that corresponds with the design base shear as computed using Equation (4.5). F_t need not exceed $0.25V$ and may be considered zero when T is 0.7 seconds or less. The remaining portion of the base shear shall be distributed over the height of the structure, including level n, according to the following formula:

$$F_x = \frac{(V - F_t)\, w_x h_x}{\sum_{i=1}^{n} w_i h_i} \tag{4.12}$$

At each level "x", the force F_x shall be applied over the area of the building in accordance with the mass distribution at that level. Stresses in each structural element shall be calculated as the effect of forces F_x and F_t applied at the appropriate levels above the base.

4.3.7 Snow Load

Snow loads are the primary roof load. ANSI Standard [8] provides isolines of ground snow corresponding to a 50-year mean recurrence interval. The ground snow is then adjusted by factors including exposure,

thermal, importance, and slope. The snow load on an unobstructed flat roof shall be calculated using the following formula:
Contiguous U.S.

$$P_f = 0.7 C_e C_t I P_g \tag{4.13}$$

Alaska

$$P_f = 0.6 C_e C_t I P_g \tag{4.14}$$

where

P_f = snow load on flat roof, lb/ft^2
C_e = exposure factor varying 0.8 to 1.2
C_t = thermal factor varying 1.0 to 1.2
I = importance factor varying 0.8 to 1.2
P_g = ground snow load varying 0 to 100 lb/ft^2 except for Alaskan location

On a sloped-roof, the snow load shall be obtained by multiplying the flat-roof load by a slope factor:

$$P_s = C_s P_f \tag{4.15}$$

where

P_s = snow load on a sloped-roof
C_s = roof slope factor varying 0 to 1.0.

It is expected that the steeper roof causes less snow to be accumulated. Partial as well as full snow loading should be considered to cause severe load effects.

4.3.8 Rain Load

Although snow loads are generally more critical than rain loads, this may not be the case for flat roofs in warm climates. Water on a flat roof ponds in low spots. Water flows into the depression deflections increase, and so the rate of accumulation rises. If the roof structure does not possess enough stiffness to resist this progression, localized failure may occur. To evaluate the effect of ponding, the deflection caused by dead and snow loads are considered. In general, roofs with a slope of 1/4 inch per foot or more are not considered susceptible to ponding instability from rain alone unless drain blockages allow deep ponds to form [8].

The amount of ponding that would result from blockage of the primary drainage system should be determined and the roof designed to withstand the ponding load that would result plus an additional 5 lb/ft^2 to account for the head needed to cause flow out of the secondary drainage system.

In some areas of the country, ordinances are in effect that limit the rate of rainwater flow from roofs into storm drains. Controlled-flow drains are often used on such roofs. Those roofs must be capable of sustaining the storm water temporarily stored on them. Many roofs designed with controlled-flow drains have a design rain load of 30 lb/ft^2 and are equipped with a secondary drainage system (for example, scuppers) that prevents ponding deeper than 3-1/2 inches on the roof [8].

4.4 Load Combinations

The load combinations in advanced analysis methods are based on the LRFD combinations [3]. Six factored combinations are provided by the LRFD Specification. One must be used to determine member sizes. Probability methods were used to determine the load combinations listed in the LRFD specification (LRFD-A4). Each factored load combination is based on the load corresponding to the 50-year recurrence as follows:

$$(1)\quad 1.4D \tag{4.16a}$$

$$(2)\quad 1.2D + 1.6L + 0.5(L_r \text{ or } S \text{ or } R) \tag{4.16b}$$

$$(3)\quad 1.2D + 1.6(L_r \text{ or } S \text{ or } R) + (0.5L \text{ or } 0.8W) \tag{4.16c}$$

$$(4)\quad 1.2D + 1.3W + 0.5L + 0.5(L_r \text{ or } S \text{ or } R) \tag{4.16d}$$

$$(5)\quad 1.2D \pm 1.0E + 0.5L + 0.2S \tag{4.16e}$$

$$(6)\quad 0.9D \pm (1.3W \text{ or } 1.0E) \tag{4.16f}$$

where

D = dead load (the weight of the structural elements and the permanent features on the structure)
L = live load (occupancy and movable equipment)
L_r = roof live load
W = wind load
S = snow load

E = earthquake load
R = rainwater or ice load

The LRFD Specification specifies an exception that the load factor on live load L in combinations (3) through (5) must be 1.0 for garages, areas designated for public assembly, and all areas where the live load is greater than 100 psf.

4.5 Resistance Factors

Based on a first-order second-moment probabilistic analysis [6], the safety of a structural member is measured by reliability or safety index [3] defined as

$$\beta = \frac{\ln\left(\bar{R}_n/\bar{Q}_n\right)}{\sqrt{V_R^2 + V_Q^2}} \tag{4.17}$$

where

$\bar{R}$ = mean resistance
$\bar{Q}$ = mean load effect
V_R = coefficient of variation of resistance $\left(= \sigma_R/\bar{R}\right)$
V_Q = coefficient of variation of load effect $\left(= \sigma_Q/\bar{Q}\right)$

in which σ equals the standard deviation.

The physical interpretation of the reliability index β is shown in Figure 1.3. The shaded area in the figure represents the probability in which $\ln(R/Q) < 0$, i.e., the probability that the resistance will be smaller than the load effect, indicating that a limit state has been exceeded. The larger the value of β, the smaller the shaded area, so that it becomes more improbable that a limit state will be exceeded. The magnitude of β reflects the safety of the member. In the development of the LRFD Specification [3] and [4], the resistance factor ϕ was determined with the objective of obtaining the β values listed in Table 4.7.

Table 4.7 Target Value of Reliability Index β

Load combination	Objective reliability index β
Dead load + live load (or snow load)	3.0 for members
	4.5 for connections
Dead load + live load + wind load	2.5 for members
Dead load + live load + earthquake load	1.75 for members

The AISC-LRFD cross-section strength equations may be written as

$$\frac{P}{\phi_c P_y} + \frac{8}{9}\frac{M}{\phi_b M_p} = 1.0 \quad \text{for} \quad \frac{P}{\phi_c P_y} \geq 0.2 \tag{4.18a}$$

$$\frac{P}{2\phi_c P_y} + \frac{M}{\phi_b M_p} = 1.0 \quad \text{for} \quad \frac{P}{\phi_c P_y} < 0.2 \tag{4.18b}$$

where

P, M = second-order axial force and bending moment
P_y = squash load
M_p = plastic moment capacity
ϕ_c, ϕ_b = resistance factors for axial strength and flexural strength

Figure 4.6 shows the cross-section strength including the resistance factors ϕ_c and ϕ_b. The resistance factors ϕ_c and ϕ_b are built into the analysis program and are thus automatically included in the calculation of the load-carrying capacity. The resistance factors are 0.85 for axial strength and 0.9 for flexural strength corresponding to AISC-LRFD Specification [3]. The ultimate connection moment capacity M_u is reduced by the resistance factor of 0.9.

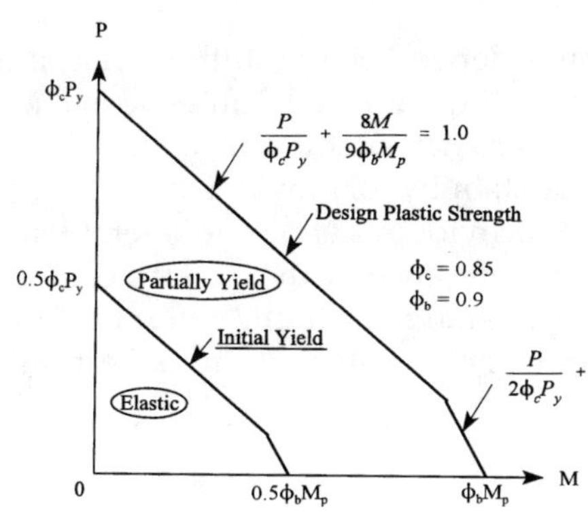

Figure 4.6 Stiffness degradation model including reduction factors.

4.6 Establishment of Structural System

Structural engineers are responsible for the safe design of structures. The design of all structural parts and components should be compatible even when the details are not designed by the same engineer. A structure should be designed to transmit dead, live, and wind loads directly to the foundation. The structural arrangement should be stable so that the structure will not overturn or collapse under external loads. Consideration should also be given to the erection procedure and stability during construction.

4.6.1 Low-Rise Structures

Lateral structural stability is usually provided in two orthogonal directions. This may be achieved by a rigid-frame system or by vertically braced bays with plan bracing or a diaphragm-braced system.

For a diaphragm-braced system, either plywood or light-gauge steel deck can be used to transmit the lateral forces; however, when the structure length-to-width ratio exceeds 3, many fasteners are required and economical concerns arise. For long buildings, plan bracing in the form of trusses is economical. The lateral force is transmitted from the side cladding through the plan bracing to the vertical braces. However, when the buildings are long and narrow or when bracing is not allowed in the vertical plane, the resistance to lateral loads is given by the rigid frame system which can be used in both directions.

4.6.2 Multistory Structures

Unbraced Frames

An unbraced frame must depend on the stiffness of its individual members for stability. Columns in unbraced frames are subject to both axial load, moments, and experience lateral translation. The columns in an unbraced frame provide stability for gravity loaded columns whose simple connections do not provide resistance to lateral load. The leaning effects generated by the presence of the gravity columns must be included in the design of the moment-frame columns. For tall buildings, unbraced rigid frames can be combined with a vertically braced truss to reduce frame drift.

Braced Frames

Lateral stability is provided by placing vertical and horizontal braces in two orthogonal directions so that the columns will not be subjected to sway moments. Bracing can be provided by structural walls enclosing the stairs, lifts, service ducts, etc. Additional lateral stiffness can be provided by bracing within external or internal walls. The bracing should be distributed throughout the structure so the shear center for lateral resistance lies approximately at the center of shear resultant in a plan. Where this is not possible, torsional forces may incur and must be considered when calculating the load effect. Braced bays should be throughout the full height of the building. If it is essential that bracing be discontinuous, provision must be made for the transfer of forces from bay to bay.

4.6.3 *Forms of Bracing*

Bracing is mainly vertical and horizontal. Vertical bracing may consist of steel trusses, moment-resisting frames, reinforced concrete shear walls (preferably not less than 5 in. thick), masonry walls (preferably not less than 6 in. thick adequately tied to the steel frames), or any combination thereof. Precautions should be taken to prevent such bracing being removed at a later stage. Temporary bracing should be provided during erection before braced systems are completed.

Horizontal bracing consists of steel trusses, concrete floors or roofs, and adequately designed and fixed profiled steel decking either in the form of roof sheeting or as formwork to a composite floor. Since the floor may not be fully loaded during construction, the profiled decking, if adequately connected, provides the diaphragm action or horizontal bracing needed during construction.

4.6.4 *Other Design Considerations*

All members of a structure should be tied together in the longitudinal, transverse, and vertical directions. Collapse of an entire structure caused by one member failure should be avoided by establishing alternative load transferring paths.

Movable joints should be provided to minimize the effect of movement arising from temperature variations and relative displacements of adjacent members. The structure should be framed on both sides of the joint, and each structure should be independent and designed to be stable without relying on the stability of an adjacent part of a structure.

Movable joints may also be required where there is a significant change in foundation, soil condition, plan configuration, or the height of the structure, to accommodate differential settlement. Where detailed calculations are not made, joints permitting movement of up to 1 in. should be provided at 170-ft centers along the length of the frame. It is also necessary to incorporate joints in the finishes and in the cladding at movable joint locations. In addition, a gap should be provided between steelworks and masonry cladding to allow for the movement of columns under loading.

4.7 Section Application

The AISC-LRFD Specification uses only one column curve for rolled and welded sections of W, WT, and HP shapes, pipe, and structural tubing. The Specification also uses some interaction equations for doubly and singly symmetric members including W, WT, and HP shapes, pipe, and structural tubing, even though the interaction equations were developed on the basis of W shapes by Kanchanalai [24].

The present advanced analysis method was developed by calibration with the LRFD column curve and interaction equations described in Chapter 3. To this end, it is concluded that the proposed methods can be used for various rolled and welded sections including W, WT, and HP shapes, pipe, and structural tubing as shown in Figure 4.7 without further modifications.

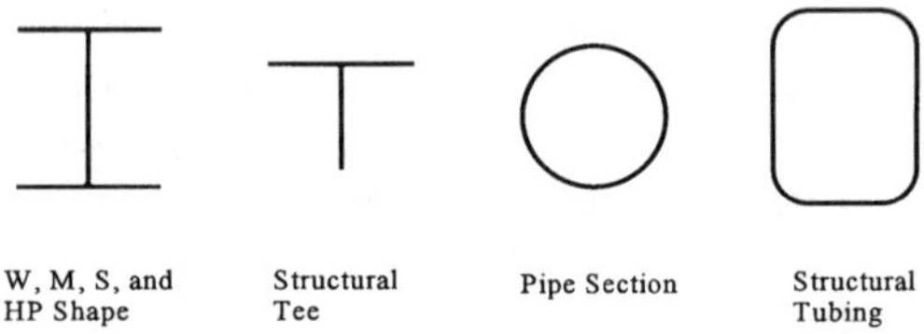

Figure 4.7 Applicable sections for advanced analysis.

4.8 Preliminary Member Sizing

To design statically indeterminate structures, it is necessary to estimate their initial member sizes before a structural analysis. Since the stiffness of individual members influences the distribution of member forces, the preliminary member sizes should be closely estimated. However, this is a challenging task as there is no systematic rule for indeterminate structures. A procedure is presented below to estimate the initial member sizes.

1. Approximate analysis with simplifying assumptions.
2. Approximate member sizing based on member forces obtained by approximate analysis.

4.8.1 Approximate Analysis

This section presents an approximate analysis of a statically indeterminate structure. Approximate analysis is used to estimate the initial member sizes. The solutions from this analysis may deviate as much as

10 or 20% from an elastic analysis. Approximate analyses make simple assumptions about structural behavior. Through these assumptions, the resulting member forces can be estimated by simple static without the use of compatibility conditions. An approximate analysis can be carried out by guessing the location of inflection points in a statically indeterminate structure. Once the locations of inflection points are assumed, the indeterminate structure can be converted into a statically determinate structure, and member forces are easily calculated.

Continuous Beams Subjected to Gravity Load

Continuous beams subject to gravity loads simulate floor systems of buildings and bridges. They are usually governed by gravity load combinations. Figure 4.8(a) shows a three-span continuous beam subject to a concentrated load. The deflection curve can be sketched schematically accounting for the fact that the deflection of the interior span is restrained by the beams of exterior spans. The inflection point is seen where the curve changes from concave to convex as shown in Figure 4.8(b). Another way to estimate the locations of inflection points is to

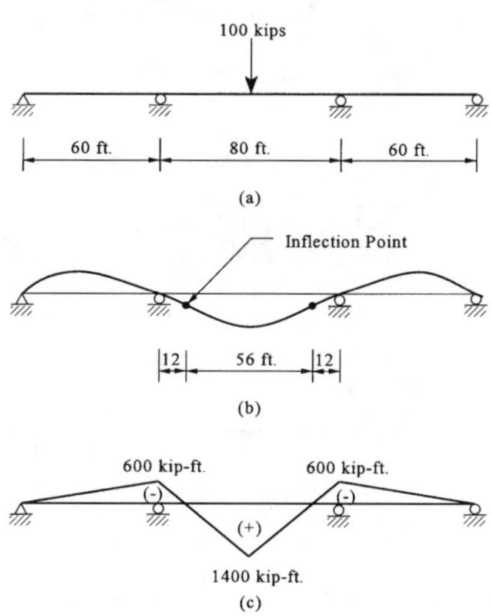

Figure 4.8 Approximation analysis of the continuous beam.

use information of the inflection points of fixed- and hinged-supported beams. Since the ends of the center span are restrained against rotation by the bending stiffness of the end spans, the inflection point of this beam falls between the two extreme cases (fixed and hinged). Figure 4.9 shows the location of the inflection point for the idealized cases to provide a guide for the approximate locations of the inflection points in continuous beams. Enough inflection points may be selected to make the statically indeterminate structure a determinate one. The inflection points with zero bending moment may be considered hinges for the static analysis. The member forces may be determined by the equilibrium equations, resulting in the approximate moment diagram shown in Figure 4.8(c).

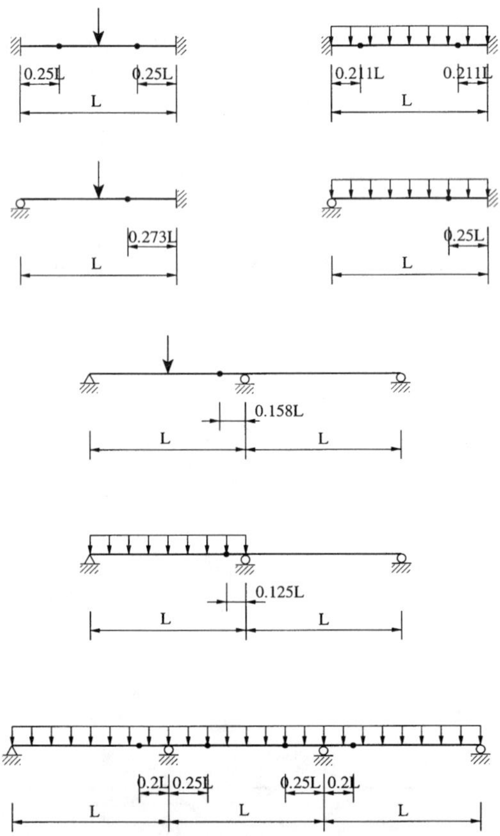

Figure 4.9 Illustration of inflection points in ideal beams.

Frame Subjected to Lateral Load

In general, frames are subject to lateral and vertical loads. Approximate analysis for framed beams is the same as that of the continuous beam described previously. The approximate analysis for framed columns (not beams) subject to lateral loads will be discussed because the column members are usually governed by the lateral load combination, while the beams are controlled by the gravity load combination. As approximate analysis for beams, it is important to estimate the locations of inflection for columns.

Figure 4.10 shows simple portal frames with rigid supports subject to a horizontal load. If the horizontal girder is rigid, it simply moves in horizontal direction experiencing no bending deformation. In this case, the inflection points occur at the mid-height in the column, and the moment produced at both supports is equal to $HL/4$. If the girder is very flexible, the inflection points in the columns move up toward the beam, and the moments at the supports become almost double those of the rigid girder. Assuming the inflection points are at mid-height of the columns is a fine estimate for practical purposes because neither extremely stiff girders nor ideally fixed supports exist and their effects on the location of the inflection point compensate each other. When a structure has pinned supports, the inflection points in the first story should be located at the supports.

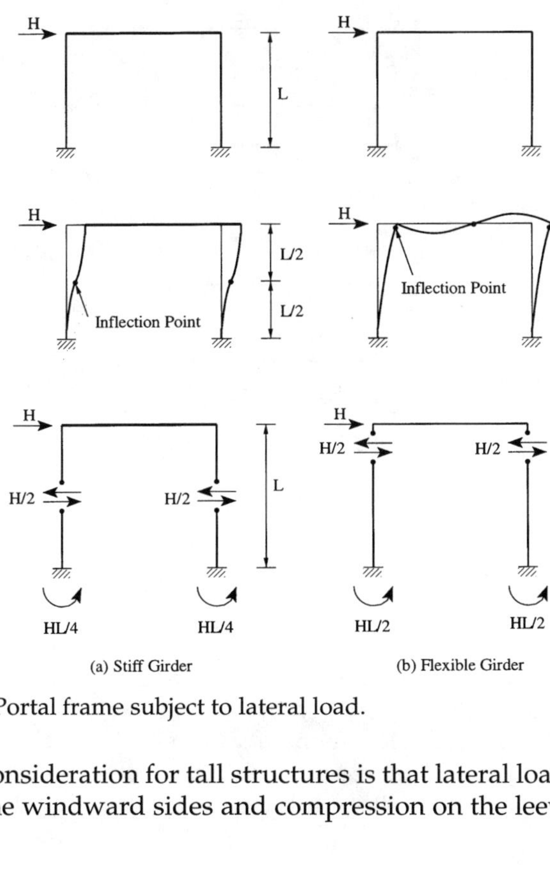

Figure 4.10 Portal frame subject to lateral load.

Another consideration for tall structures is that lateral loads produce tension on the windward sides and compression on the leeward sides.

The change in the axial forces may be estimated as a proportion of their distance from the centroid of the column group. Norris and Wilbur [31] discuss the details of the portal and cantilever method for approximate analysis of frames.

Figure 4.11(a) shows a two-bay, four-story frame to illustrate approximate analysis. First, the inflection points are assumed to be at mid-height of the column shown in Figure 4.11(a). Second, the story shear

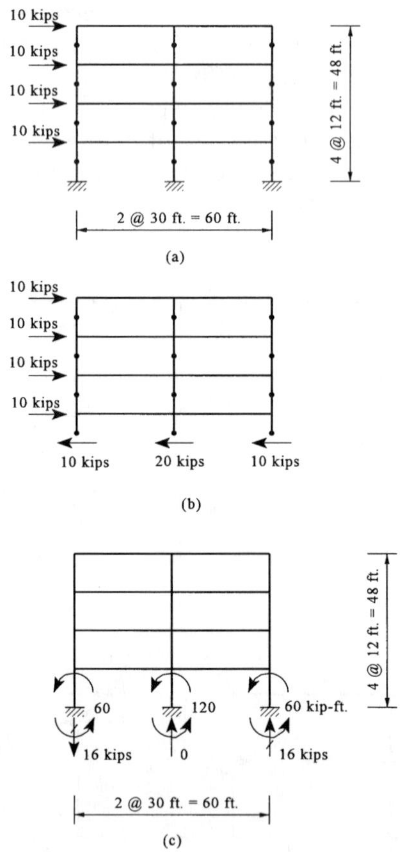

Figure 4.11 Approximate analysis of frame.

at the first story is distributed to the columns as shown in Figure 4.11(b) corresponding to the assumption that interior column carries twice the shear of exterior ones. The moments in the columns of the first story [Figure 4.11(c)] are calculated by multiplying the shear forces by a moment arm equal to one half the column height. Finally, the overall bending moment (with respect to the mid-height of the first story column) is calculated as 960kips-ft[= (10kips) × (42ft + 30ft + 18ft + 6ft)], and the axial force due to the overturning moment in exterior columns is 16 kips (= 960kips-ft/60ft). The upper stories may be treated in the same manner.

Herein, we are not concerned with the force in the beams because these are usually governed by the gravity load combination. This allows the simple implementation of the conventional portal or cantilever methods.

4.8.2 Approximate Member Sizing

Once the approximate member forces are estimated, the selection of member sizes is straightforward for the beams but not the columns. This is because axial compression and bending moments are coupled in the columns. One way to overcome this difficulty is to convert the bending moment to an equivalent axial compression load, and then to add this equivalent load to the original compression forces. The AISC-LRFD Specification provides an efficient procedure for the selection of trial section, originally developed by Burgett [11]. The total equivalent axial load is written as:

$$P_{u,eq} = P_u + M_{ux}m + M_{uy}mu \tag{4.19}$$

where

P_u = actual factored axial load, kips
M_{ux} = factored bending moment about the strong axis, k-ft
M_{uy} = factored bending moment about the weak axis, k-ft
m = factor taken from Table 4.8
u = factor taken from the Column Design Table in the AISC-LRFD Specification

The selection of a column is based on a trial-and-error process as follows:

1. Select the first trial value m corresponding to the effective length KL and let $u = 2$.
2. Compute an equivalent axial compression load $P_{u,eq} = P_u + M_{ux}m + M_{uy}mu$.
3. Select a trial section from the Column Design Tables in the AISC-LRFD.
4. Find the second trial m and u corresponding to the trial section.

5. Compute the updated $P_{u,eq}$.
6. Repeat Steps 3 through 5 until there is no change in $P_{u,eq}$.

Table 4.8 The m Values for Preliminary Beam-Column Design

F_y	36 ksi							50 ksi						
KL(ft)	10	12	14	16	18	20	22 and over	10	12	14	16	18	20	22 and over
First approximation														
All shapes	2.0	1.9	1.8	1.7	1.6	1.5	1.3	1.9	1.8	1.7	1.6	1.4	1.3	1.2
Subsequent approximation														
W4	3.1	2.3	1.7	1.4	1.1	1.0	0.8	2.4	1.8	1.4	1.1	1.0	0.9	0.8
W6	3.2	2.7	2.1	1.7	1.4	1.2	1.0	2.8	2.2	1.7	1.4	1.1	1.0	0.9
W8	2.8	2.5	2.1	1.8	1.5	1.3	1.1	2.5	2.2	1.8	1.5	1.3	1.2	1.1
W8	2.5	2.3	2.2	2.0	1.8	1.6	1.4	2.4	2.2	2.0	1.7	1.5	1.3	1.2
W10	2.1	2.0	1.9	1.8	1.7	1.6	1.4	2.0	1.9	1.8	1.7	1.5	1.4	1.3
W12	1.7	1.7	1.6	1.5	1.5	1.4	1.3	1.7	1.6	1.5	1.5	1.4	1.3	1.2
W14	1.5	1.5	1.4	1.4	1.3	1.3	1.2	1.5	1.4	1.4	1.3	1.3	1.2	1.2

Adopted from Uang et al.,1990.

An example of the initial member sizing is illustrated with the following data:

$$P_u = 600\text{kips}$$

$$M_{ux} = 100\text{k-ft}$$

$$M_{uy} = 50\text{k-ft}$$

$$KL_x = KL_y = 12\text{ft}$$

$$F_y = 50\text{ksi}$$

The procedures for the preliminary member sizing are as follows:

1. For $KL = 12$ ft, select a first trial value of $m = 1.8$, from Table 4.8 and let $u = 2$.
2. $P_{u,eq} = P_u + M_{ux}m + M_{uy}mu = 600 + (100)(1.8) + (50)(1.8)(2) = 960$ kips.

3. Select W14x90 ($\phi_c P_n = 1,010$) from the Column Design Table in the AISC-LRFD Specification.

4. Find $m = 1.4$ from Table 4.8 and $u = 1.94$ from the Column Design Table.

5. Compute the updated $P_{u,eq}$

$$P_{u,eq} = 600 + (100)(1.4) + 50(1.4)(1.94) = 876\text{kips}$$

6. Select W14x90 ($\phi_c P_n$ =1,010).

Yura [41] suggested a simpler approach for the initial member sizing compared to Equation (4.19). The equivalent axial load proposed by Yura is written as:

$$P_{u,eq} = P_u + \frac{2M_{ux}}{d} + \frac{7.5M_{uy}}{b} \tag{4.20}$$

where

d = beam depth
b = beam width

Herein, the previous example is used to compare with the results of Equation (4.19). When a W14 shape for the beam-column with the width b of 14 in. is assumed, the equivalent axial force is

$$P_{u,eq} = 600 + \frac{(2)(100)(12)}{14} + \frac{(7.5)(50)(12)}{14} = 1,093\text{kips}$$

From the Column Design Tables, W14x99 ($\phi_c P_n = 1,110$) may be selected. Yura's equation predicts one size larger member compared to Equation (4.19).

The equivalent axial load approaches given by Equations (4.19 and 4.20) are useful in preliminary sizing of a member under gravity load combinations. However, when the bending moment is dominant, those equations may incur considerable error. In this case, the equivalent moment approach, rather than the equivalent axial force approach, is preferable. Yura proposed an equation for estimating the equivalent moment:

$$M_{u,eq} = M_{ux} + \frac{P_u d}{2} \tag{4.21}$$

This equation is applicable for the initial column sizing in unbraced frames under lateral load combination.

4.9 Modeling of Structural Members

Different types of advanced analysis are: (1) Plastic-zone method, (2) Quasi-plastic hinge method, (3) Elastic-plastic hinge method, and (4) Refined-plastic hinge method. An important consideration in making these advanced analyses practical is the required number of elements for a member in order to predict realistically the behavior of frames.

In the plastic-zone method, frame members are discretized into several finite elements, and the cross-section of each finite element is further subdivided into many fibers. Although the plastic-zone solution is regarded as an "exact solution", it may not be used in engineering design because it is too computationally intensive [14].

A quasi-plastic hinge approach for capturing spread of plasticity effects was proposed by Attalla et al. [10]. The method enables a more realistic structural analysis than the conventional plastic hinge approaches, with less computational effort than that of spread-of-plasticity. However, this approach still requires many elements to model one member.

A simple and efficient alternative is to use the elastic-plastic hinge analysis. However, for stocky members with significant yielding, it over-predicts the strength and stiffness of members.

In the recent works by Liew [28] and Kim [25] among others, the practical advanced analysis using refined-plastic hinge concept has been proposed for two-dimensional frame analyses. This method is adequate with only one or two elements per member.

A sensitivity study of advanced analysis is performed on the required number of elements for a beam member subject to distributed transverse loads. Two-element model adequately predicts the strength of a member. To model a parabolic out-of-straightness in a beam-column, a two-element model with a maximum initial deflection at the mid-height of a member adequately captures imperfection effects. The required number of elements in modeling each member to provide accurate predictions of the strengths is summarized in Table 4.9. It is concluded that practical advanced analysis is computationally efficient. The sensitivity study will be presented in detail in what follows.

Table 4.9 Necessary Number of Elements

Member	Number of element
Beam member subjected to uniform loads	2
Column member of braced frame	2
Column member of unbraced frame	1

4.9.1 Number of Elements for a Beam Subjected to Distributed Transverse Loads

Isolated Beams

Case studies are performed to determine the appropriate number of elements needed to model beams subject to distributed loads with various boundaries and end moments. Predicted beam strengths according to refined-plastic hinge analysis are summarized in Table 4.10. Two- or four-element models predict the strength of a member with errors less than 3% compared with exact solutions.

Table 4.10 Strength Ratio Corresponding to Number of Elements in Modeling the Beam Member

Case No	Boundary Condition		Applied End Mom.		Ratio of Predicted Load-Carrying Capacity to Accurate Strength (Total Number of Element)					Accurate Strength w kip/in
	A	B	A	B	2	3	4	6	12	
1	Hinge	Roll.	0.0	0.0	0.99	1.12	0.99	0.99	0.99	2.43
2	Fix	Fix	0.0	0.0	1.00	1.12	1.00	1.00	0.99	4.86
3	Fix	Roll.	0.0	0.0	1.03	1.03	1.03	1.03	1.00	3.54
4	Hinge	Roll.	$0.4M_p$	0.0	1.01	1.04	0.99	1.01	0.99	1.92
5	Hinge	Roll.	$0.4M_p$	$0.2M_p$	1.03	1.02	1.00	1.03	0.99	2.13
Figure	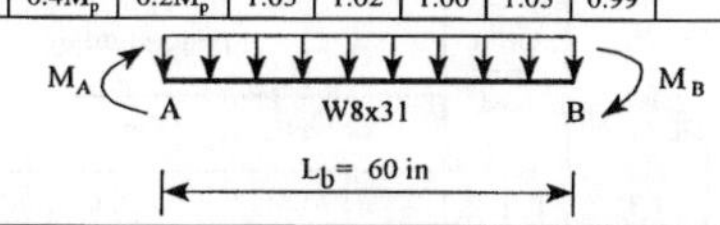									

Framed Beams

The first case study is extended to describe an unbraced frame subject to distributed gravity loads and horizontal loads shown in Figure 4.12. A 12-element model predicts a load-carrying capacity of 48.93 kips (in terms of total gravity load). Two- or four-element models predict almost the identical strength with an error of less than 1% when compared to the 12-element model. As a result, a two-element model of a beam with distributed loads is adequate for practical design.

4.9.2 Number of Elements for a Column Without Geometric Imperfections

Braced Column

For braced frames, clearly more than one element in modeling a column is necessary because plastic hinges in braced frames may form between the ends. Strengths of braced columns with various boundary conditions and end moments are evaluated in Tables 4.11 and 4.12. The two-element model predicts the strengths with errors of less than 2% when compared to those predicted by the four-element model.

Table 4.11 Strength Ratio Corresponding to Number of Elements in Modeling the Braced Column Member Hinged at Top and Bottom

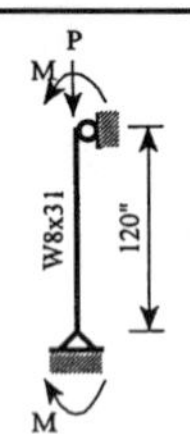

Applied Incr. Load		Predicted Strength						Strength Ratio	
		1 Element		2 Element		4 Element			
P	M	P_1/P_y	M_1/M_p	P_2/P_y	M_2/M_p	P_4/P_y	M_4/M_p	P_1/P_4	P_2/P_4
3	0	0.995	0.0	0.908	0.0	0.911	0.0	1.09	1.00
3	5	0.689	0.345	0.641	0.321	0.644	0.322	1.07	1.00
1	10	0.271	0.814	0.257	0.772	0.256	0.768	1.06	1.01

Table 4.12 Strength Ratio Corresponding to Number of Elements in Modeling the Braced Column Member Hinged at Top and Fixed at Bottom

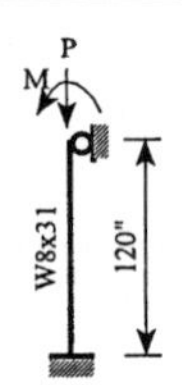

Applied Incr. Load		Predicted Strength						Strength Ratio	
		1 Element		2 Element		4 Element			
P	M	P_1/P_y	M_1/M_p	P_2/P_y	M_2/M_p	P_4/P_y	M_4/M_p	P_1/P_4	P_2/P_4
3	0	0.964	0.0	0.929	0.0	0.938	0.0	1.03	0.99
3	5	0.696	0.348	0.694	0.347	0.694	0.347	1.00	1.00
1	10	0.274	0.824	0.274	0.822	0.272	0.818	1.01	1.01

Columns in Braced Frame

Figure 4.13 shows a braced frame consisting of one-column and one-beam subject to a concentrated gravity load. The four-element model predicts a load-carrying capacity of 215.6 kips (in terms of the concentrated gravity load). The one-element model predicts the strength 5% above the four-element model. Two-elements provide almost the same strength (error 1%). Thus, the two-element model for the braced column is considered to adequately predict the strength.

Unbraced Column

For an unbraced frame, one element per column is adequate to predict the strength because the maximum second-order moments in a column usually occur at the ends. The one-element model of an unbraced col-

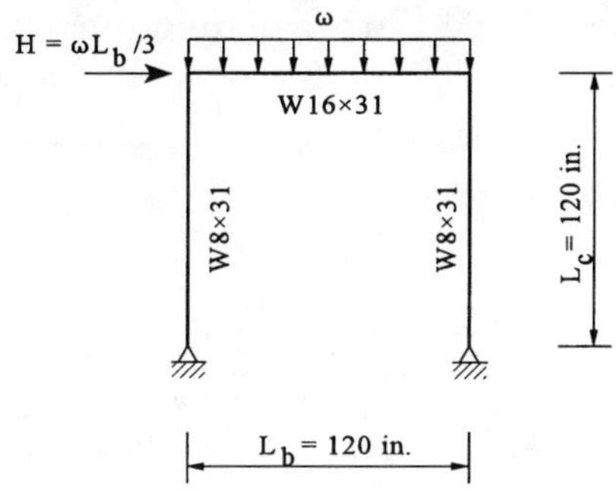

Figure 4.12 Configuration and load condition of unbraced portal frame.

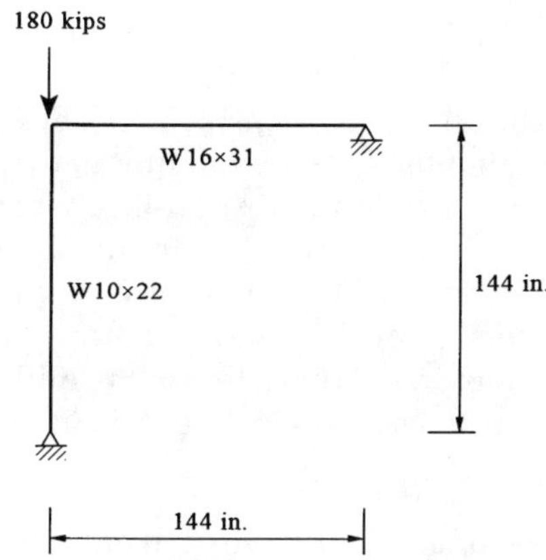

Figure 4.13 Configuration and load condition of braced frame.

umn predicts the strength with the error less than 2%, as compared to the four-element model and summarized in Table 4.13.

Columns in Unbraced Frame

The case study is further extended to unbraced frames subject to uniform gravity loads and horizontal loads shown in Figure 4.12. The four-element model predicts a load-carrying capacity of 16.53 kips (in terms of the horizontal load). The one- or two-element model predicts strength accurately with less than 1% error when compared with that of the four-element model. As a result, the one-element model is recommended for an unbraced column.

4.9.3 Number of Elements for a Column with Geometric Imperfections

The Canadian Standard [18] and the AISC Code of Standard Practice [3] recommend an out-of-plumbness equal to $L_c/500$ accounting for the

Table 4.13 Strength Ratio Corresponding to Number of Elements in Modeling the Unbraced Column Member Guided at Top and Fixed at Bottom

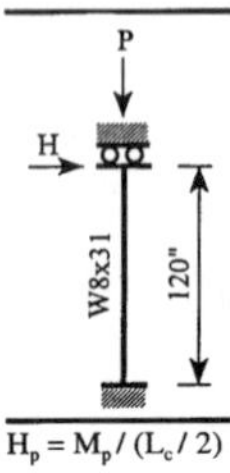

Applied Incr. Load		Predicted Strength						Strength Ratio	
		1 Element		2 Element		4 Element			
P	H	P_1	H_1/H_p	P_2	H_2/H_p	P_4	H_4/H_p	P_1/P_4	P_2/P_4
3	0	0.881	0.000	0.908	0.000	0.911	0.000	0.97	1.00
3	0.1	0.596	0.356	0.615	0.367	0.622	0.373	0.96	0.99
1	0.2	0.225	0.811	0.229	0.828	0.228	0.822	0.99	1.00

erection tolerance between stories in unbraced frames. Out-of-plumbness can be easily modeled with only one element. For braced frames, the AISC-LRFD [3], ECCS [19] [20], AS [35], and CSA [18] Specifications recommend that the out-of-straightness vary parabolically with a maximum in-plane deflection at the mid-height equal to 1/1,000 times the column length. However, they do not provide guidance on how to model the parabolic imperfection explicitly in an advanced analysis. Many elements would be ideal, but not practical. In this study, two elements with a maximum initial deflection at the mid-height of a member capture imperfection effects. This is described in what follows.

Axially Loaded Columns

In practical advanced analysis, the two-element column has an initial geometric imperfection of $L_c/1,000$ at mid-height. The column strengths predicted by the refined-plastic hinge analysis are summarized in Table 4.14 and compared with the LRFD curve in Figure 4.14. The errors are less than 5% for slenderness ratios of L/r up to 140, ($\lambda_c = 1.57$), a practical limit for almost all columns.

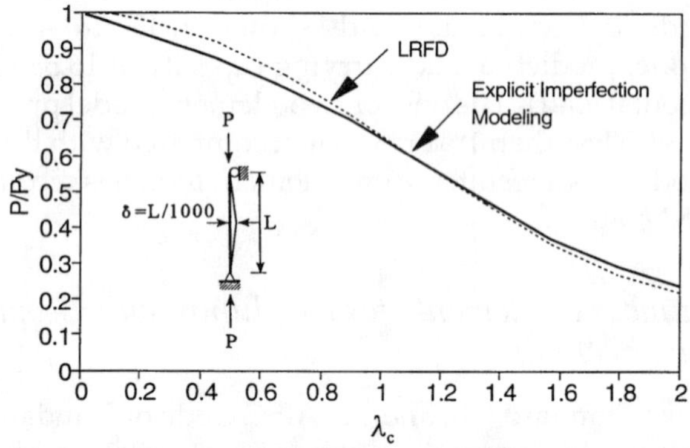

Figure 4.14 Comparison of strength curves for column.

Table 4.14 Comparison of Strength for the Axially Loaded Pin-ended Column

L/r	λ_c	LRFD P_L/P_y	Proposed method P_{AA}/P_y	Error P_{AA}/P_L
1	0.0112	1.000	0.994	0.994
10	0.1122	0.995	0.969	0.974
20	0.2243	0.979	0.938	0.958
40	0.4486	0.919	0.871	0.948
60	0.6729	0.827	0.792	0.958
80	0.8972	0.714	0.700	0.980
100	1.1215	0.590	0.590	1.000
120	1.3458	0.468	0.475	1.015
140	1.5701	0.356	0.369	1.037
160	1.7944	0.272	0.293	1.077
180	2.0187	0.215	0.235	1.093

P_{AA} = ultimate strength by the advanced analysis; P_L = ultimate strength by the LRFD method; P_y = squash load.

Isolated Beam-Columns

Galambos and Ketter [23] developed the interaction curves for pin-ended beam-columns subject to axial thrust in combination with bending moments using the plastic-zone theory. The stress-strain relationship was assumed to be elastic-perfectly plastic with a 33 ksi yield stress and a 30,000 ksi elastic modulus. Galambos and Ketter's interaction curves account for residual stresses ($0.3F_y$) but not geometric imperfections. An adjustment can be made to these interaction curves to consider these imperfections [39] [41].

In the refined-plastic hinge analysis, the two-element column is assumed to have an initial geometric imperfection of $L_c/1,000$ at mid-height. The strength curves granted by the refined-plastic hinge analysis compare well with the LRFD and the adjusted plastic-zone curves. The errors are no more than 5% as shown in Figure 4.15.

Rigidly Jointed Truss

Clarke et al. [16] [17] studied a rigidly jointed roof truss subject to gravity loads between and at the joints of the top chord members (Figure 4.16). The chords have both axial and flexural forces since it is rigidly jointed. The W-sections are oriented such that the weak-axis is subjected to bending. The section properties are summarized in Table 4.15. In the plastic-zone analysis, the chords were assumed to have a sinusoidal out-of-straightness with a maximum in-plane imperfection $\delta_0 = L/1,000$ at mid-height. All cross-sections in the chords were assumed to have

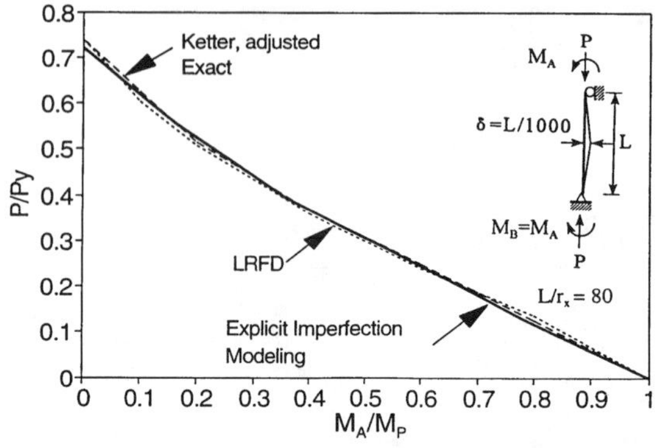

Figure 4.15 Comparison of strength curves for beam-column.

compressive residual stresses. Strain hardening was included according to the stress-strain curve in Figure 4.17. Twenty elements were used for each chord between the joints. The load factor determined by the plastic-zone method is 1.93.

Table 4.15 Section Properties of Rigid-Jointed Truss Members

Member	Location	Section	$A(mm^2)$	$I_y(mm^4)$	$S_y(mm^3)$	$F_y(MPa)$
A, D	Top chord	250UC72.9	9290	38.7x10^6	462x10^3	250
B, C	Top chord	200UC52.2	6640	17.7x10^6	264x10^3	250
E, F	Bot. chord	100UC14.8	1890	$1.14x10^6$	$35.3x10^3$	260
G, I	Web chord	150UC23.4	2980	$4.03x10^6$	$80.9x10^3$	260
H	Web chord	100UC14.8	1890	$1.14x10^6$	$35.3x10^3$	260

In the refined plastic analysis, each chord is modeled by two elements with an initial out-of-straightness of $\delta_0 = L/1,000$ at mid-length shown in Figure 4.18. A load factor of 1.88 is predicted by this method. The proposed method with two elements per member predicts well the ultimate load with conservative errors of less 5%. The load-deflection behavior of the truss is compared between the plastic-zone and the refined-plastic hinge analysis as shown in Figure 4.19.

Braced Frame

The plastic-zone solutions reported by Chen and Zhou (see [14]) are used here as the benchmark check for the braced frame. The configuration of the portal frame is shown in Figure 4.20. It is prevented from side-sway and its beams and columns are rigidly connected. All members are fully braced from out-of-plane displacement, and their cross-sections are compact. The beam is 210 in. long, and the columns are 140 in. for

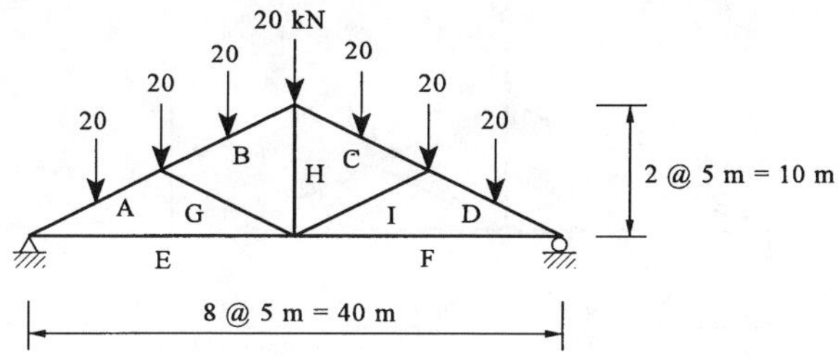

Figure 4.16 Configuration of Clarke's rigidly jointed truss [16] [17].

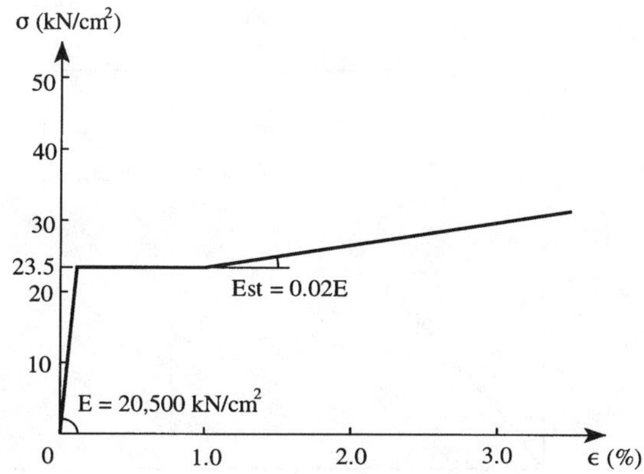

Figure 4.17 Stress-strain relationships of rigidly jointed truss.

case 1 and 280 in. for case 2. The frame is subjected to uniform gravity loads only.

In the plastic-zone analysis, the columns were given a sinusoidal out-of-straightness with a maximum in-plane deflection $\delta_0 = L/1,000$ at mid-height. All cross-sections were assumed to have residual stresses, the maximum compressive stress being $0.3F_y$. An elastic-perfectly plastic stress-strain relationship was assumed.

In the refined plastic analysis, each column is modeled by two elements with an initial out-of-straightness of $\delta_0 = L/1,000$ at mid-eight. The beam is modeled with eight elements. The results of the proposed method are compared with those of the LRFD and plastic-zone methods and are summarized in Table 4.16. The proposed method predicts well the ultimate load, and the conservative errors against the plastic-zone solution are less than 6%. On the contrary, the error of the LRFD method is found to be more than 20%. Since the proposed method accounts for moment redistribution caused by inelasticity, it should require more economic member sizes than the conventional LRFD method.

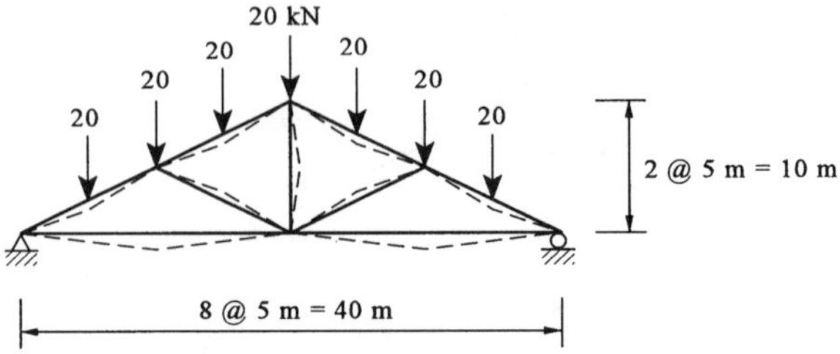

Figure 4.18 Explicit imperfection modeling with two elements per member of rigidly jointed truss.

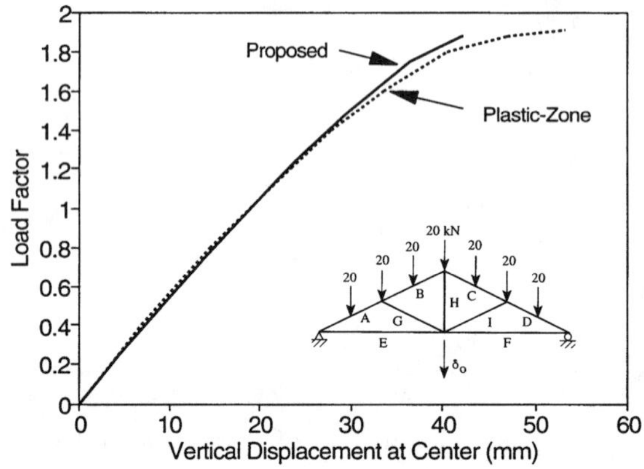

Figure 4.19 Comparison of load-displacement relationships of rigidly jointed truss.

4.10 *Modeling of Geometric Imperfection*

Geometric imperfection modeling is required to account for fabrication and erection tolerances. The imperfection modeling methods used here are the explicit imperfection, the equivalent notional load, and the further reduced tangent modulus models. Users may choose one of these three models in an advanced analysis.

Geometric imperfection modeling is required for a frame but not a truss element because the program computes the axial strength of a truss member using the LRFD column strength equations which account for geometric imperfections.

4.10.1 *Explicit Imperfection Modeling*

Braced Frame

The ECCS [19] [20], AS [35], and CSA [12] Specifications recommend an initial crookedness of 1/1,000 the column length. The AISC Code

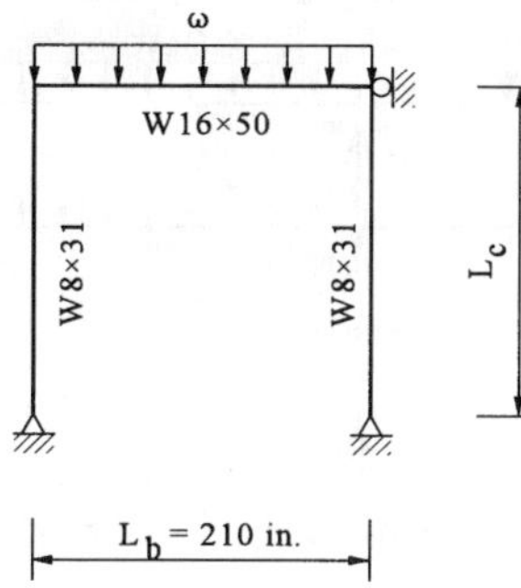

Figure 4.20 Configuration and load condition of braced portal frame.

Table 4.16 Comparison of Ultimate Strength of Braced Portal Frame

Col. length m (in.)	Proposed method P_{AA}/P_y	LRFD P_L/P_y	Plastic-zone P_{PZ}/P_y	Error	
				P_{AA}/P_{PZ}	P_L/P_{PZ}
3.6 m (140 in.)	0.243	0.220	0.246	0.99	0.89
7.1 m (280 in.)	0.234	0.192	0.243	0.96	0.79

P_{AA} = ultimate strength by the advanced analysis; P_L = ultimate strength by the LRFD method; P_{PZ} = ultimate strength by the plastic-zone method; P_y = squash load.

recommends a maximum fabrication tolerance of $L_c/1,000$ for member out-of-straightness. In this study, a geometric imperfection of $L_c/1,000$ is used. The ECCS [19] [20], AS [35], and CSA [12] Specifications recommend that the out-of-straightness varies parabolically with the maximum in-plane deflection at the mid-height. They do not, however, specify how the parabolic imperfection should be modeled in an analysis. Herein, two elements with a maximum initial deflection at the mid-height are shown to adequately capture imperfection effects.

Note that an explicit model of a braced frame requires inconvenient simulation of an imperfection at mid-height of the column. The direction of geometric imperfections should coincide with the deflection direction caused by the bending moment. Using the wrong direction of geometric imperfection in braced frames with rigid joints may improve the bending stiffness of the columns rather than reduce it. Unfortunately, the determination of the direction of geometric imperfection is not a simple task. Figure 4.21 shows a two-bay two-story braced frame subject to concentrated and distributed gravity loads. It is difficult to determine the direction of the geometric imperfection for an interior column. To bypass this difficulty in modeling braced frames, it is recommended that the further reduced tangent modulus model be used.

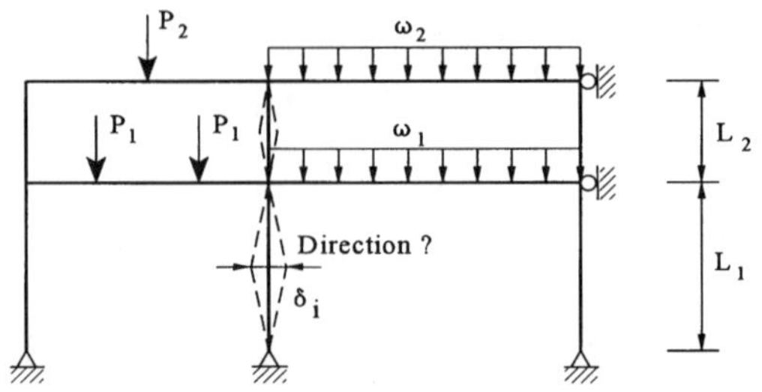

Figure 4.21 Difficulty in determining direction of geometric imperfection.

Unbraced Frame

The Canadian Standard [12] and the AISC Code of Standard Practice [3] limit the erection out-of-plumbness to $L_c/500$ per story. The maximum erection tolerances in the AISC are limited to 1 in. toward the exterior of buildings and 2 in. toward the interior of buildings less than 20 stories. Considering the maximum permitted average lean of 1.5 in. in the same direction of a story, the geometric imperfection of $L_c/500$ can be used for buildings up to six stories with each story measuring approximately 10 ft. For taller buildings, this imperfection value ($L_c/500$) is conservative because the accumulated geometric imperfection calculated by 1/500 times building height is greater than the maximum permitted erection tolerance.

In this study, $L_c/500$ is the out-of-plumbness used. No modifications are needed because the system strength is usually governed by a weak story with an out-of-plumbness of $L_c/500$ [29]. The constant imperfection is a simple practical design approach.

4.10.2 Equivalent Notional Loads Modeling

Braced Frame

For braced frames, an equivalent notional load may be applied at column mid-heights since the ends of the column are braced. An equivalent notional load factor equal to 0.004 equivalent to an out-of-straightness of $L_c/1,000$ is used here.

A drawback of this method when used for braced frames is that it requires the input of notional loads. Another difficulty is that the axial forces must be known before the analysis to determine the notional loads. The estimation of these axial forces is often difficult for large structures subject to lateral wind loads such as as those shown in Figure 4.22. The lateral loads change the axial forces in the columns from those simply caused by gravity loads. The notional load method also introduces the difficulty of determining the direction in which the notional loads should be supplied (Figure 4.22) just as the explicit imperfection model does. The use of the further reduced tangent modulus model bypasses these difficulties in braced frames.

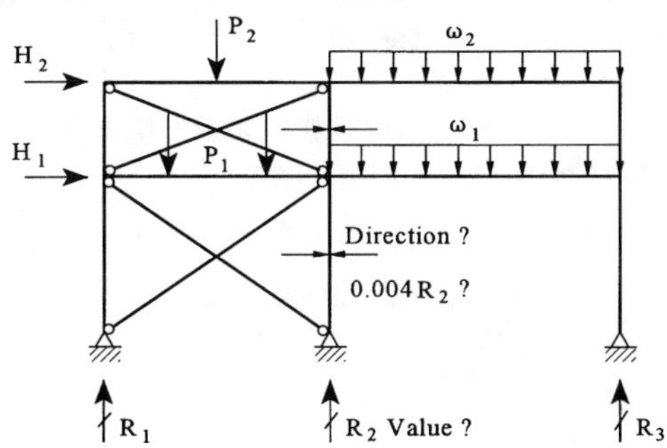

Figure 4.22 Difficulty in determining the axial forces and direction of geometric imperfection.

Unbraced Frame

Herein, the equivalent notional load factor (0.002) is used. The notional load should be applied laterally at the top of each story. For sway frames subject to combined gravity and lateral loads, the notional loads should be added to the lateral loads.

4.10.3 Further Reduced Tangent Modulus Modeling

Geometric imperfections may be simulated by applying an "equivalent reduction" of member stiffness. This may be achieved by a further reduction of tangent modulus as [25] [26]:

$$E_t' = 4\frac{P}{P_y}\left(1 - \frac{P}{P_y}\right) E\xi_i \quad \text{for } P > 0.5P_y \tag{4.22a}$$

$$E_t' = E\xi_i \quad \text{for } P \leq 0.5P_y \tag{4.22b}$$

where

E_t' = reduced E_t
ξ_i = reduction factor for geometric imperfection

Herein, the reduction factor of 0.85 is used and built into the computer program. This 0.85 reduction factor may be used for braced and unbraced frames. The advantage of this approach over the other two models is its convenience and simplicity. It eliminates the inconveniences of both the explicit imperfection and the equivalent notional load methods. Another advantage of this technique is that it does not require guessing as to the direction of the geometric imperfections. The magnitudes of geometric imperfections are summarized in Table 4.17. Of the three geometric imperfection models, the further reduced tangent modulus has been demonstrated as the simplest and is, therefore, the one recommended for general use.

Table 4.17 Magnitude of Geometric Imperfection

Geometric imperfection method	Magnitudes
(1) Explicit imperfection modeling method	$\Psi = 2/1,000$ for unbraced frames $\Psi = 1/1,000$ for braced frames
(2) Equivalent notional load method	$\alpha = 2/1,000$ for unbraced frames $\alpha = 4/1,000$ for braced frames
(3) Further reduced tangent modulus method	$E_t' = 0.85E_t$

4.11 *Load Application*

4.11.1 *Proportional Loading*

In advanced analysis, the gravity and lateral loads are simultaneous and do not account for unloading. This is illustrated in what follows. The force-state parameter of the proposed methods including resistance factors can be written as:

$$\alpha = \frac{P}{\phi_c P_y} + \frac{8}{9}\frac{M}{\phi_b M_p} \quad \text{for} \quad \frac{P}{\phi_c P_y} \geq \frac{2}{9}\frac{M}{\phi_b M_p} \tag{4.23a}$$

$$\alpha = \frac{P}{2\phi_c P_y} + \frac{M}{\phi_b M_p} \quad \text{for} \quad \frac{P}{\phi_c P_y} < \frac{2}{9}\frac{M}{\phi_b M_p} \tag{4.23b}$$

For illustration, if a gravity load P equal to $0.7\phi_c P_y$ is applied to the cantilever (Figure 4.23), the parameter α is 0.7 and the inelastic stiffness reduction factor η is 0.85 (Equation 3.8). When a bending moment is applied to the column in addition to P for α equal to 0.9, the reduction factor η is reduced to 0.36. In this method of stiffness reduction, the proposed methods treat only the resulting forces but not the stress-state of fibers in the cross-section, and consequently the methods do not account for the unloading effect.

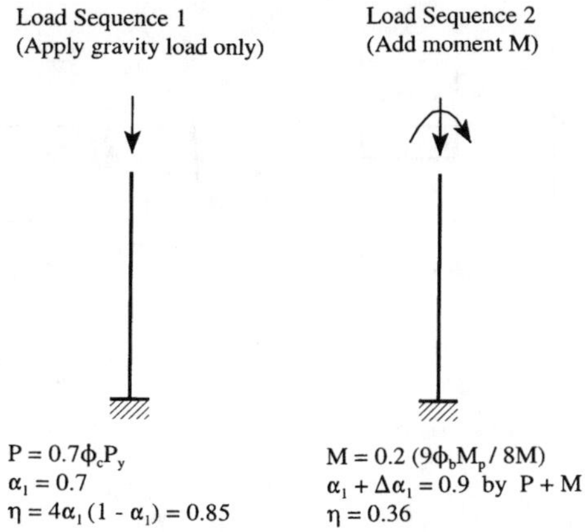

Figure 4.23 Stiffness degradation process in the proposed methods.

On the contrary, the plastic-zone analysis can account for the unloading because the stress-strain in each fiber in the cross-section can be traced in the analysis. The cantilever in Figure 4.24 is a good example illustrating this effect. When the moment is added with the force P, the strain on the concave side of the column is increased and on the convex side it is reversed.

As a result, the proposed method under-predicts the strength of frames subject to sequential loads, large gravity loads first and then lateral loads. It is, however, justified for the practical design because the development of the LRFD interaction equations was also based on strength curves subject to simultaneous loading [24] and the current LRFD elastic analysis uses the proportional loading rather than the sequential loading.

4.11.2 *Incremental Loading*

It is necessary, in an advanced analysis, to input each increment load (not the total loads) to trace nonlinear load-displacement behavior. The incremental loading process can be achieved by scaling down the combined factored loads by a number between 10 and 50. For a highly redundant structure (such as one greater than six stories), dividing by about 10 is recommended and for a nearly statically determinate structure (such as a portal frame), the incremental load may be factored down by 50. One may choose a number between 10 and 50 to reflect the redundancy of a particular structure. Since a highly redundant structure has the potential to form many plastic hinges and the applied load increment is automatically reduced as new plastic hinges form, the larger incremental load (i.e., the smaller scaling number) may be used.

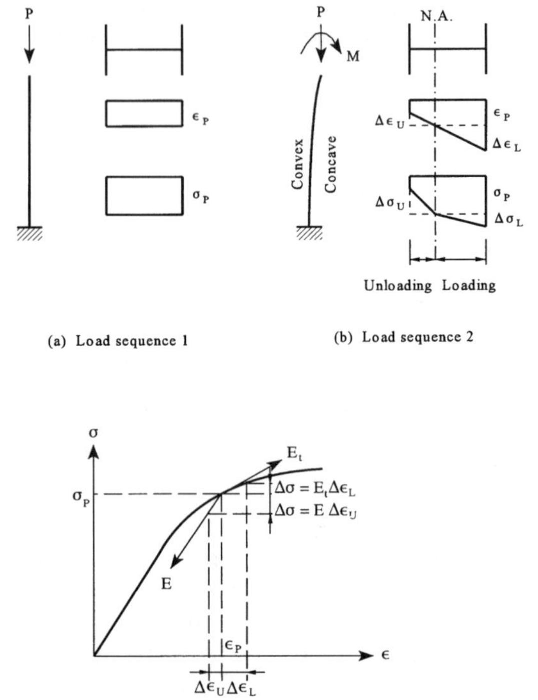

(c) Stress-strain relationship

Figure 4.24 Stiffness degradation process in the plastic-zone analysis.

The simple incremental solution method is implemented for practical use. The errors in using the simple incremental solution method are minimized by automatically scaling the incremental loads when changes in the element stiffness parameter exceed a predefined tolerance.

4.12 *Analysis*

Analysis is important in the proposed design procedures because the advanced analysis method captures key behaviors including second-order and inelasticity in its analysis program. Advanced analysis does not require separate member capacity checks by the specification equations. On the other hand, the conventional LRFD method accounts for inelastic second-order effects in its design equations (not in analysis). The LRFD method requires tedious separate member capacity checks. Input data used for advanced analysis is easily accessible to users, and the input format is similar to the conventional linear elastic analysis. The format will be described in detail in Chapter 5. Analyses can be simply carried out by executing the program described in Chapter 5. This program continues to analyze with increased loads, and stops when a structural system reaches its ultimate state.

4.13 Load-Carrying Capacity

The LRFD method does not capture the inelastic redistribution of internal forces throughout a structural system, since the first-order forces, even with the B_1 and B_2 factors, account for the second-order geometric effect but not the inelastic redistributions of internal forces. The LRFD provides a conservative estimation of the ultimate load-carrying capacity due to material yielding, but advanced analysis directly considers moment redistribution and thus allows smaller member sizes to be selected. This is particularly beneficial in highly indeterminate steel frames. Because consideration at moment redistribution may not always be desirable, the two approaches (including and excluding inelastic moment redistribution) are presented. First, the load-carrying capacity, including the effect of inelastic moment redistribution, is obtained from the final loading step (limit state) given by the computer program. Second, the load-carrying capacity without the inelastic moment redistribution is obtained by extracting that force sustained when the first plastic hinge formed. Generally, advanced analysis predicts the same member size as the LRFD method when moment redistribution is not considered. Further illustrations on these two choices will be presented in Chapters 5 and 6.

4.14 Serviceability Limits

The serviceability conditions specified by the LRFD consist of five limit states: (1) deflection, vibration, and drift; (2) thermal expansion and contraction; (3) connection slip; (4) camber; and (5) corrosion. The most common parameter affecting the design serviceability of steel frames is the deflections. The deflection limits are specified by the code. The advanced analysis does not set any specific limitations of serviceability, but the usual serviceability requirements are presented in what follows.

Frame deflections are usually measured in terms of a non-dimensional parameter called "drift". Inter-story drift is the ratio of inter-story deflection to story height. Frame drift is the ratio of the lateral deflection at the top of the frame to the frame height. Deflection limits are always investigated at service load conditions.

According to the ASCE Ad Hoc Committee report [7], the accepted range of drift limits by first-order analyses is $1/750$ to $1/250$ times the building height H with a typical value of $H/400$. The typical limits on the interstory drift are $1/500$ to $1/200$ times the story height. Another parameter related to the calculation of building drift is the second-order $P - \Delta$ effect. The $P - \Delta$ effect occurs due to the presence of gravity loads acting on the displaced structure. A steel building with standard dimensions and designed to a first-order drift limit of 1/500 may experience additional second-order deflections in the range of 10 to 30% of the first-order ones. Therefore, if a second-order analysis is used to directly assess the serviceability of a building frame, the limit for lateral drift and interstory deflection should be relaxed.

Service beam live load deflections may be limited to $L/360$ when the beam carries brittle finishers such as plaster ceilings; otherwise, the limit can be increased to $L/240$, where L is the span length of the beam. A maximum absolute deflection limit must be prescribed depending on the construction material. This absolute limit should be within the range of 3/4 to 1 in. [22].

Based on the studies by the Ad Hoc Committee [7] and Ellingwood [21], the deflection limits recommended (Table 4.18) were proposed for general use. For some multistory frames, values other than those in Table 4.18 may be more appropriate; for instance, $H/500$ may be more suitable when the cladding cannot accommodate larger movements. At service load levels, no plastic hinges are permitted anywhere in the structure to avoid permanent deformation under service loads.

Table 4.18 Deflection Limitations of Frame

Item	Deflection ratio
(1) Floor girder deflection for service live load	$L/360$
(2) Roof girder deflection	$L/240$
(3) Lateral drift for service wind load	$H/400$
(4) Interstory drift for service wind load	$H/300$

The first-order displacements at service loads can be obtained by either using the proposed analysis program or a usual commercial program. The proposed analysis program is recommended since the input data prepared for the ultimate strength checks may also be used for the serviceability checks without major change. Using the proposed analysis program, the first-order displacements can be obtained by setting the number of load increments equal to 1 (NINCRE = 1), and then by multiplying the resulting displacements by the scaling number. This procedure is based on the fact that the structural system behaves linearly and elastically under a small incremental load.

4.15 Ductility Requirements

Adequate rotation capacity is required for members to develop their full plastic moment capacity. This is achieved when members are adequately braced and their cross-sections are compact. The limits for compact sections and lateral unbraced lengths described in the following section lead to an inelastic rotation capacity of at least three times the elastic rotation for non-seismic regions and seven times for seismic regions.

4.15.1 Compactness

In the advanced analysis of steel structures, it is assumed that a section will sustain its plastic moment capacity until sufficient plastic hinges develop to form a failure mechanism. However, if the section is made of thin plate elements, premature local buckling may occur and the section can neither attain a plastic moment nor sustain the plastic moment up to the desired rotation capacity after attaining the plastic moment. For example, if the width-to-thickness ratios of the flanges and web are too high, they may buckle locally, thus violating the basic assumptions underlying the advanced analysis. Figure 4.25 shows possible local buckling modes for the compression flanges of two beams: one under moment gradient and the other under a uniform moment.

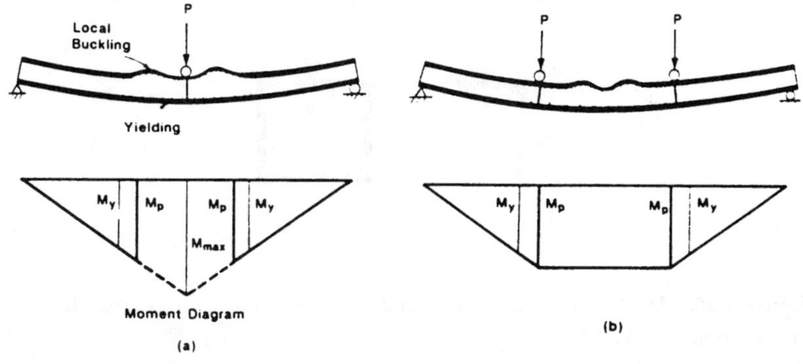

Figure 4.25 Local buckling of flanges of beams.

Definition of Compact Section

To ensure compactness, the LRFD Specification defines two sets of limiting width-to-thickness ratios ($\lambda = b/t$) : λ_p and λ_r (Table 4.19). On the basis of these limiting values, steel sections are subdivided into three categories: compact, noncompact, and slender sections. If the width-to-thickness ratios for all the elements in a cross-section are less than λ_p, then it is compact. A compact section is capable of developing a fully plastic stress distribution (plastic moment), and can sustain rotations approximately three times beyond the yield before the possibility of local buckling arises. If the width-to-thickness ratios are between λ_r and λ_p, the section is noncompact. Noncompact sections can develop the yield stress before local buckling occurs. They may not, however, resist local buckling at the strain levels required to develop the fully plastic stress distribution. If the width-to-thickness ratio of one or more elements of a section is greater than λ_r, then the section is slender and may locally buckle under elastic stresses before the yielding is achieved. For determining the width-to-thickness ratio (b/t) of a plate element in a thin-walled section, the width b is explicitly defined in Figure 4.26. Further discussion of Table 4.19 is forthcoming.

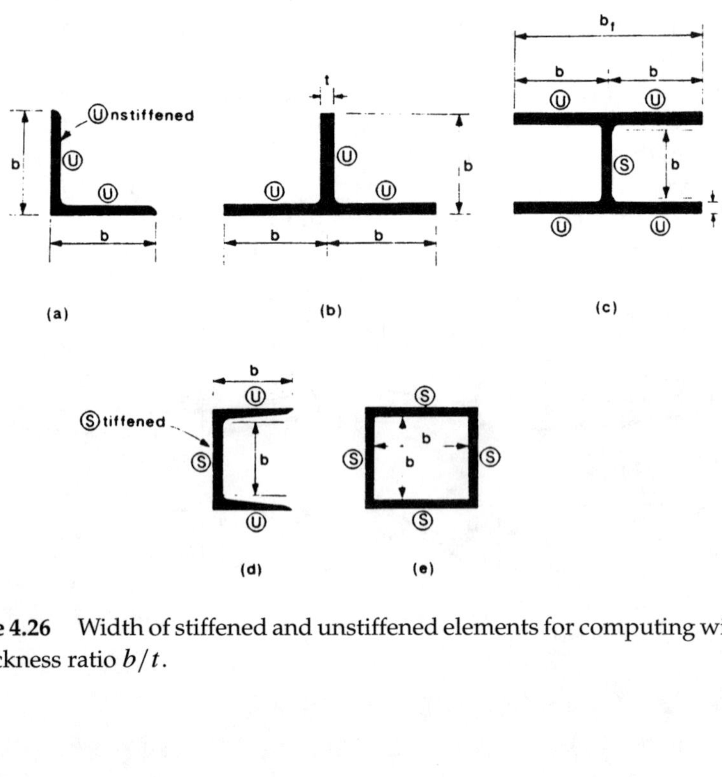

Figure 4.26 Width of stiffened and unstiffened elements for computing width-to-thickness ratio b/t.

Table 4.19 Limiting Values of Width-to-Thickness Ratio ($M = b/t$) to Avoid Premature Local Buckling

Type of element	$k_{\min}$	λ_p (compact)	λ_r (noncompact)
Unstiffened			
Single angles	0.425	NA	$76/\sqrt{F_y}$
Flanges of I-shaped rolled beams and channel section in flexure	0.7	$65/\sqrt{F_y}$	$141/(\sqrt{F_y} - 10)$
Flanges of I-shaped hybrid or welded beams in flexure	0.7	$65/\sqrt{F_{yf}}$	$162/\sqrt{[F_{yf} - 16.5)/k_c]}$
Flanges of channels or I-shapes in pure compression	0.7	NA	$95/\sqrt{F_y}$
Stems of tees	1.277	NA	$127/\sqrt{F_y}$
Stiffened			
Uniform thickness flanges of tubular sections and flange cover plates	4.7	$190/\sqrt{F_y}$	$238/\sqrt{F_y}$
Webs in flexural compression	—	$640/\sqrt{F_y}$	$970/\sqrt{F_y}$
Perforated cover plates	6.97	NA	$317/\sqrt{F_y}$
All other uniformly stressed stiffened elements, i.e., supported along two edges	5.0	NA	$253/\sqrt{F_y}$
Circular hollow sections, D/t, in axial compression, in flexure	—	$2070/F_y$	$3300/F_y$, $8970/F_y$

F_{yf} = yield stress of the flange, ksi; h = clear distance between flanges when welds are used; $k_c = 4/\sqrt{(k/t_w)}$ but not less than $0.35 \le k_c \le 0.763$; D = diameter.

Adopted from AISC-LRFD, 1994.

Buckling Strength of Plate Elements

The recommended λ_p and λ_r values are based on the buckling strength plate elements with various boundary conditions. Formal mathematics provides the elastic buckling strength of an axially loaded steel plate element as:

$$F_{cr} = \frac{k\pi^2 E}{12\left(1 - \nu^2\right)\left(\frac{b}{t}\right)^2} = 26,210\frac{k}{\left(\frac{b}{t}\right)^2} \tag{4.24}$$

where

ν = Poisson ratio of the material
b/t = width-to-thickness ratio of the plate
k = buckling coefficient

For elements in most structural members, the value of k depends mainly on the boundary conditions (or edge conditions) of the plate element.

On the basis of the boundary conditions, plate elements in structural members can be divided into two categories: unstiffened elements and stiffened elements. Unstiffened elements are supported along one edge parallel to the axial stress; for example, legs of single angles, flanges of wide-flange sections, and stems of tees. Stiffened elements are supported along both the edges parallel to the axial stress; for example, flanges of square and rectangular tubular sections, perforated cover plates, and webs of wide-flange and channel sections. The recommended values of the buckling coefficient k for several types of thin elements are listed in Table 4.19.

A plate element can develop the full yield stress without buckling only if its width-to-thickness ratio does not exceed a certain value. For a plate without residual stresses, this value can be obtained by rearranging Equation (4.24)

$$\frac{b}{t} = 162\sqrt{\frac{k}{F_y}} \tag{4.25}$$

where

F_y = yield strength of steel in ksi.

Limiting Width-to-Thickness Ratios

To attain the full yielding, elements with residual stresses must undergo strains larger than the elements without residual stresses. Thus, the LRFD recommends a conservative version of Equation (4.25) by limiting the b/t ratio denoted as λ_r to:

$$\lambda_r = 113\sqrt{\frac{k}{F_y}} \tag{4.26}$$

In order to attain a fully plastic stress distribution over the entire section, the plate elements have to undergo even higher strains. Thus, λ_p in LRFD is based on even more strict requirements approximately given by the following:

for unstiffened elements

$$\lambda_p = 75\sqrt{\frac{k}{F_y}} \tag{4.27}$$

and for stiffened elements

$$\lambda_p = 94\sqrt{\frac{k}{F_y}} \tag{4.28}$$

The values of λ_r and λ_p recommended by the LRFD [3] in Table B5.1 can approximately be obtained by using the appropriate value of k in Equations (4.26 through 4.28). These value of λ_r and λ_p are given in Table 4.19. Since webs of sections under flexural compression are partially under tension, the recommended values of λ_r and λ_p are much higher than those for the uniformly stressed stiffened elements (Table 4.19). For webs under a combined flexural and axial compression, λ_p is recommended:

for $P/\left(\phi_b P_y\right) \leq 0.125$

$$\lambda_p = \frac{640}{\sqrt{F_y}}\left(1 - 2.75\frac{P}{\phi_b P_y}\right) \tag{4.29}$$

and for $P/\left(\phi_b P_y\right) > 0.125$

$$\lambda_p = \frac{191}{\sqrt{F_y}}\left[2.33 - \frac{P}{\phi_b P_y}\right] \geq \frac{253}{\sqrt{F_y}} \tag{4.30}$$

where

P = required axial load capacity
P_y = yield axial load
ϕ_b = resistance factor for bending, i.e., 0.9

Note that limiting b/t ratio to λ_p is to ensure that the member can be designed using the advanced analysis methods. In areas of high seismicity, however, sections must also be able to develop a higher ductility/rotation capacity (seven to nine times the elastic rotation). In these regions, the limiting values of λ_p must be further reduced as recommended by LRFD. For flanges of I-shaped and channel sections, λ_p for seismic areas is:

$$\lambda_p = \frac{52}{\sqrt{F_y}} \tag{4.31}$$

For webs under a combined flexural and axial compression, λ_p in seismic area is:

for $P/\left(\phi_b P_y\right) \leq 0.125$

$$\lambda_b = \frac{520}{\sqrt{F_y}}\left[1 - 1.54\frac{P}{\phi_b P_y}\right] \tag{4.32}$$

for $P/\left(\phi_b P_y\right) \geq 0.125$

$$\lambda_p = \frac{191}{\sqrt{F_y}}\left[2.33 - \frac{P}{\phi_b P_y}\right] \geq \frac{253}{\sqrt{F_y}} \tag{4.33}$$

In the advanced analysis presented here, all sections are assumed to be compact. The limiting width-to-thickness ratios for compact sections are summarized in Table 4.20. Sections that do not meet these width-to-thickness requirements may be strengthened/stiffened in the region of the plastic hinge. Figure 4.27 suggests some methods by which this may be accomplished.

4.15.2 Lateral Torsional Buckling

I-sections bent about the strong axis are susceptible to lateral torsional buckling depending on the distance between the points of lateral support [30] [34] [36] [41]. Lateral-torsional buckling (Figure 4.28) of a beam is very similar to the effect of local buckling of a cross-section in detrimental consequences (Section 4.15.1). Both may prevent the beam from attaining its full plastic moment capacity, and even if it is attained, the desired rotation capacity may not be. Local buckling occurs when the high width-to-thickness ratios of the elements of a cross section are high. Lateral-torsional buckling occurs when the unbraced length of a narrow beam is too long.

The effect of the unbraced length on the moment-carrying and rotation capacity is illustrated schematically in Figure 4.29. When the unbraced

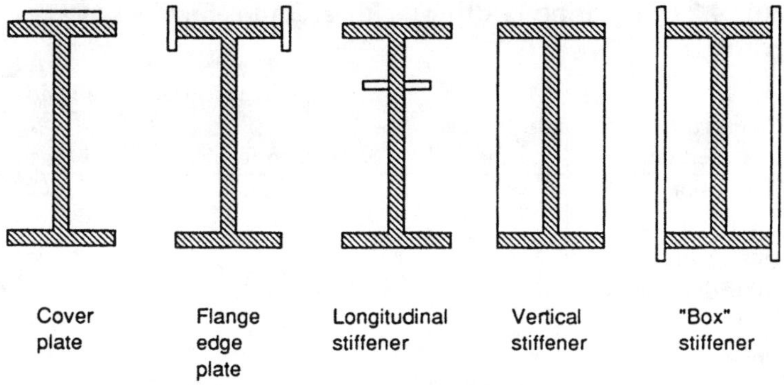

Figure 4.27 Methods of stiffening wide-flange shapes to prevent local buckling of compression elements.

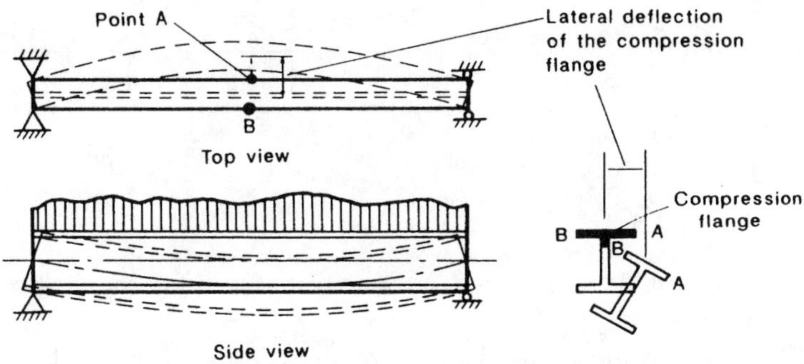

Figure 4.28 Lateral torsional buckling of a wide-flange.

length is long, the member will likely fail by elastic lateral-torsional buckling and the rotation capacity will be small. When the unbraced length is short, the full plastic moment will be attained or exceeded since the section can deliver a large rotation capacity. For intermediate lengths, inelastic buckling occurs and the rotation capacity is higher than that for elastic buckling. For satisfactory performance, the rotation capacity of a beam (or a member of a structure) should be equal to or greater than the angle required to form a mechanism.

To ensure that the proper moment and rotation capacities are provided, the LRFD specification has defined three sets of limiting unbraced lengths: L_{pd}, L_p, and L_r. When unbraced length $L_b \leq L_{pd}$, the member is suitable for advanced analysis, i.e., it will be able to sustain plastic moment up to the rotation capacity necessary to form a failure mechanism (taken as three times the yielding rotation by LRFD). When $L_{pd} < L_b \leq L_p$, the member will attain the plastic moment but may not provide the desired rotation capacity. When $L_p < L_b \leq L_r$, the member will fail by inelastic lateral torsional buckling and never attain the

Table 4.20 Limiting Width-Thickness Ratios for Compact Section

Description of element	Width-to thickness ratio	Limiting width-to-thickness ratios λ_p Non-seismic	Seismic
Flanges of I-shaped sections (including hybrid sections) and channels in flexure[a]	b/t	$65/\sqrt{F_y}$	$52/\sqrt{F_y}$
Webs in combined flexural and axial compression	h/t_w	For $P_u/\phi_b P_y \leq 0.125$	
		$\frac{640}{\sqrt{f_y}}\left(1 - \frac{2.75P_u}{\phi_b P_y}\right)$	$\frac{520}{\sqrt{f_y}}\left(1 - \frac{1.54P_u}{\phi_b P_y}\right)$
		For $P_u/\phi_b P_y > 0.125$	
		$\frac{191}{\sqrt{F_y}}\left(2.33 - \frac{P_u}{\phi_b P_y}\right) \geq \frac{253}{\sqrt{F_y}}$	

[a] For hybrid beams, use F_{yf} in place of F_y.
Adopted from AISC-LRFD, 1994.

plastic moment. When $L_b > L_r$, the member will fail by elastic lateral torsional buckling.

The elastic limiting length L_r is obtained from the elastic lateral torsional buckling moment M_{cr} for beams with various cross-sectional shapes. For example, I-sections under a uniform moment can sustain a critical moment given by the elastic solution of the governing differential equation [13]:

$$M_{cr} = \frac{\pi}{L_b}\sqrt{EI_y GJ + \left(\frac{\pi E}{L_b}\right)^2 I_y C_w} \tag{4.34}$$

where

L_b = unbraced length
E = Young's modulus in ksi
G = shear modulus in ksi
I_y = the moment of inertia about the weak axis in in.^{4}
J = torsional constant in in.^{4}
C_w = warping constant in in.^{6}

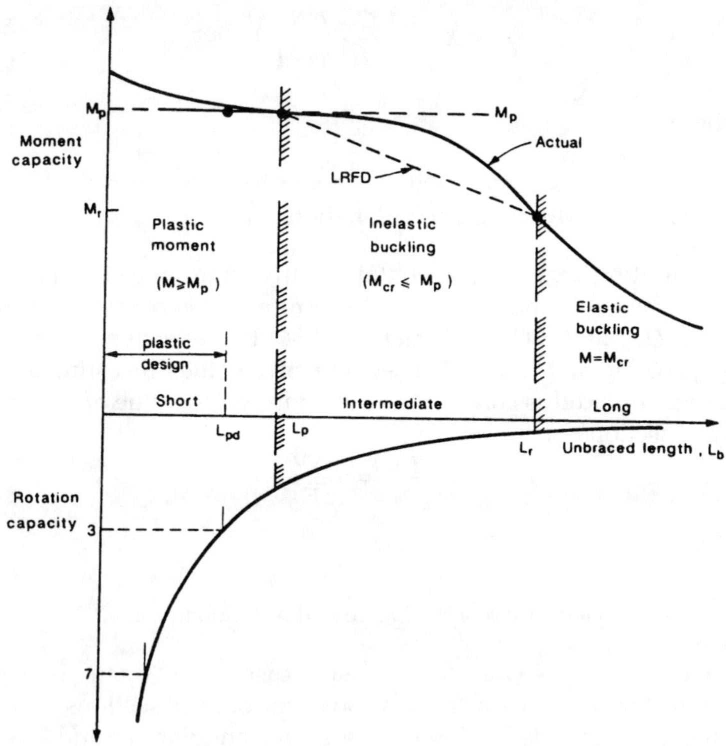

Figure 4.29 Moment and rotation capacity of a beam with its unbraced length.

By equating $M_{cr} = \left(F_y - F_r\right) S_x$ in this equation and rearranging, $L_r (= L_b)$ is obtained as

$$L_r = \frac{r_y X_1}{\left(F_y - F_r\right)} \sqrt{1 + \sqrt{1 + X_2 \left(F_y - F_r\right)^2}} \tag{4.35}$$

where

F_y = yield stress in ksi
F_r = the maximum compressive residual stress in ksi
r_y = the radius of gyration of the section about the minor axis in in.

$$X_1 = \frac{\pi}{S_x} \sqrt{\frac{EAGJ}{2}} \tag{4.36}$$

$$X_2 = 4\frac{C_w}{I_y}\left(\frac{S_x}{GJ}\right)^2 \tag{4.37}$$

where

S_x = elastic section modulus about the strong axis in in.^{3}
A = cross-sectional area of the beam in in.^{2}

The limiting length L_p is obtained from the inelastic critical moment for beams. The inelastic critical moment is obtained by replacing EI_y, GJ, and EC_w in Equation (4.34) by "effective" counterparts $[(EI_y)_e, (GJ)_e$, and $(EC_w)_e]$. These effective values are estimated using the tangent modulus concept. The recommended value of L_p for I and channel sections is:

$$\frac{L_p}{r_y} = \frac{300}{\sqrt{F_{yf}}} \tag{4.38}$$

where

F_{yf} = yield stress of the flange of the section in ksi.

Equation (4.38) is valid for the usual case of an "open" I-shape but becomes conservative for "closed" box-type or solid sections. For example, the recommended value for a solid rectangular bar and box-beam is:

$$\frac{L_p}{r_y} = \frac{3,750}{M_p}\sqrt{JA} \tag{4.39}$$

where

M_p = plastic moment in kip-in.
A = cross-sectional area in in.^{2}
J = torsional constraint in in.^{4}

Tests on I-shapes have shown that the maximum effective spacing between points of lateral support adjacent to a plastic hinge is affected by several factors, the most important of which is the moment gradient, i.e., the change in moment over this distance. The limiting unbraced length L_{pd} for advanced analysis can be obtained from a beam model in which the beam is partly elastic and partly strain hardened. The solution of the resulting differential equation, with some simplification, can be expressed in terms of the moment ratio M_1/M_p as:

$$\frac{L_{pd}}{r_y} = \frac{3,600 + 2,200M_1/M_p}{F_y} \tag{4.40}$$

where

M_1 = smaller moment at the end of unbraced length of beam in kip-in.

M_1/M_p = positive when the moments cause reverse curvature

In Equation (4.40), L_{pd} is measured from a plastic hinge location to the adjacent bracing point, and r_y is the radius of gyration with respect to the weak axis. Equation (4.40) assumes that the moment diagram within the unbraced length is linear. For nonlinear moment distribution between the braces, judgment should be used to choose a representative ratio. All plastic hinge locations associated with the failure mechanism shall be braced to resist lateral and torsional displacements. Equation (4.40) is developed to provide a rotation capacity at least three times that of the initial yield curvature. This is sufficient for most applications; however, in areas of high seismicity, rotations of seven to nine times the yield curvature may be required. In these cases, LRFD recommends

$$\frac{L_{pd}}{r_y} = \frac{2500}{F_y} \tag{4.41}$$

Adequate lateral support to beams or structural members may be provided at intervals by cross-beams, cross-frames, struts, and roof purlins [Figure 4.30(a-d)]. It may also be provided by embedment of a compression flange in a concrete floor slab [Figure 4.30(e and f)].

Note that at the region of the last plastic hinge, a large rotation capacity is not required. Therefore, the lateral support in this region can be provided at a spacing of more than L_{pd}. The lateral support requirements at the last hinge, and for regions not adjacent hinges, are no different from those of elastic design. The flexural design strength is determined according to the unbraced length and the limits: L_{pd}, L_p, or L_r (Section F1, LRFD).

The AISC-LRFD Specification specifies limitations for beam members (as described above), but does not directly address beam-columns. More studies are necessary to determine reasonable limits which will lead to reliable beam-column rotation capacities. Based on a study [39], the limitations for beam members are used for beam-column members until the Specification provides the specific values for beam-column members.

4.16 Adjustment of Member Sizes

If one of three conditions (strength, serviceability, or ductility) is not satisfied, appropriate adjustments of the member sizes should be made. This can be done by referring to the sequence of plastic hinge formation

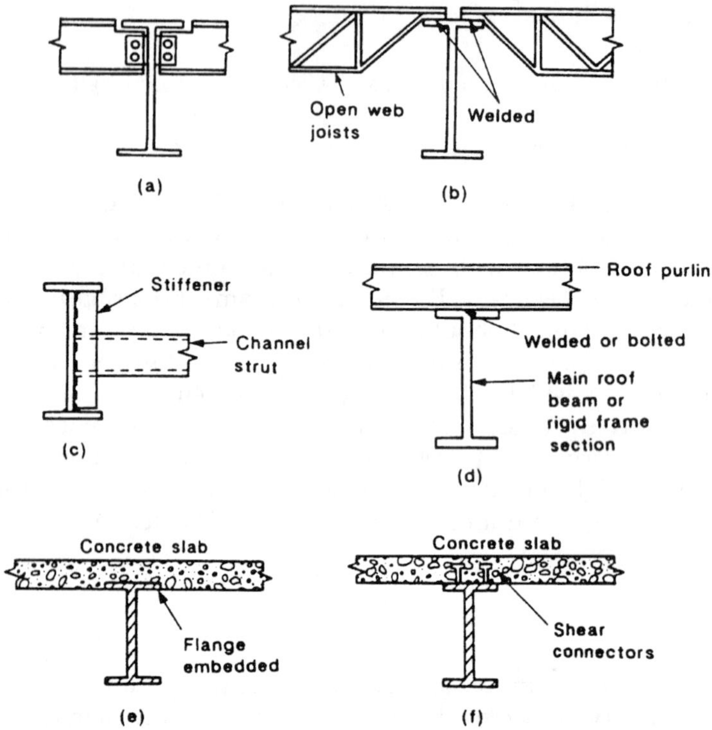

Figure 4.30 Types of lateral supports for beams.

shown in the "P.OUT". For example, if the load-carrying capacity of a structural system is less than the factored load effect, the member with the first plastic hinge should be replaced with a stronger member. On the other hand, if the load-carrying capacity exceeds the factored load effect significantly, members without plastic hinges may be replaced with lighter members. If lateral drift exceeds drift requirements, columns or beams should be sized-up, or a braced structural system should be considered instead to meet this serviceability limit.

In semi-rigid frames, behavior is influenced by the combined effects of members and connections. As an illustration, if an excessive lateral drift occurs in a structural system, the drift may be reduced by increasing member sizes or using more rigid connections. If the strength of a beam exceeds the required strength, it may be adjusted by reducing the beam size or using more flexible connections. Once the member and connection sizes are adjusted, the iteration leads to an optimum design. Figure 4.31 shows a flow chart of analysis and design procedure in the use of advanced analysis.

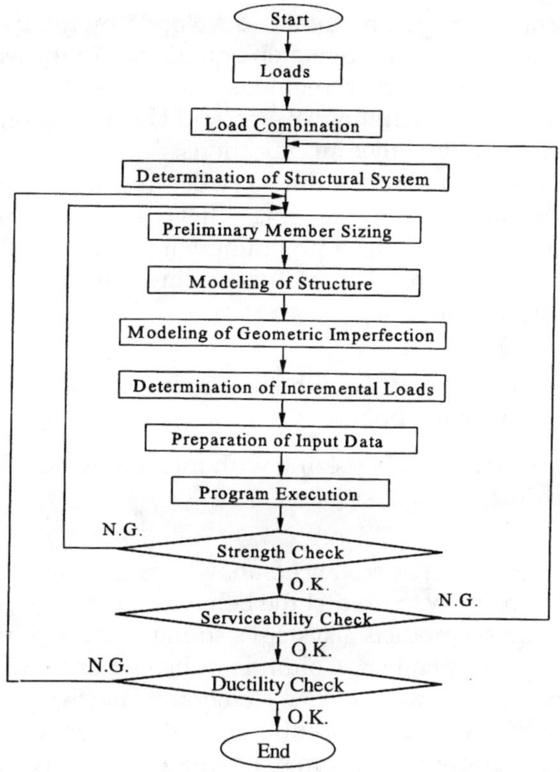

Figure 4.31 Analysis and design procedure.

4.17 Summary

The proposed advanced analysis procedure follows the LRFD design format except that the resistance R_n refers to a structural system, rather than a simple member. The load combinations in advanced analysis are the same as those used by the LRFD [3].

A structural system must be designed to prevent failure or excessive deformation due to various loads including dead, live, impact, wind, earthquake, snow, and rain load. The member sizes are determined based on combined factored loads. The live load reduction is given by the ASCE 7-88. It is important to use the live load reduction factor in analyzing a structural system because the influence area for each beam and column is different, and this results in different reduction factors.

The strength reduction factors ϕ_c and ϕ_b are built in the analysis program and thus are automatically included in the resulting load-carrying capacity. The strength reduction factors are selected as 0.85 for axial strength and 0.9 for flexural strength corresponding to AISC-LRFD Specification [3].

The advanced analysis method is developed by calibration with the LRFD column curve and interaction equations. To this end, it may be concluded that the proposed methods may be used for various rolled and welded sections including W, WT, and HP shapes, pipe, and structural tubing without further modifications.

To design statically indeterminate structures, it is necessary to estimate member sizes before analyzing. This is a challenging task and no systematic rule is available for preliminary member sizing of a statically indeterminate structure. Approximate sizing can be achieved through the following two steps:

1. Approximate analysis with simplifying assumption of the locations of inflection points.
2. Approximate member sizing with member forces obtained by an approximate analysis.

An important issue in advanced analysis is the number of elements required per member to predict the behavior of frames. The proposed advanced analysis predicts a system's strength accurately using one or two elements per member. Columns in braced and unbraced frames may be modeled by two and one element, respectively. Nodal points should be provided where concentrated loads are applied. When a beam is subject to uniform loads, it may be modeled by two elements. The uniform loads should be converted into these concentrated loads.

The magnitudes of geometric imperfections are selected as follows:

1. Explicit imperfection modeling

   : $\Psi = 2/1{,}000$ for unbraced frames

   : $\Psi = 1/1{,}000$ for braced frames

2. Equivalent notional load modeling

   : $\alpha = 2/1{,}000$ for unbraced frames

   : $\alpha = 4/1{,}000$ for braced frames

3. Further reduced tangent modulus modeling

   : $E_t' = 0.85E_t$

The explicit imperfection model and the equivalent notional load model require the user to input either imperfections or notional loads, respectively. The further reduced tangent modulus approach includes imperfection effects implicitly in the computer program.

In using the proposed method, the gravity and lateral loads should be applied at the same time (not sequentially), since it does not account for unloading. The input of incremental loads, rather than the total load is required so nonlinear load-displacement behavior can be traced. The incremental loads can be determined by dividing the combined factored loads using the recommended scaling number between 10 and 50. The errors in using the simple incremental solution method can be minimized by an automatic scaling of the incremental loads when changes in an element stiffness parameter exceed a predefined tolerance in the program.

The input data format is similar to that of the linear elastic analysis with proper considerations of geometric imperfections and incremental loads. Analysis can be simply carried out by the execution of the analysis program described in detail in Chapter 5. The program continues the analysis for increased loads and stops when it reaches its ultimate state.

From the output file, the ultimate load-carrying capacity, i.e., the ultimate strength ϕR_n of a structure can be obtained. The adequacy of the structural system can be directly evaluated by comparing the resulting load-carrying capacity ϕR_n with the applied factored loads $\sum \gamma_i Q_i$. Separate member capacity checks are not necessary because the predicted ultimate load of a structural system will not violate individual member capacity encompassed by the AISC-LRFD equations.

The serviceability limits depend on specifications. The proposed methods do not set any specific serviceability limit, but the usual serviceability requirements may be checked:

1. Floor girder deflection for service live load : $L/360$
2. Roof girder deflection : $L/240$
3. Lateral drift for service wind load : $H/400$
4. Interstory drift for service wind load : $H/300$

At service load levels, no plastic hinges may occur to avoid permanent deformations under service loads.

Adequate inelastic rotation capacity is required for members in order to develop their full plastic moment capacity. The required rotation capacity can be achieved when members are adequately braced and their cross-sections are compact. The limitations of compact sections and lateral unbraced length are given by the LRFD and lead to at least three times an inelastic rotation capacity for non-seismic regions and seven times for seismic regions.

If initial member sizes are inadequate, adjustments should be made. Adjustments can be carried out referring to the sequence of plastic hinge formation shown in the output. For example, if the resulting load-carrying capacity of a structural system is less than the applied factored loads, the member containing the first plastic hinge should be first replaced with a stronger member. On the contrary, if the load-carrying capacity of a structural system exceeds the applied factored load significantly, members without plastic hinges may be replaced with lighter members. If the lateral drift of a structural system does not meet the drift requirement, column sizes should be increased for serviceability, or a braced structural system may be considered instead. In the case of semi-rigid frames, the system behavior is influenced by the combined effects of members and connections. As an illustration, if an excessive lateral drift occurs in a structural system, the drift may be reduced by either increasing member size or by using more rigid connections. If the strength of a beam exceeds the required strength, it may be adjusted by either reducing the beam size or using more flexible connections.

References

[1] AASHTO (1989) *Standard Specifications for Highway Bridges*, 14th ed. Washington, DC, The American Association of State Highway and Transportation Officials.

[2] AISC (1978) *Specification for the Design, Fabrication and Erection of Structural Steel for Buildings*, American Institute of Steel Construction, Chicago, IL.

[3] AISC (1994) *Load and Resistance Factor Design Specification*, American Institute of Steel Construction, 2nd ed., Chicago, IL.

[4] AISC (1986) *Load and Resistance Factor Design Specification for Structural Steel Buildings*, American Institute of Steel Construction, Chicago, IL.

[5] AISC (1989) *Allowable Stress Design Specification for Structural Steel Buildings*, American Institute of Steel Construction, Chicago, IL.

[6] Ang, A. H-S. and Cornell, C.A. (1974) Reliability bases of structural safety and design, *J. Struct. Div.*, ASCE, 100(ST9), 1755-1769.

[7] Ad Hoc Committee on Serviceability (1986) Structural serviceability: a critical appraisal and research needs, *J. Struct. Eng.*, ASCE, 112(12), 2646-2664.

[8] ANSI (1988) *Building Code Requirements for Minimum Design Loads in Buildings and Other Structures*, ANSI A58.1-1982, American National Standards Institute, New York, 94 pp.

[9] ASCE (1993) *Minimum Design Loads for Buildings and Other Structures*, ASCE Standard, Revision of ANSI A58.1-1988, American Society of Civil Engineers, New York, 94 pp.

[10] Attala, M.N., Deierlein, G.G., and McGuire, W. (1994) Spread of plasticity: Quasi-plastic-hinge approach, *J. Struct. Eng.*, 120(8), 2451-2473.

[11] Burgett, Lewis B. (1973) Selection of a "Trial" Column Section. *Eng. J.*, AISC, 10(2), 54-59.

[12] Canadian Standard Association (1989) *Limit States Design of Steel Structures*, CAN/CSA-S16.1-M89.

[13] Chen, W.F. and Lui, E.M. (1986) *Structural Stability - Theory and Implementation*, Elsevier, New York, 490 pp.

[14] Chen, W.F. and Toma, S. (1994) *Advanced Analysis of Steel Frames*, CRC Press, Boca Raton, FL, 383 pp.

[15] Cheong-Siat-Moy, F. (1972) Consideration of secondary effects in frame design, *J. Struct. Div.*, 103(ST10), 2005-2019.

[16] Clarke, M.J., Bridge, R.Q., Hancock, G.J., and Trahair, N.S. (1993) Benchmarking and Verification of Second-Order Elastic and Inelastic Frame Analysis Programs, in SSRC TG 29 *Workshop and Nomograph on Plastic Hinge Based Methods for Advanced Analysis and Design of Steel Frames*, D.W. White and W.F. Chen, Eds., SSRC, Lehigh University, Bethlehem, PA.

[17] Clarke, M. J., Bridge, R. Q., Hancock, G. J., and Trahair, N. S. (1992) Advanced analysis of steel building frames, *J. Constr. Steel Res.*, 23(1-3), 1-29.

[18] CSA (1994) *Limit States Design of Steel Structures*, CAN/CSA-S16.1-M94, Canadian Standards Association.

[19] ECCS (1984) Ultimate Limit State Calculations of Sway Frames with Rigid Joints, Technical Committee 8 - Structural Stability Technical Working Group 8.2 - System, Publication No. 33, 20 pp.

[20] ECCS (1991) *Essentials of Eurocode 3 Design Manual for Steel Structures in Buildings*, ECCS-Advisory Committee 5, No. 65, 60 pp.

[21] Ellingwood (1989) Serviceability Guidelines for Steel Structures, *Eng. J.*, AISC, 26, 1st Quarter, pp. 1-8.

[22] Fisher, J.M. and West, M.A. (1990), Serviceability Design Considerations for Low-Rise Buildings, Steel Design Guide Series 3, American Institute of Steel Construction, 21pp.

[23] Galambos, J. M., and Ketter, R. L. (1959) Columns under combined bending and thrust, *J. Eng. Mech. Div., Proc.*, ASCE, 85(EM2), 1-30.

[24] Kanchanalai, T. (1977) The Design and Behavior of Beam-Columns in Unbraced Steel Frames, AISI Project No. 189, Report No. 2, Civil Engineering/Structures Research Lab., University of Texas at Austin, 300 pp.

[25] Kim, S.E. (1996) Practical Advanced Analysis for Steel Frame Design, Ph.D. dissertation, School of Civil Engineering, Purdue University, West Lafayette, IN, 271 pp.

[26] Kim, S.E. and Chen, W.F. (1996) Practical Advanced Analysis for Steel Frame Design, The ASCE Structural Congress XIV Special Proceedings Volume on Analysis and Computation, Chicago, IL, April, 19-30.

[27] Kim, S.E. and Chen, W.F. (1996) A sensitivity study on number of elements in refined plastic hinge analysis, *Comp. Struct.*, (To appear).

[28] Liew, J.Y.R. (1992) Advanced Analysis for Frame Design, Ph.D. dissertation, School of Civil Engineering, Purdue University, West Lafayette, IN, May, 393 pp.

[29] Maleck, A. E., White, D. W., and Chen, W. F. (1995) "Practical Application of Advanced Analysis in Steel Design," Proceedings of the 4th Pacific Structural Steel Conference, Singapore, Vol. 1, 119-126.

[30] Nethercot, D.A. and Trahair, N.S. (1976) Inelastic lateral buckling of determinate beams, *J. Struct. Div.*, ASCE, 102(ST4), 701-717.

[31] Norris, C. H., Wilbur, J. B., and Utku, S. (1976) *Elementary Structural Analysis*, 3rd ed., New York, McGraw-Hill.

[32] Parikh, B. P. (1966) Elastic-Plastic Analysis and Design of Unbraced Multi-Story Steel Frames, Ph.D. thesis, Lehigh University, Bethlehem, PA.

[33] Pesquera, C. I. (1984) Integrated Analysis and Design of Steel Frames with Interactive Computer Graphics, Ph.D. thesis, Cornell University, Ithaca, NY.

[34] Rosenblueth, E. (1965) Slenderness effects in buildings, *J. Struct. Div.*, ASCE, 91(ST1), 229-252.

[35] Standards Australia (1990) AS4100-1990, *Steel Structures*, Sydney, Australia.

[36] Stevens, L. K. (1967) Elastic stability of practical multistory frames, *Proc. Civil Eng.*, 36, London, England.

[37] Uang, C.M., Wattar, S.W., and Leet, K.M. (1990) Proposed revision of the equivalent axial load method for LRFD steel and composite beam-column design, *Eng. J.*, 1st Qtr., AISC, Chicago, IL.

[38] *Uniform Building Code* (1991) Whittier, CA: International Conference of Building Officials.

[39] White, D. W. (1993) Plastic hinge methods for advanced analysis of steel frames, *J. Constr. Steel Res.*, 24(2), 121-152.

[40] White, D. W. and Hajjar, J. F. (1990) Application of second-order elastic analysis in design-research to practice, AISC, National Steel Construction Conference, Kansas City, MO, March, pp. 11.1-11.22.

[41] Yura, J. A. (1988) Elements for teaching load and resistance factor design. University of Texas at Austin, TX, pp. 55-71.

[42] Ziemian, R. D. and McGuire, W. (1992) A method for incorporating live load reduction provisions in frame analysis, *Eng. J.*, AISC, 29, 1st Quarter, pp. 1-3.

chapter five

Computer Program

5.1 Introduction

This chapter describes the Practical Advanced Analysis Program (PAAP) for two-dimensional steel frame design [5] [7]. The program integrates the methods and techniques developed in Chapters 2 and 3. The names of variables and arrays correspond as closely as possible to those used in theoretical derivations. The main objective of this chapter is to present an "educational version" of software to enable engineers and graduate students to perform planar frame analysis for a more realistic prediction of the strength and behavior of structural systems.

The instructions necessary for user input into PAAP are presented in Section 5.3. Except for the requirement to input geometric imperfections and incremental loads, the input data format of the program is basically the same as that of the usual linear elastic analysis program. The user is advised to read all the instructions, paying particular attention to the appended notes to achieve an overall view of the data required for a specific PAAP analysis. Readers having no previous experience with nonlinear analyses will benefit from a careful study of the example in Section 5.4. It is intended to clarify the data-generation process. The reader should recognize that no system of units is assumed in the program, and take the responsibility to make units consistent. Mistaken unit conversion and input are a common source of erroneous results. The complete source code and the input data of the example in Section 5.4 are provided on the attached diskette. Lastly, this chapter will discuss a possible modification procedure for an in-house program in order to obtain an equivalent program of PAAP.

5.2 Program Overview

5.2.1 Nonlinear Analysis Routines

The nonlinear analysis routines in PAAP are based on the matrix stiffness method and the following incremental force-displacement simultaneous equations are satisfied:

$$[K_t]\{\dot{d}\} = \{\dot{f}\} \tag{5.1}$$

where

$\{\dot{d}\}$ = vector of unknown incremental nodal displacements
$\{\dot{f}\}$ = vector of known equivalent incremental nodal loads
$[K_t]$ = global tangent-stiffness matrix, reflecting the effective stiffness of deformed structure

The matrix $[K_t]$ is the assembly of the transformed element tangent-stiffness matrices discussed in Section 2.4.

5.2.2 Organization of Computer Program

This FORTRAN program is divided into three programs: DATAGEN, INPUT, and PAAP. The first program, DATAGEN, reads an input data file P.DAT and generates a modified data file INFILE. The second program, INPUT, rearranges INFILE into three working data files, DATAO, DATA1, and DATA2. The third program, PAAP, reads the working data files and provides two output files named P.OUT1 and P.OUT2. P.OUT1 contains an echo of the information from the input data file P.DAT. This file may be used to check for numerical and incompatibility errors in input data. P.OUT2 contains the load and displacement information for various joints in the structure as well as the element joint forces for all types of elements at every load step. The load-displacement results are presented at the end of every load increment. The sign conventions for loads and displacements should follow the frame degrees of freedom as shown in Figures 5.1 and 5.2.

The element joint forces are obtained by summing the product of the element incremental displacements at every load step. The element joint forces act in the global coordinate system and must be in equilibrium with applied forces. After the output files are generated, the user can view these files on the screen or print them with the MS-DOS PRINT command. The schematic diagram in Figure 5.3 sets out the operation procedure used by PAAP and its supporting programs [5].

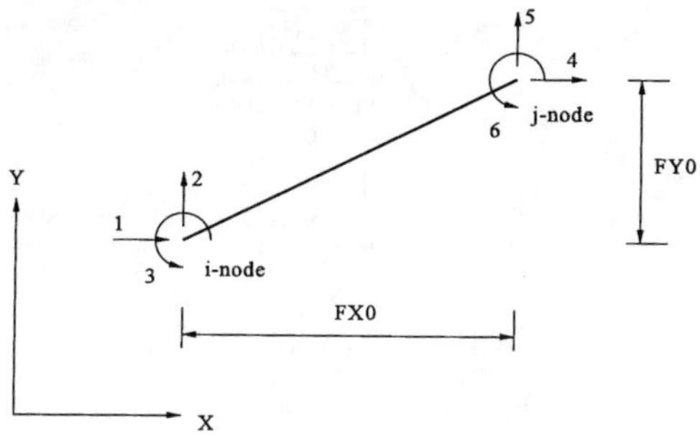

Figure 5.1 Degrees of freedom numbering for the frame element.

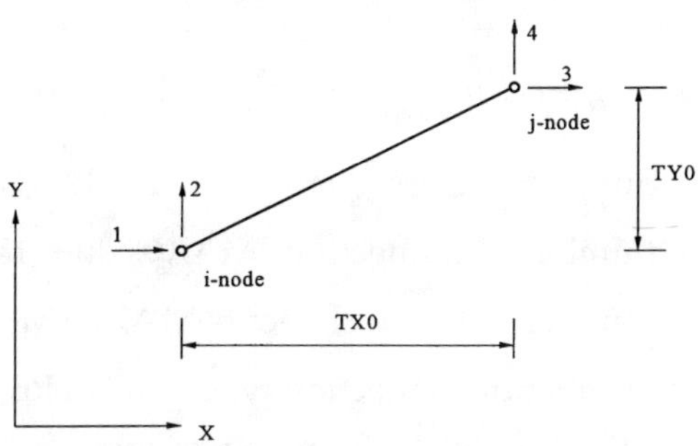

Figure 5.2 Degrees of freedom numbering for the truss element.

5.2.3 Hardware Requirements

This program has been tested in two computer processors. First was on an IBM 486 or equivalent personal computer system using Microsoft's FORTRAN 77 compiler v1.00 and Lahey's FORTRAN 77 compiler v5.01. Second, its performance in the workstation environment was tested on a Sun 5 using a Sun FORTRAN 77 compiler. The program sizes of DATAGEN, INPUT, and PAAP are 8 KB, 9 KB, and 94 KB, respectively. The total size of the three programs is small as 111 KB (= 0.111 MB), and so the 3.5" high density diskette (1.44MB) can accommodate the three programs and several example problems.

The memory required to run the program depends on the size of the problem. A computer with minimum 640K of memory and a 30MB hard disk is generally required. For the PC applications, the array sizes are restricted as follows:

1. Maximum total degrees of freedom MAXDOF = 300
2. Maximum translational degrees of freedom MAXTOF = 300
3. Maximum rotational degrees of freedom MAXROF = 100

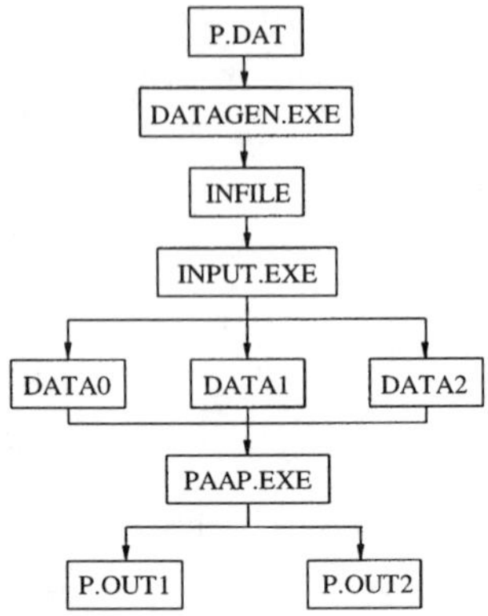

Figure 5.3 Operating procedures of the proposed program.

4. Maximum number of truss elements MAXTRS = 150
5. Maximum number of connections MAXCNT = 150
6. Maximum number of frame elements MAXFRM = 100
7. Maximum number of connection types MCTYPE = 30
8. Maximum number of frame types MFTYPE = 30
9. Maximum number of truss types MTTYPE = 30

It is possible to run bigger jobs by modifying the above values in the PARAMETER statements in the source codes. INPUT (line no. 22 and 23) and PAAP (line no. 96, 1147, 1810, 2232, 2336, and 2497).

5.2.4 Execution of Program

A computer diskette is provided, containing four directories with the following files.

1. Directory PSOURCE
   - DATAGEN.FOR
   - INPUT.FOR
   - PAAP.FOR
2. Directory PTEST
   - DATAGEN.EXE
   - INPUT.EXE
   - PAAP.EXE

- RUN.BAT (batch file)
- P.DAT (input data for a test-run)
- P.OUT1 (output for a test-run)
- P.OUT2 (output for a test-run)

3. Directory PEXAMPLE
   - All input data for the example problems presented in Chapter 6
4. Directory CONNECT
   - 3PARA.FOR (program for semi-rigid connection parameters)
   - 3PARA.EXE
   - CONN.DAT (input data)
   - CONN.OUT (output for three parameters)

For execution of the programs, first they must be copied onto the hard disk, i.e., copy DATAGEN.EXE, INPUT.EXE, PAAP.EXE, RUN.BAT, and P.DAT from the directory PTEST on the diskette to the hard disk. Before launching into the program, the user should test the system by running the sample example provided in the directory. The input data P.DAT contains the information presented in the example in Section 5.4. The programs are executed by issuing the command "RUN". The batch file, RUN.BAT executes DATAGEN, INPUT, and PAAP in sequence. The output files produced are P.OUT1 and P.OUT2. These files should be compared with those provided in Section 5.4 and in the directory PTEST of the diskette. The programs are compiled by Microsoft's FORTRAN 77 compiler v1.00. The user may recompile the source programs in the directory PSOURCE by using a different compiler.

The input data for all the problems in Chapter 6 are provided in the directory PEXAMPLE. The user may use the input data for his or her reference and confirmation of the results presented in Chapter 6. It should be noted that RUN is a batch command to facilitate the execution of PAAP. Entering the command RUN will write the new files including DATA0, DATA1, DATA2, P.OUT1, and P.OUT2 over the old ones. Therefore, output files should be renamed before running a new problem.

The program can generate the output files P.OUT1 and P.OUT2 in a reasonable time period described in the following. The run time on an IBM 486 PC with memory of 640K to get the output files for the eight-story frame shown in Figure 6.38 is taken as 4 min 10 s and 2 min 30 s in real time rather than CPU time by using Microsoft FORTRAN and Lahey

FORTRAN, respectively. In the Sun 5, the run time varies approximately 2 to 3 min depending on the degree of occupancy by users.

The directory CONNECT contains the program that computes the three parameters needed for semi-rigid connections. The operation procedure of the program, the input data format, and two examples have been presented in Section 2.5.5.

5.3 User's Manual

5.3.1 General Rules

Analysis Options

PAAP was developed on the basis of the theory presented in Chapter 2. While the purpose of the program is basically for advanced analysis using second-order inelastic concept, the program can be used also for first- and second-order elastic analysis. For a first-order elastic analysis, the total factored load should be applied in one load increments to suppress numerical iteration in the nonlinear analysis algorithm. For a second-order elastic analysis, a yield strength of an arbitrarily large value should be assumed for all members to prevent yielding.

Coordinate System

A two-dimensional (x, y) global coordinate system is used for the generation of all the input and output data associated with the joints. The following input and output data are prepared with respect to the global coordinate system.

1. Input data
   - joint coordinates
   - point restraints
   - joint load
2. Output data
   - joint displacement
   - member forces

Type of Elements

The analysis library consists of three elements: (1) a plane frame, (2) a plane truss, and (3) a connection element. The connection is represented by a zero-length rotational spring element with a user-specified nonlinear moment-rotation curve. Loading is allowed only at nodal points. Geometric and material nonlinearities can be accounted for by using an iterative load-increment scheme. Zero-length plastic hinges are lumped at the element ends.

Locations of Nodal Points

The geometric dimensions of the structures are established by placing joints (or nodal points) on the structures. Each joint is given an identification number and is located in a plane associated with a global two-dimensional coordinate system. The structural geometry is completed by connecting the predefined joints with structural elements that may be a frame, a truss, or a connection. Each element also has an identification number.

The following are some of the factors that need to be considered in placing joints in a structure:

1. The number of joints should be sufficient to describe the initial geometry and the response behavior of the structures.
2. Joints need to be located at points and lines of discontinuity, e.g., at changes in material properties or section properties.
3. Joints should be located at points on the structure where forces and displacements need to be evaluated.
4. Joints should be located at points where concentrated loads will be applied. The applied loads should be concentrated and act on the joints.
5. Joints should be located at all support points. Support conditions are represented in the structural model by restricting the movement of the specific joints in specific directions.
6. Second-order inelastic behavior can be captured by the use of one or two elements per member corresponding to the following guidelines:
   - Beam member subjected to uniform loads: 2 elements
   - Column member of braced frame: 2 elements
   - Column member of unbraced frame: 1 element

Degrees of Freedom

A two-joint frame element has six displacement components as shown in Figure 5.1. Each joint can translate in the global x- and y-directions, and rotate about the global z-axis. The directions associated with these displacement components are known as degrees of freedom of the joint. A two-joint truss element has four degrees of freedom as shown in Figure 5.2. Each joint has two translational degrees of freedom and no rotational component.

If the displacement of a joint corresponding to any one of its degrees of freedom is known to be zero (such as at a support), then it is labeled an inactive degree of freedom. Degrees of freedom where the displacements are not known are termed active degrees of freedom. In general, the displacement of an inactive degree of freedom is usually known, and the purpose of the analysis is to find the reaction in that direction. For an active degree of freedom, the applied load is known (it could be zero), and the purpose of the analysis is to find the corresponding displacement.

Units

There are no "built-in" units in PAAP. The user must prepare the input in a consistent set of units. The output produced by the program will conform to the same set of units. Therefore, if the user chooses to use kips and inches as the input units, all the dimensions of the structure must be entered in inches and all the loads in kips. The material properties should also conform to these units. The output units will then be in kips and inches, so that the frame member axial force will be in kips, bending moments will be in kip-inches, and displacements will be in inches. Joint rotations, however, are in radians, irrespective of units.

5.3.2 Input Instructions

Described here is the input sequence and data structure used to create an input file called P.DAT. The analysis program, PAAP, can analyze any structures with up to 165 degrees of freedom, but it is possible to recompile the source code to accommodate more degrees of freedom by changing the size of the arrays in the PARAMETER and COMMON statements. The limitation of degree of freedom can be solved by using dynamic storage allocation. This procedure is common in finite element programs [3] [4], and it will be used in the next release of the program.

The input data file is prepared in a specific format. The input data consists of 13 data sets including 5 control data, 3 section property data, 3 element data, 1 boundary condition, and 1 load data set as follows.

1. Title
2. Analysis and design control
3. Job control
4. Total number of element types
5. Total number of elements
6. Connection properties
7. Frame element properties
8. Truss element properties

9. Connection element data
10. Frame element data
11. Truss element data
12. Boundary conditions
13. Incremental loads

Input of all data sets are mandatory, but some of the data associates with elements (data set No. 6 through 11) may be skipped depending on the use of the element. The order of data lines in the input file must be strictly maintained. Instructions for inputting data are described in what follows and summarized in Table 5.1 to provide a quick reference [5]. The mark "*" in the columns in what follows indicates that the real value ("F" or "E" format) should be entered, otherwise input the integer value ("I" format).

Data Set 1: Title

Note	Columns	Variable	Description
1	1-70	TITLE	A comment card

Note:

1. This title line provides a description of the input data and identifies different job names.

Data Set 2: Analysis and Design Control

Note	Column	Variable	Description
1	1-5	IGEOIM	Geometric imperfection method 0: No geometric imperfection (default) 1: Explicit imperfection modeling 2: Equivalent notional load 3: Further reduced tangent modulus
2	6-10	ILRFD	Strength reduction factor $\phi_c = 0.85$, $\phi_b = 0.9$ 0: No reduction factors considered (default) 1: Reduction factors considered

Note:

1. If IGEOIM = 0 is specified, geometric imperfections are not considered. If IGEOIM = 1, explicit imperfections must be input in the element data set No. 10 and 11. If IGEOIM = 2, notional loads

must be input in the incremental load data set No. 13. If IGEOIM = 3, further reduced tangent modulus is automatically considered by specifying column members in the element property data set No. 7 and 8.

2. For analysis, no input is required. For design, strength reduction factors $\phi_c = 0.85$ and $\phi_b = 0.9$ are automatically considered by the program if ILRFD = 1.

Data Set 3: Job Control

Note	Columns	Variable	Description
1	1-5	NNODE	Total number of nodal points of the structure
2	6-10	NBOUND	Total number of supports
3	11-15	NINCRE	Allowable number of load increments (default = 100), at least two or three times larger than the scaling number

Note:

1. The total number of nodal points of the structure is required. NNODE should not be larger than 100 (= Max. D.O.F. 300/3) in the frame element and 150 (= 300/2) in the truss element. If this value is exceeded, the user needs to recompile the source code by setting a larger value for "MAXDOF" in the PARAMETER statements.
2. The total number of supports is required. For example, a simply supported one-span beam has NBOUND = 2.
3. The total load incremental steps are required. The analysis is terminated when the number of the total load incremental steps is reached. NINCRE must be two or three times larger than the scaling number. For first-order analysis, the total factored load should be applied in one load increment to suppress numerical iteration in the nonlinear analysis algorithm, i.e., let NINCRE = 1.

Data Set 4: Total Number of Element Type

Note	Columns	Variable	Description
1	1-5	NCTYPE	Number of connection types (≤ 30)
1	6-10	NFTYPE	Number of frame element types (≤ 30)
1	11-15	NTTYPE	Number of truss element types (≤ 30)

Note:

1. If the limits specified in the above are exceeded, the user needs to recompile the program by specifying suitable values for the variables in the PARAMETER statements of the source code.

Data Set 5: Total Number of Element

Note	Columns	Variable	Description
1	1-5	NUMCNT	Number of connection elements (≤ 150)
1	6-10	NUMFRM	Number of frame elements (≤ 100)
1	11-15	NUMTRS	Number of truss elements (≤ 150)

Note:

1. If the limits specified in the above are exceeded, the user needs to recompile the program by specifying suitable values for the variables in the PARAMETER statements of the source code.

Data Set 6: Connection Property

If NCTYPE $\neq 0$, this line must be supplied NCTYPE times. Otherwise, this line should be omitted.

Note	Columns	Variable	Description
1	1-5	ICTYPE	Connection type identifier (an integer)
1	6-15*	MU	Connection ultimate moment capacity
1	16-25*	RKI	Connection initial stiffness
1	26-35*	N	Connection shape parameter

Note:

1. Connection property sets must be specified in the order as shown. Users should refer to Section 2.5.2 for the definition of the connection parameters, M_u, R_{ki}, and n, and Section 2.5.5 for the estimation of the three parameters. The connection property set is stored in an array CTYPE (ICTYPE, J $= 1, 3$).

Data Set 7: Frame Element Property

If NFTYPE $\neq 0$, this line must be supplied NFTYPE times. Otherwise, this line should be omitted.

Note	Columns	Variable	Description
	1-5	IFTYPE	Frame type identifier (an integer)
1	6-15*	A	Cross-sectional area
1	16-25*	I	Moment of inertia
1	26-35*	Z	Plastic section modulus
1	36-45*	E	Modulus of elasticity
1, 2	46-55*	FY	Material yield strength
3	56-60	IFCOL	Identification of column element, IFCOL=1 for column when IGEOIM = 3 (default = 0)

Notes:

1. Material property sets for the frame elements must be specified in the order shown. The material set is stored in an array FTYPE (IFTYPE, J = 1, 5).
2. For elastic analysis, a material yield strength of an arbitrarily large value (ex. FY = 1E+10) should be assumed for all elements to prevent the formation of plastic hinges.
3. If further reduced tangent modulus is used for the geometric imperfection (IGEOIM = 3), IFCOL must be equal to 1 for column members subjected to dominant axial forces. Leave it blank for beam members and for the cases of IGEOIM = 1 or 2.

Data Set 8: Truss Element Property

If NFTYPE $\neq 0$, this line must be supplied NTTYPE times. Otherwise, this line should be omitted.

Note	Columns	Variable	Description
	1-5	ITTYPE	Truss type identifier (an integer)
1	6-10*	A	Cross-sectional area
1	11-25*	I	Moment of inertia
1	26-35*	E	Modulus of elasticity
1, 2	36-45*	FY	Yield stress
3	46-50	ITCOL	Identification of column member, ITCOL = 1 for column when IGEOIM = 3 (default = 0)

Note:

1. Material property sets for the truss elements must be specified in the order shown. For truss elements in axial compression, the

analysis is terminated once the axial force in the truss element exceeds the column strength given by the LRFD column strength equations. Therefore, the moment of inertia about the weak-axis should be specified for a pinned-ended truss element. For an element subject to a tensile force, the cross-sectional effective area (with allowance for bolt holes) may be specified. The material set is stored in an array TTYPE (ITTYPE, J = 1, 4).

2. For elastic analysis, a material yield strength of an arbitrarily large value (ex. FY = 1E+10) should be assumed for all elements to prevent the formation of plastic hinges.
3. If the further reduced tangent modulus is used for geometric imperfections (IGEOIM = 3), ITCOL must be equal to 1 for column members subjected to dominant axial forces. Leave it blank for beam members and for the cases of IGEOIM = 1 or 2.

Data Set 9: Connection Element Data

This line should be omitted if NUMCNT = 0. Otherwise, this line must be supplied NUMCNT times.

Note	Columns	Variable	Description
	1-5	LCNT	Connection element number
	6-10	IFMCNT	Frame element number containing the connection
1	11-15	IEND	Identification of element ends containing the connection 1: Connection attached at element end *i* 2: Connection attached at element end *j*
	16-20	JDCNT	Connection type number
2	21-25	NOSMCN	Number of same elements for automatic generation (default = 1)
3	26-30	NELINC	Element number (IFMCNT) increment of automatically generated elements (default = 1)

Note:

1. See Figure 5.1 for the identification of element ends.
2. The same connections may be automatically generated (leave blank if the connection is not automatically generated).
3. If NOSMCN $\neq$ 1, NELINC should be input except NELINC = 1.

Data Set 10: Frame Element Data

This line should be omitted if NUMFRM = 0. Otherwise, this line must be entered NUMFRM times.

Note	Column	Variable	Description
	1-5	LFRM	Frame element number
1	6-15*	FXO	Horizontal projected length; positive for i-j direction in global x-direction
1	16-25*	FYO	Vertical projected length; positive for i-j direction in global y-direction
	25-30	JDFRM	Frame type number
	31-35	IFNODE	Number of node i
	36-40	JFNODE	Number of node j
2	41-45	NOSMFE	Number of same elements for automatic generation (default = 1)
3	46-50	NODINC	Node number increment of automatically generated elements (default = 1)

Note:

1. The projected lengths for frame elements are shown in Figure 5.1.
2. The same elements may be automatically generated. Leave this blank if the element is not automatically generated.
3. If NOSMFE $\neq$ 1, NODINC should be input except NODINC = 1.

Data Set 11: Truss Element Data

Note	Column	Variable	Description
	1-5	LTRS	Truss element number
1	6-15*	TXO	Horizontal projected length; positive for i-j direction in global x-direction
1	16-25*	TYO	Vertical projected length; positive for i-j direction in global y-direction
	26-30	JDTRS	Truss type number
	31-35	ITNODE	Number of node i
	36-40	JTNODE	Number of node j
2	51-55	NOSMTE	Number of same elements for automatic generation (default = 1)
3	56-60	NODINC	Node number increment of automatically generated elements (default = 1)

Note:

1. The projected lengths for truss elements are shown in Figure 5.2.

2. The same elements may be automatically generated. Leave this blank if the element is not automatically generated.
3. If NOSMTE $\neq 1$, NODINC should be input except NODINC = 1.

Data Set 12: Boundary Condition

Note	Column	Variable	Description
1	1-5	NODE	Node number of support
	6-10	XFIX	XFIX = 1 for restrained in global x-direction
	11-15	YFIX	YFIX = 1 for restrained in global y-direction
	16-20	RFIX	RFIX = 1 for restrained in rotation
2	21-25	NOSMBD	Number of same boundary conditions for automatic generation (default = 1)
3	26-30	NODINC	Node number increment of automatically generated supports (default = 1)

Note:

1. Total number of boundary data must be equal to NBOUND of the data set 3.
2. The same supports may be automatically generated. Leave this blank if the support is not automatically generated.
3. If NOSMBD $\neq 1$, NODINC should be input except NODINC = 1.

Data Set 13: Incremental Loads

Note	Column	Variable	Description
1	1-5	NODE	Node number where a load is applied
2	6-15*	XLOAD	Incremental load in the global x-direction
3	16-25*	YLOAD	Incremental load in the global y-direction
4	26-35*	RLOAD	Incremental moment in counter-clockwise
5	36-40	NOSMLD	Number of same loads for automatic generation (default = 1)
6	41-45	NODINC	Node number increment of automatically generated loads (default = 1)

Note:

1. All concentrated forces must be applied at nodal points in the global coordinates system.
2. Right direction is positive.

3. Upward direction is positive.
4. Counter-clockwise is positive.
5. The same loads may be automatically generated. Leave this blank if the loads are not automatically generated. If NOSMLD $\neq 1$, NODINC should be input except NODINC $= 1$.

5.4 Example

An example is provided in this section for two reasons: one, to test the computer system and the program installation, and two, to gain an overall view of data preparation, execution of the program, and output interpretation. Complete input and output data are provided in this section and stored in the directory PTEST on the attached diskette.

5.4.1 Frame Configuration and Load Condition

Figure 5.4 shows a rectangular portal frame subjected to concentrated gravity loads of 80λ and 60λ kips acting on one-third points of the beam. The question here is what the collapse load factor λ should be. The beam and column members are W21x83 and W18x50, respectively. A36 steel is used. The frame has a fixed support on the left and a hinged support on the right. The beam and the columns are laterally braced.

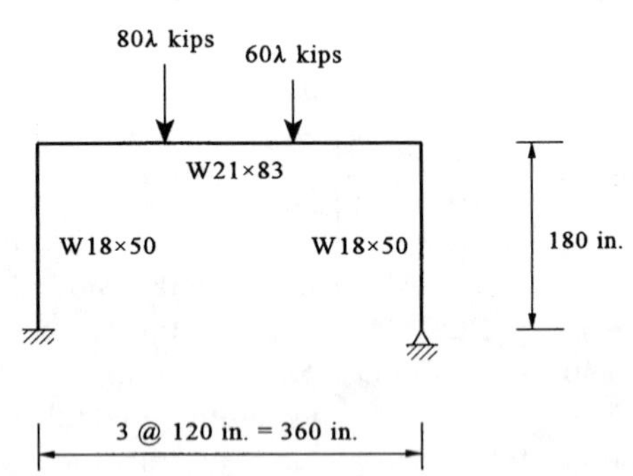

Figure 5.4 Portal frame subjected to two concentrated loads.

5.4.2 Input Data Preparation

The beam is discretized into three elements because two nodal points are needed for the loads shown in Figure 5.5. The columns are repre-

Table 5.1 Input Data Format for the Program PAAP

Data set	Column	Variable	Description
1. Title	A70	—	Job title and general comments
2. Analysis and design control	1-5	IGEOIM	Geometric imperfection method 0: No geometric imperfection (default) 1: Explicit imperfection modeling 2: Equivalent notional load 3: Further reduced tangent modulus
	6-10	ILRFD	Strength reduction factor $\phi_c = 0.85$, $\phi_b = 0.9$ 0: No reduction factors considered (default) 1: Reduction factors considered
3. Job control	1-5	NNODE	Total number of nodal points of the structure
	6-10	NBOUND	Total number of supports
	11-15	NINCRE	Allowable number of load increments (default = 100); at least two or three times larger than the scaling number
4. Total number of element type	1-5	NCTYPE	Number of connection types (1-30)
	6-10	NFTYPE	Number of frame types (1-30)
	11-15	NTTYPE	Number of truss types (1-30)
5. Total number of element	1-5	NUMCNT	Number of connection elements (1-150)
	6-10	NUMFRM	Number of frame elements (1-100)
	11-15	NUMTRS	Number of truss elements (1-150)
6. Connection property	1-5	ICTYPE	Connection type number
	6 – 15*	M_u	Ultimate moment capacity of connection
	16 – 25*	R_{ki}	Initial stiffness of connection
	26 – 35*	N	Shape parameter of connection
7. Frame element property	1-5	IFTYPE	Frame type number
	6 – 15*	A	Cross-section area
	16 – 25*	I	Moment of inertia
	26 – 35*	Z	Plastic section modulus
	36 – 45*	E	Modulus of elasticity
	46 – 55*	FY	Yield stress
	55-60	IFCOL	Identification of column member, IFCOL = 1 for column (default = 0)
8. Truss element property	1-5	ITYPE	Truss type number
	6 – 15*	A	Cross-section area
	16 – 25*	I	Moment of inertia
	26 – 35*	E	Modulus of elasticity
	36 – 45*	FY	Yield stress
	46-50	ITCOL	Identification of column member, ITCOL = 1 for column (default = 0)

Table 5.1 Input Data Format for the Program PAAP (cont.)

Data set	Column	Variable	Description
9. Connection element data	1-5	LCNT	Connection element number
	6-10	IFMCNT	Frame element number containing the connection
	11-15	IEND	Identification of element ends containing the connection 1: Connection attached at element end i 2: Connection attached at element end j
	16-20	JDCNT	Connection type number
	21-25	NOSMCN	Number of same elements for automatic generation (default = 1)
	26-30	NELINC	Element number (IFMCNT) increment of automatically generated elements (default = 1)
10. Frame element data	1-5	LFRM	Frame element number
	6 – 15*	FXO	Horizontal projected length; positive for i-j direction in global x-direction
	16 – 25*	FYO	Vertical projected length; positive for i-j direction in global y-direction
	26-30	JDFRM	Frame type number
	31-35	IFNODE	Number of node i
	36-40	JFNODE	Number of node j
	41-45	NOSMFE	Number of same elements for automatic generation (default = 1)
	46-50	NODINC	Node number increment of automatically generated elements (default = 1)
11. Truss element data	1-5	LTRS	Truss element number
	6 – 15*	TXO	Horizontal projected length; positive for i-j direction in global x-direction
	16 – 25*	TYO	Vertical projected length; positive for i-j direction in global y-direction
	26-30	JDTRS	Truss type number
	31-35	ITNODE	Number of node i
	36-40	JTNODE	Number of node j
	51-55	NOSMTE	Number of same elements for automatic generation (default = 1)
	56-60	NODINC	Node number increment of automatically generated elements (default = 1)

Table 5.1 Input Data Format for the Program PAAP (cont.)

Data set	Column	Variable	Description
12. Boundary condition	1-5	NODE	Node number of support
	6-10	XFIX	XFIX = 1 for restrained in global x-direction (default = 0)
	11-15	YFIX	YFIX = 1 for restrained in global y-direction (default = 0)
	16-20	RFIX	RFIX = 1 for restrained in rotation (default = 0)
	21-25	NOSMBD	Number of same boundary conditions for automatic generation (default = 1)
	26-30	NODINC	Node number increment of automatically generated supports (default = 1)
13. Incremental loads	1-5	NODE	Node number where a load applied
	6 – 15*	XLOAD	Incremental load in global x-direction (default = 0)
	16 – 25*	YLOAD	Incremental load in global y-direction (default = 0)
	26 – 35*	RLOAD	Incremental moment in global θ-direction (default = 0)
	36-40	NOSMLD	Number of same loads for automatic generation (default = 1)
	41-45	NODINC	Node number increment of automatically generated loads (default = 1)

* indicates the real value ("F" or "E" format) to be input, otherwise the integer value ("I" format).

sented by one element per member as discussed in Section 5.3.1. All members are modeled with the frame elements. The scaling number to determine the incremental load is selected as 20 so that the incremental loads are 4 kips (= 80/20) and 3 kips (=60/20). Kip-inch units are used. Since this example is an analysis rather than a design, geometric imperfections are not considered (IGEOIM=0), and no strength reduction factors are accounted for (ILRFD=0). The input data P.DAT may be prepared in Table 5.2 by simply following the input data format instructed in Section 5.3.2.

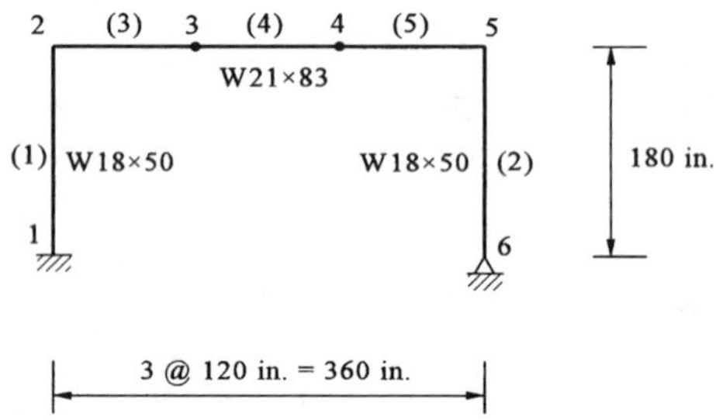

Figure 5.5 Structural modeling of the portal frame.

Table 5.2 Input Data P.DAT for the Portal Frame

```
Portal frame subjected to two concentrated loads
   0    0
   6    2   100
   0    2     0
   0    5     0
   1       14.7      800.0       101.0   29000.0       36.0
   2       24.3     1830.0       196.0   29000.0       36.0
   1        0.0      180.0    1     1     2
   2        0.0      180.0    1     6     5
   3      120.0        0.0    2     2     3     3
   1    1     1    1
   6    1     1
   3                  -4.0
   4                  -3.0
```

5.4.3 *Program Execution*

The program is executed by simply entering the batch file command "RUN" on the screen, or issuing DATAGEN, INPUT, and PAAP in sequence. The screen shows the incremental loading steps as listed (also in Table 5.3) which correspond to the progress of analysis. After execution, several files including INFILE, DATA0, DATA1, DATA2, P.OUT1, and P.OUT2 are generated. Among these, P.OUT1, and P.OUT2 are of interest. P.OUT1, listed in Table 5.4, contains an echo of the information of the input data P.DAT. P.OUT2, in Table 5.5, contains the incremental displacement and member forces.

5.4.4 Output Interpretation

The collapse load results in 93.24 kips ($\lambda = 1.166$) on Node 3 shown in Table 5.5. At this load, three plastic hinges have formed at Nodes 5 and 3. Figures 5.6 and 5.7 show the load-displacement curve for Node 3 and the force distribution at the collapse state.

The load applied at Node 3 corresponding to the formation of the first plastic hinge is 87.47 kips ($\lambda = 1.093$) at Load Step 24. The inelastic moment redistribution is not yet significant up to Load Step 24. If one selects this load rather than the collapse load as a nominal strength of the frame, the same member sizes will be provided as those by conventional LRFD methods because it does not consider moment redistribution either.

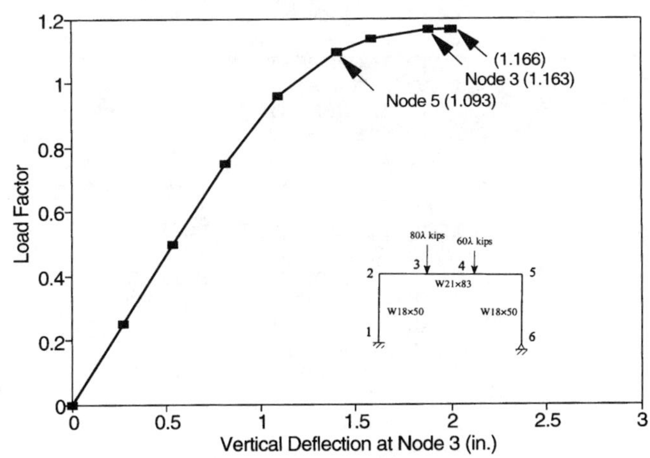

Figure 5.6 Load-displacement curves for the portal frame.

To take advantage of inelastic moment redistribution, the collapse load may be chosen to define the system strength, and smaller member sizes are obtained by advanced analysis compared to the LRFD. The load-carrying capacity of the structural system may be taken either at the formation of the first plastic hinge or at the ultimate state.

Table 5.3 Incremental Loading Steps Shown on the Screen Corresponding to the Progress of Analysis

```
LOAD STEP =  1
LOAD STEP =  1
LOAD STEP =  1
LOAD STEP =  1
LOAD STEP =  1
LOAD STEP =  1
LOAD STEP =  1
LOAD STEP =  2
LOAD STEP =  3
LOAD STEP =  4
LOAD STEP =  5
LOAD STEP =  6
LOAD STEP =  7
LOAD STEP =  8
LOAD STEP =  9
LOAD STEP =  10
LOAD STEP =  11
LOAD STEP =  12
LOAD STEP =  13
LOAD STEP =  14
LOAD INCREMENT IS SCALED
LOAD STEP =  15
LOAD INCREMENT IS SCALED
LOAD STEP =  16
LOAD INCREMENT IS SCALED
LOAD STEP =  17
LOAD INCREMENT IS SCALED
LOAD STEP =  18
LOAD INCREMENT IS SCALED
LOAD STEP =  19
LOAD INCREMENT IS SCALED
LOAD STEP =  20
LOAD INCREMENT IS SCALED
LOAD INCREMENT IS SCALED
LOAD STEP =  21
LOAD INCREMENT IS SCALED
LOAD STEP =  22
LOAD INCREMENT IS SCALED
LOAD INCREMENT IS SCALED
LOAD STEP =  23
LOAD INCREMENT IS SCALED
LOAD INCREMENT IS SCALED
LOAD STEP =  24
LOAD STEP =  25
LOAD STEP =  26
LOAD STEP =  27
LOAD STEP =  28
LOAD STEP =  29
LOAD STEP =  30
LOAD STEP =  31
LOAD INCREMENT IS SCALED
LOAD STEP =  32
LOAD STEP =  33
LOAD STEP =  34
```

Table 5.4 Output P.OUT1 for Portal Frame

```
**********************************************
*         NO GEOMETRIC IMPERFECTION          *
**********************************************

**********************************************
*      ANALYSIS (NO STRENGTH REDUCED)        *
**********************************************

**********************************************
* SECOND-ORDER REFINED PLASTIC HINGE ANALYSIS *
**********************************************

=================================
= CRC Tangent Modulus model     =
= Parabolic stiffness reduction =
=================================

==================
= STRUCTURE DATA =
==================

    NUMBER OF DEGREES OF FREEDOM OF THE STRUCTURE =  13
    NUMBER OF LOAD INCREMENTS                     = 100
    NUMBER OF LOAD SEQUENCES                      =   1

    NUMBER OF CONNECTION TYPES                    =   0
    NUMBER OF FRAME TYPES                         =   2
    NUMBER OF TRUSS TYPES                         =   0

    NUMBER OF CONNECTION ELEMENTS                 =   0
    NUMBER OF FRAME ELEMENTS                      =   5
    NUMBER OF TRUSS ELEMENTS                      =   0



======================
= FRAME ELEMENT DATA =
======================


  --------------------
  - ELEMENT NUMBER  1 -
  --------------------
      XO    = 0.00000D+00    YO   = 0.18000D+03    LO   = 0.18000D+03
      A     = 0.14700D+02    I    = 0.80000D+03    ZP   = 0.10100D+03
      E     = 0.29000D+05    FY   = 0.36000D+02
      ICNT  =       0        IEND =       0        IDCNT=       0

      ELEMENT    DOF :   1    2    3    4    5    6
      ----------------------------------------------
      STRUCTURE DOF :    0    0    0    1    2    3
```

Table 5.4 Output P.OUT1 for Portal Frame (cont.)

```
- ELEMENT NUMBER  2 -
---------------------
    XO    = 0.00000D+00     YO   = 0.18000D+03     LO   = 0.18000D+03
    A     = 0.14700D+02     I    = 0.80000D+03     ZP   = 0.10100D+03
    E     = 0.29000D+05     FY   = 0.36000D+02
    ICNT  =         0       IEND =         0       IDCNT=         0

    ELEMENT   DOF :   1    2    3    4    5    6
    --------------------------------------------
    STRUCTURE DOF :   0    0   13   10   11   12


---------------------
- ELEMENT NUMBER  3 -
---------------------
    XO    = 0.12000D+03     YO   = 0.00000D+00     LO   = 0.12000D+03
    A     = 0.24300D+02     I    = 0.18300D+04     ZP   = 0.19600D+03
    E     = 0.29000D+05     FY   = 0.36000D+02
    ICNT  =         0       IEND =         0       IDCNT=         0

    ELEMENT   DOF :   1    2    3    4    5    6
    --------------------------------------------
    STRUCTURE DOF :   1    2    3    4    5    6


---------------------
- ELEMENT NUMBER  4 -
---------------------
    XO    = 0.12000D+03     YO   = 0.00000D+00     LO   = 0.12000D+03
    A     = 0.24300D+02     I    = 0.18300D+04     ZP   = 0.19600D+03
    E     = 0.29000D+05     FY   = 0.36000D+02
    ICNT  =         0       IEND =         0       IDCNT=         0

    ELEMENT   DOF :   1    2    3    4    5    6
    --------------------------------------------
    STRUCTURE DOF :   4    5    6    7    8    9


---------------------
- ELEMENT NUMBER  5 -
---------------------
    XO    = 0.12000D+03     YO   = 0.00000D+00     LO   = 0.12000D+03
    A     = 0.24300D+02     I    = 0.18300D+04     ZP   = 0.19600D+03
    E     = 0.29000D+05     FY   = 0.36000D+02
    ICNT  =         0       IEND =         0       IDCNT=         0

    ELEMENT   DOF :   1    2    3    4    5    6
    --------------------------------------------
    STRUCTURE DOF :   7    8    9   10   11   12


================================
= STRUCTURE DEGREES OF FREEDOM =
================================

  NUMBER OF TRANSLATIONAL DOF =    8
```

Table 5.4 Output P.OUT1 for Portal Frame (cont.)

```
DOF NUMBERING :   1  2 10 11  4  5  7  8
NUMBER OF ROTATIONAL DOF     =   5
DOF NUMBERING :   3 13 12  6  9
```

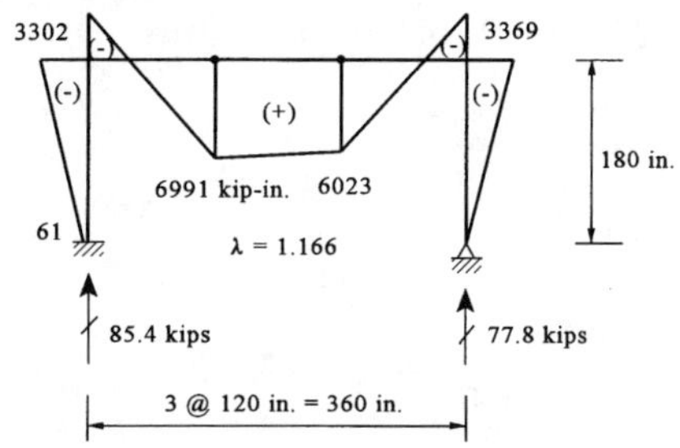

Figure 5.7 Force distribution at the collapse.

The member forces must not exceed the plastic strength of the cross-section given by Equation (2.38) where the plastic hinge forms. To prove this, the member forces at Node 5 (where the first plastic hinge forms) are checked in what follows. Note that the plastic hinge is formed at the end of Member 2 rather than Member 5 since the plastic strength of Member 2 is smaller than that of Member 5. The squash load and plastic moment capacity of the section W18x50 are

$$P_y = (A_g)(F_y) = (14.7)(36) = 529.2 \text{ kips}$$

$$M_p = (Z)(F_y) = (101)(36) = 3,636 \text{ kip-in.}$$

The member forces on the j-end of Element 2 at Load Step 34 from P.OUT2 are:

$$P = 77.8 \text{ kips}$$

$$M = 3,369 \text{ kip-in.}$$

and so

$$P/P_y = 77.8/529.2 = 0.147 < 0.2$$

The value of the interaction equations may be calculated as [Equation (2.38b)]:

$$P/2P_y + M/M_p = 0.147/2 + 3,369/3,636 = 1.00$$

As a result, the internal forces computed by PAAP are confirmed to satisfy the AISC-LRFD cross-section plastic strength at the location of the plastic hinge.

Table 5.5 Output P.OUT2 for Portal Frame

```
============================
= LOAD-DEFLECTION BEHAVIOR =
============================


  -----------------
  - LOAD STEP   1 -
  -----------------
      LOAD SEQUENCE  1
      DETERMINANT OF STIFFNESS MATRIX = 0.524 TIMES 10 TO THE POWER  56
      TRACE OF STIFFNESS MATRIX       = 0.122 TIMES 10 TO THE POWER   8

           NODE      ITEM              X             Y             R

             1    NODAL FORCE     0.0000D+00    0.0000D+00    0.0000D+00
                  DISPLACEMENT    0.0000D+00    0.0000D+00    0.0000D+00

             2    NODAL FORCE     0.0000D+00    0.0000D+00    0.0000D+00
                  DISPLACEMENT    0.1814D-01   -0.1509D-02   -0.4333D-03

             3    NODAL FORCE     0.0000D+00   -0.4000D+01    0.0000D+00
                  DISPLACEMENT    0.1798D-01   -0.5384D-01   -0.2773D-03

             4    NODAL FORCE     0.0000D+00   -0.3000D+01    0.0000D+00
                  DISPLACEMENT    0.1781D-01   -0.5099D-01    0.3054D-03

             5    NODAL FORCE     0.0000D+00    0.0000D+00    0.0000D+00
                  DISPLACEMENT    0.1764D-01   -0.1447D-02    0.3653D-03

             6    NODAL FORCE     0.0000D+00    0.0000D+00    0.0000D+00
                  DISPLACEMENT    0.0000D+00    0.0000D+00   -0.3296D-03

      ELEMENT FORCES IN GLOBAL COORDINATE
      -----------------------------------

        FRAME ELEMENT :                    ELEMENT FORCES
        ELEMENT NUMBER        Fx                Fy                Mz

               1   i     0.9955897D+00    0.3572976D+01   -0.3372871D+02
                   j    -0.9955897D+00   -0.3572976D+01   -0.1454126D+03
               2   i    -0.9948938D+00    0.3427024D+01    0.0000000D+00
                   j     0.9948938D+00   -0.3427024D+01    0.1791413D+03
               3   i     0.9952295D+00    0.3572542D+01    0.1454126D+03
                   j    -0.9952295D+00   -0.3572542D+01    0.2833445D+03
               4   i     0.9952296D+00   -0.4270006D+00   -0.2833445D+03
                   j    -0.9952296D+00    0.4270006D+00    0.2321016D+03
               5   i     0.9952295D+00   -0.3426614D+01   -0.2321016D+03
                   j    -0.9952295D+00    0.3426614D+01   -0.1791413D+03


  -----------------
  - LOAD STEP   2 -
  -----------------
      LOAD SEQUENCE  1
      DETERMINANT OF STIFFNESS MATRIX = 0.523 TIMES 10 TO THE POWER  56
      TRACE OF STIFFNESS MATRIX       = 0.122 TIMES 10 TO THE POWER   8
```

Table 5.5 Output P.OUT2 for Portal Frame (cont.)

```
   NODE       ITEM                  X                  Y                  R

     1   NODAL FORCE       0.0000D+00        0.0000D+00        0.0000D+00
         DISPLACEMENT      0.0000D+00        0.0000D+00        0.0000D+00

     2   NODAL FORCE       0.0000D+00        0.0000D+00        0.0000D+00
         DISPLACEMENT      0.3634D-01       -0.3019D-02       -0.8669D-03

     3   NODAL FORCE       0.0000D+00       -0.8000D+01        0.0000D+00
         DISPLACEMENT      0.3598D-01       -0.1077D+00       -0.5547D-03

     4   NODAL FORCE       0.0000D+00       -0.6000D+01        0.0000D+00
         DISPLACEMENT      0.3564D-01       -0.1020D+00        0.6110D-03

     5   NODAL FORCE       0.0000D+00        0.0000D+00        0.0000D+00
         DISPLACEMENT      0.3528D-01       -0.2896D-02        0.7308D-03

     6   NODAL FORCE       0.0000D+00        0.0000D+00        0.0000D+00
         DISPLACEMENT      0.0000D+00        0.0000D+00       -0.6595D-03


ELEMENT FORCES IN GLOBAL COORDINATE
-----------------------------------

  FRAME ELEMENT :                   ELEMENT FORCES
  ELEMENT NUMBER          Fx                 Fy                 Mz

        1   i     0.1990783D+01     0.7145699D+01    -0.6733366D+02
            j    -0.1990783D+01    -0.7145699D+01    -0.2907476D+03
        2   i    -0.1989390D+01     0.6854303D+01    -0.1421085D-13
            j     0.1989390D+01    -0.6854303D+01     0.3583321D+03
        3   i     0.1988510D+01     0.7144729D+01     0.2907476D+03
            j    -0.1988510D+01    -0.7144729D+01     0.5668277D+03
        4   i     0.1990060D+01    -0.8543533D+00    -0.5668277D+03
            j    -0.1990060D+01     0.8543533D+00     0.4642940D+03
        5   i     0.1988654D+01    -0.6853577D+01    -0.4642940D+03
            j    -0.1988654D+01     0.6853577D+01    -0.3583321D+03
```

```
-----------------
- LOAD STEP  24 -
-----------------
    LOAD SEQUENCE  1
    DETERMINANT OF STIFFNESS MATRIX = 0.197 TIMES 10 TO THE POWER  54
    TRACE OF STIFFNESS MATRIX       = 0.665 TIMES 10 TO THE POWER   7

   NODE       ITEM                  X                  Y                  R

     1   NODAL FORCE       0.0000D+00        0.0000D+00        0.0000D+00
         DISPLACEMENT      0.0000D+00        0.0000D+00        0.0000D+00

     2   NODAL FORCE       0.0000D+00        0.0000D+00        0.0000D+00
         DISPLACEMENT      0.5796D+00       -0.3435D-01       -0.1157D-01

     3   NODAL FORCE       0.0000D+00       -0.8747D+02        0.0000D+00
         DISPLACEMENT      0.5692D+00       -0.1402D+01       -0.6925D-02

     4   NODAL FORCE       0.0000D+00       -0.6560D+02        0.0000D+00
         DISPLACEMENT      0.5660D+00       -0.1304D+01        0.7588D-02

     5   NODAL FORCE       0.0000D+00        0.0000D+00        0.0000D+00
```

Table 5.5 Output P.OUT2 for Portal Frame (cont.)

```
          DISPLACEMENT     0.5565D+00    -0.3199D-01     0.1013D-01

     6    NODAL FORCE      0.0000D+00     0.0000D+00     0.0000D+00
          DISPLACEMENT     0.0000D+00     0.0000D+00    -0.7484D-02


 * PLASTIC HINGE FORMS AT THE B-TH END OF FRAME  ELEMENT  2 *

   ELEMENT FORCES IN GLOBAL COORDINATE
   -----------------------------------

     FRAME ELEMENT :                  ELEMENT FORCES
     ELEMENT NUMBER        Fx               Fy                 Mz

            1   i     0.1856462D+02   0.7922421D+02  -0.2746162D+03
                j    -0.1856462D+02  -0.7922421D+02  -0.3021068D+04
            2   i    -0.1856347D+02   0.7384624D+02  -0.3028688D-12
                j     0.1856347D+02  -0.7384624D+02   0.3382509D+04
            3   i     0.1806836D+02   0.7917971D+02   0.3021068D+04
                j    -0.1806836D+02  -0.7917971D+02   0.6504591D+04
            4   i     0.1855154D+02  -0.8282966D+01  -0.6504591D+04
                j    -0.1855154D+02   0.8282966D+01   0.5508822D+04
            5   i     0.1812797D+02  -0.7375928D+02  -0.5491170D+04
                j    -0.1812797D+02   0.7375928D+02  -0.3382509D+04



 -----------------
 - LOAD STEP  31 -
 -----------------
    LOAD SEQUENCE  1
    DETERMINANT OF STIFFNESS MATRIX = 0.509 TIMES 10 TO THE POWER  48
    TRACE OF STIFFNESS MATRIX       = 0.487 TIMES 10 TO THE POWER   7

        NODE      ITEM               X              Y              R

          1    NODAL FORCE     0.0000D+00     0.0000D+00     0.0000D+00
               DISPLACEMENT    0.0000D+00     0.0000D+00     0.0000D+00

          2    NODAL FORCE     0.0000D+00     0.0000D+00     0.0000D+00
               DISPLACEMENT    0.7784D+00    -0.3756D-01    -0.1554D-01

          3    NODAL FORCE     0.0000D+00    -0.9307D+02     0.0000D+00
               DISPLACEMENT    0.7618D+00    -0.1883D+01    -0.8612D-02

          4    NODAL FORCE     0.0000D+00    -0.6980D+02     0.0000D+00
               DISPLACEMENT    0.7584D+00    -0.1628D+01     0.9709D-02

          5    NODAL FORCE     0.0000D+00     0.0000D+00     0.0000D+00
               DISPLACEMENT    0.7452D+00    -0.3432D-01     0.1301D-01

          6    NODAL FORCE     0.0000D+00     0.0000D+00     0.0000D+00
               DISPLACEMENT    0.0000D+00     0.0000D+00    -0.8517D-02


 * PLASTIC HINGE FORMS AT THE B-TH END OF FRAME  ELEMENT  3 *

   ELEMENT FORCES IN GLOBAL COORDINATE
   -----------------------------------
```

Table 5.5 Output P.OUT2 for Portal Frame (cont.)

```
 FRAME ELEMENT :                       ELEMENT FORCES
 ELEMENT NUMBER          Fx               Fy                Mz

          1    i     0.1840759D+02   0.8510455D+02    0.2361571D+02
               j    -0.1840759D+02  -0.8510455D+02   -0.3270691D+04
          2    i    -0.1839058D+02   0.7776625D+02   -0.8662957D+00
               j     0.1839058D+02  -0.7776625D+02    0.3369108D+04
          3    i     0.1758502D+02   0.8524858D+02    0.3270691D+04
               j    -0.1758502D+02  -0.8524858D+02    0.6990376D+04
          4    i     0.1838319D+02  -0.8098971D+01   -0.6990376D+04
               j    -0.1838319D+02   0.8098971D+01    0.6013824D+04
          5    i     0.1776205D+02  -0.7730906D+02   -0.5935473D+04
               j    -0.1776205D+02   0.7730906D+02   -0.3369108D+04


> MOMENT AT THE B-TH END OF FRAME ELEMENT  1 IS GREATER THAN
> THE PLASTIC MOMENT. THEREFORE, THE LOAD INCREMENT IS SCALED
> DOWN BY A FACTOR OF 0.48171D+00


> DETERMINANT OF THE TANGENT STIFFNESS MATRIX IS NEGATIVE AT LOAD
> STEP  32 WITH THE PRESENT LOAD INCREMENT.  THUS, LOAD INCREMENT
> IS REDUCED TO 1/5 OF PREVIOUS VALUE FOR THIS LOAD STEP.


-----------------
- LOAD STEP  32 -
-----------------
    LOAD SEQUENCE  1
    DETERMINANT OF STIFFNESS MATRIX = 0.274 TIMES 10 TO THE POWER  48
    TRACE OF STIFFNESS MATRIX       = 0.484 TIMES 10 TO THE POWER   7

       NODE      ITEM                 X              Y              R

          1   NODAL FORCE       0.0000D+00     0.0000D+00     0.0000D+00
              DISPLACEMENT      0.0000D+00     0.0000D+00     0.0000D+00

          2   NODAL FORCE       0.0000D+00     0.0000D+00     0.0000D+00
              DISPLACEMENT      0.8023D+00    -0.3777D-01    -0.1649D-01

          3   NODAL FORCE       0.0000D+00    -0.9323D+02     0.0000D+00
              DISPLACEMENT      0.7839D+00    -0.1999D+01    -0.8141D-02

          4   NODAL FORCE       0.0000D+00    -0.6992D+02     0.0000D+00
              DISPLACEMENT      0.7804D+00    -0.1688D+01     0.1020D-01

          5   NODAL FORCE       0.0000D+00     0.0000D+00     0.0000D+00
              DISPLACEMENT      0.7665D+00    -0.3442D-01     0.1351D-01

          6   NODAL FORCE       0.0000D+00     0.0000D+00     0.0000D+00
              DISPLACEMENT      0.0000D+00     0.0000D+00    -0.8633D-02

    ELEMENT FORCES IN GLOBAL COORDINATE
    -----------------------------------

 FRAME ELEMENT :                       ELEMENT FORCES
 ELEMENT NUMBER          Fx               Fy                Mz

          1    i     0.1838871D+02   0.8534250D+02    0.5936591D+02
               j    -0.1838871D+02  -0.8534250D+02   -0.3300817D+04
          2    i    -0.1838491D+02   0.7780854D+02   -0.6901127D-01
               j     0.1838491D+02  -0.7780854D+02    0.3368970D+04
          3    i     0.1748324D+02   0.8549509D+02    0.3301198D+04
               j    -0.1748324D+02  -0.8549509D+02    0.6991137D+04
          4    i     0.1836043D+02  -0.8029784D+01   -0.6991137D+04
```

Table 5.5 Output P.OUT2 for Portal Frame (cont.)

```
                  j   -0.1836043D+02    0.8029784D+01    0.6021851D+04
              5   i    0.1770466D+02   -0.7736269D+02   -0.5942950D+04
                  j   -0.1770466D+02    0.7736269D+02   -0.3368970D+04


 > DETERMINANT OF THE TANGENT STIFFNESS MATRIX IS NEGATIVE AT LOAD
 > STEP  33 WITH THE PRESENT LOAD INCREMENT.  THUS, LOAD INCREMENT
 > IS REDUCED TO 1/5 OF PREVIOUS VALUE FOR THIS LOAD STEP.


 -----------------
 - LOAD STEP  33 -
 -----------------
     LOAD SEQUENCE  1
     DETERMINANT OF STIFFNESS MATRIX = 0.273 TIMES 10 TO THE POWER  48
     TRACE OF STIFFNESS MATRIX       = 0.484 TIMES 10 TO THE POWER   7

          NODE      ITEM                X              Y              R

            1    NODAL FORCE      0.0000D+00     0.0000D+00     0.0000D+00
                 DISPLACEMENT     0.0000D+00     0.0000D+00     0.0000D+00

            2    NODAL FORCE      0.0000D+00     0.0000D+00     0.0000D+00
                 DISPLACEMENT     0.8024D+00    -0.3777D-01    -0.1648D-01

            3    NODAL FORCE      0.0000D+00    -0.9323D+02     0.0000D+00
                 DISPLACEMENT     0.7840D+00    -0.1999D+01    -0.1311D-01

            4    NODAL FORCE      0.0000D+00    -0.6993D+02     0.0000D+00
                 DISPLACEMENT     0.7805D+00    -0.1688D+01     0.1020D-01

            5    NODAL FORCE      0.0000D+00     0.0000D+00     0.0000D+00
                 DISPLACEMENT     0.7665D+00    -0.3442D-01     0.1351D-01

            6    NODAL FORCE      0.0000D+00     0.0000D+00     0.0000D+00
                 DISPLACEMENT     0.0000D+00     0.0000D+00    -0.8633D-02

     ELEMENT FORCES IN GLOBAL COORDINATE
     -----------------------------------

       FRAME ELEMENT :                    ELEMENT FORCES
       ELEMENT NUMBER          Fx               Fy               Mz

              1   i    0.1838790D+02    0.8534869D+02    0.5949963D+02
                  j   -0.1838790D+02   -0.8534869D+02   -0.3300795D+04
              2   i   -0.1838510D+02    0.7781355D+02   -0.8379172D-02
                  j    0.1838510D+02   -0.7781355D+02    0.3368953D+04
              3   i    0.1748234D+02    0.8549179D+02    0.3300796D+04
                  j   -0.1748234D+02   -0.8549179D+02    0.6991140D+04
              4   i    0.1835961D+02   -0.8021957D+01   -0.6991140D+04
                  j   -0.1835961D+02    0.8021957D+01    0.6022797D+04
              5   i    0.1770378D+02   -0.7736824D+02   -0.5943633D+04
                  j   -0.1770378D+02    0.7736824D+02   -0.3368953D+04


 -----------------
 - LOAD STEP  34 -
 -----------------
     LOAD SEQUENCE  1
     DETERMINANT OF STIFFNESS MATRIX = 0.264 TIMES 10 TO THE POWER  48
     TRACE OF STIFFNESS MATRIX       = 0.484 TIMES 10 TO THE POWER   7

          NODE      ITEM                X              Y              R

            1    NODAL FORCE      0.0000D+00     0.0000D+00     0.0000D+00
```

Table 5.5 Output P.OUT2 for Portal Frame (cont.)

```
               DISPLACEMENT     0.0000D+00     0.0000D+00     0.0000D+00

          2    NODAL FORCE      0.0000D+00     0.0000D+00     0.0000D+00
               DISPLACEMENT     0.8033D+00    -0.3778D-01    -0.1655D-01

          3    NODAL FORCE      0.0000D+00    -0.9324D+02     0.0000D+00
               DISPLACEMENT     0.7848D+00    -0.2007D+01    -0.1310D-01

          4    NODAL FORCE      0.0000D+00    -0.6993D+02     0.0000D+00
               DISPLACEMENT     0.7813D+00    -0.1692D+01     0.1023D-01

          5    NODAL FORCE      0.0000D+00     0.0000D+00     0.0000D+00
               DISPLACEMENT     0.7673D+00    -0.3443D-01     0.1354D-01

          6    NODAL FORCE      0.0000D+00     0.0000D+00     0.0000D+00
               DISPLACEMENT     0.0000D+00     0.0000D+00    -0.8637D-02

     ELEMENT FORCES IN GLOBAL COORDINATE
     -----------------------------------

       FRAME ELEMENT :                 ELEMENT FORCES
       ELEMENT NUMBER          Fx                Fy                 Mz

             1   i     0.1838746D+02    0.8535819D+02    0.6091188D+02
                 j    -0.1838746D+02   -0.8535819D+02   -0.3302039D+04
             2   i    -0.1838476D+02    0.7781527D+02   -0.2569555D-02
                 j     0.1838476D+02   -0.7781527D+02    0.3368947D+04
             3   i     0.1747656D+02    0.8550181D+02    0.3302062D+04
                 j    -0.1747656D+02   -0.8550181D+02    0.6991186D+04
             4   i     0.1835893D+02   -0.8019304D+01   -0.6991186D+04
                 j    -0.1835893D+02    0.8019304D+01    0.6023094D+04
             5   i     0.1770089D+02   -0.7736990D+02   -0.5943896D+04
                 j    -0.1770089D+02    0.7736990D+02   -0.3368947D+04

 >> SOLUTION DOES NOT CONVERGE, A LIMIT STATE MAY HAVE BEEN REACHED <<

========================================
= SEQUENCE OF PLASTIC HINGE FORMATION =
========================================

     PLASTIC HINGE              LOCATION

            1           B-TH END OF FRAME ELEMENT  2
            2           B-TH END OF FRAME ELEMENT  3
```

5.5 Modification of In-House Program

The present analysis program may be directly used for practical design. Its concepts, however, may be implemented by simple modifications imposed on in-house elastic analysis programs. The modifications may be made by incorporating key behavioral effects of a steel frame including second-order, gradual yielding associated with residual stresses and flexure, geometric imperfections, and connection nonlinearity into an in-house program. Herein, the concept of the modification will be briefly presented.

5.5.1 Stability Function

To capture second-order effects, stability functions are recommended because they lead to large savings in modeling and solution efforts by using one or two elements per member. The simplified stability functions reported by Chen and Lui [2] or an alternative may be used. Considering the prismatic beam-column element shown in Figure 2.16(a), the incremental force-displacement relationship of this element may be written as

$$\begin{bmatrix} \dot{M}_A \\ \dot{M}_B \\ \dot{P} \end{bmatrix} = \frac{EI}{L} \begin{bmatrix} S_1 & S_2 & 0 \\ S_2 & S_1 & 0 \\ 0 & 0 & A/I \end{bmatrix} \begin{bmatrix} \dot{\theta}_A \\ \dot{\theta}_B \\ \dot{e} \end{bmatrix} \tag{5.2}$$

where

S_1, S_2	=	stability functions
$\dot{M}_A, \dot{M}_B$	=	incremental end moment
$\dot{P}$	=	incremental axial force
$\dot{\theta}_A, \dot{\theta}_B$	=	incremental joint rotation
$\dot{e}$	=	incremental axial displacement
A, I, L	=	area, moment of inertia, and length of beam-column element
E	=	modulus of elasticity

In this formulation, all members are assumed to be adequately braced to prevent out-of-plane buckling, and their cross-sections are compact to avoid local buckling.

5.5.2 Cross-Section Plastic Strength

Based on the AISC-LRFD bilinear interaction equations [1], the cross-section plastic strength may be expressed as Equations (5.3a and 5.3b). These AISC-LRFD cross-section plastic strength curves may be adopted for both strong and weak-axis bending (Figure 2.20).

$$\frac{P}{\phi_c P_y} + \frac{8}{9}\frac{M}{\phi_b M_p} = 1.0 \quad \text{for} \quad \frac{P}{\phi_c P_y} \geq 0.2 \tag{5.3a}$$

$$\frac{P}{2\phi_c P_y} + \frac{M}{\phi_b M_p} = 1.0 \quad \text{for} \quad \frac{P}{\phi_c P_y} < 0.2 \tag{5.3b}$$

where

P, M = second-order axial force and bending moment
P_y = squash load
M_p = plastic moment capacity
ϕ_c, ϕ_b = resistance factors for axial strength and flexural strength

The reduction factors ϕ_c and ϕ_b are built in the analysis program and are automatically included in the load-carrying capacity. The reduction factors are selected as 0.85 for axial strength and 0.9 for flexural strength just as the AISC-LRFD Specification does [1].

5.5.3 CRC Tangent Modulus

The CRC tangent modulus concept is employed to account for the gradual yielding effect due to residual stresses along the length of members under axial loads between two plastic hinges. In this concept, the elastic modulus E instead of moment of inertia I is reduced to account for the reduction of the elastic portion of the cross-section because the reduction of elastic modulus is easier to implement than that of moment of inertia for different sections. The reduction rate in stiffness between weak-and strong-axis is different, but this is not considered here. This is because rapid degradation in stiffness in the weak-axis strength is compensated well by the stronger weak-axis plastic strength. As a result, this simplicity will make the present methods practical. From Chen and Lui [2], the CRC E_t is written as (Figure 2.23):

$$E_t = 1.0E \quad \text{for} \quad P \leq 0.5P_y \tag{5.4a}$$

$$E_t = 4\frac{P}{P_y}E\left(1 - \frac{P}{P_y}\right) \quad \text{for} \quad P > 0.5P_y \tag{5.4b}$$

5.5.4 Parabolic Function

The tangent modulus model in Equations (5.4a and 5.4b) is suitable for $P/P_y > 0.5$, but it is not sufficient to represent the stiffness degradation for cases with small axial forces and large bending moments. A gradual stiffness degradation of plastic hinge is required to represent the distributed plasticity effects associated with bending actions. We shall introduce the softening plastic hinge model to represent the gradual transition from elastic stiffness to zero stiffness associated with a fully developed plastic hinge. When the softening plastic hinges are present at both ends of an element, the incremental force-displacement relationship may be expressed as [7]:

$$\begin{bmatrix} \dot{M}_A \\ \dot{M}_B \\ \dot{P} \end{bmatrix} = \frac{E_t I}{L} \begin{bmatrix} \eta_A \left[S_1 - \frac{S_2^2}{S_1}(1-\eta_B) \right] & \eta_A \eta_B S_2 & 0 \\ \eta_A \eta_B S_2 & \eta_B \left[S_1 - \frac{S_2^2}{S_1}(1-\eta_A) \right] & 0 \\ 0 & 0 & A/I \end{bmatrix} \begin{bmatrix} \dot{\theta}_A \\ \dot{\theta}_B \\ \dot{e} \end{bmatrix} \tag{5.5}$$

where

$\dot{M}_A, \dot{M}_B, \dot{P}$ = incremental end moments and axial force, respectively
S_1, S_2 = stability functions
E_t = tangent modulus
η_A, η_B = element stiffness parameters

The parameter η represents a gradual stiffness reduction associated with flexure at sections. The partial plastification at cross-sections in the end of elements is denoted by $0 < \eta < 1$. The η may be assumed to vary according to the parabolic expression (Figure 2.27):

$$\eta = 4\alpha(1-\alpha) \quad \text{for } \alpha > 0.5 \tag{5.6}$$

where α is the force-state parameter obtained from the limit state surface corresponding to the element end (Figure 2.26):

$$\alpha = \frac{P}{P_y} + \frac{8}{9}\frac{M}{M_p} \quad \text{for } \frac{P}{P_y} \geq \frac{2}{9}\frac{M}{M_p} \tag{5.7a}$$

$$\alpha = \frac{P}{2P_y} + \frac{M}{M_p} \quad \text{for } \frac{P}{P_y} < \frac{2}{9}\frac{M}{M_p} \tag{5.7b}$$

where

P, M = second-order axial force and bending moment at the cross-section
M_p = plastic moment capacity

5.5.5 Geometric Imperfection

For geometric imperfections, the explicit imperfection modeling method, the equivalent notional load method, or the further reduced tangent method may all be used. Since the first two methods require the imperfections to be explicitly input, the further reduced tangent method will be used to model the imperfections in code to be implemented in "in-house" program.

The reduced tangent modulus method further reduces the tangent modulus E_t to account for the degradation of stiffness due to geometric imperfections [5] [6]:

$$E_t' = 4\tfrac{P}{P_y}\left(1 - \tfrac{P}{P_y}\right) E\xi_i \quad \text{for} \quad P > 0.5P_y \tag{5.8a}$$

$$E_t' = E\xi_i \quad \text{for} \quad P \leq 0.5P_y \tag{5.8b}$$

where

E_t' = reduced E_t
ξ_i = reduction factor for geometric imperfection

With the reduction factor 0.85, the further reduced tangent modulus curve for the CRC E_t with geometric imperfections is shown in Figure 2.52. The advantage of this method over the other two is its convenience in design because it completely eliminates the inconvenience of the explicit imperfection modeling or the equivalent notional load.

5.5.6 Semi-Rigid Connection

The connection may be modeled as a rotational spring in the moment-rotation relationship represented by Equation (2.9). Figure 2.45 shows a beam-column element with semi-rigid connections at both ends. If the effect of connection flexibility is incorporated into the member stiffness, the incremental element force-displacement relationship of Equation (5.1) is modified as [2]:

$$\begin{bmatrix} \dot{M}_A \\ \dot{M}_B \\ \dot{P} \end{bmatrix} = \frac{E_t I}{L} \begin{bmatrix} S_{ii}^* & S_{ij}^* & 0 \\ S_{ij}^* & S_{jj}^* & 0 \\ 0 & 0 & A/I \end{bmatrix} \begin{bmatrix} \dot{\theta}_A \\ \dot{\theta}_B \\ \dot{e} \end{bmatrix} \tag{5.9}$$

where

$$S_{ii}^* = \left(S_{ii} + \frac{E_t I S_{ii} S_{jj}}{L R_{ktB}} - \frac{E_t I S_{ij}^2}{L R_{ktB}} \right) / R^* \tag{5.10a}$$

$$S_{jj}^{*} = \left(S_{jj} + \frac{E_t I S_{ii} S_{jj}}{L R_{ktA}} - \frac{E_t I S_{ij}^2}{L R_{ktA}} \right) / R^{*} \tag{5.10b}$$

$$S_{ij}^{*} = S_{ij} / R^{*} \tag{5.10c}$$

$$R^{*} = \left(1 + \frac{E_t I S_{ii}}{L R_{ktA}}\right)\left(1 + \frac{E_t I S_{jj}}{L R_{ktB}}\right) - \left(\frac{E_t I}{L}\right)^2 \frac{S_{ij}^2}{R_{ktA} R_{ktB}} \tag{5.10d}$$

where

R_{ktA}, R_{ktB} = tangent stiffness of connections A and B
S_{ii}, S_{ij} = generalized stability functions
S_{ii}^{*}, S_{jj}^{*} = modified stability functions that account for the presence of end connections.

5.6 Summary

In this chapter, the analysis program PAAP for a two-dimensional steel frame design has been discussed. The program has been developed for practical design use based on the refined-plastic hinge concept [5] [7]. The program is divided into three FORTRAN programs including DATAGEN, INPUT, and PAAP. The first program, DATAGEN, reads an input file P.DAT and produces the generated data file INFILE. The second program, INPUT, rearranges INFILE into three data files including DATA0, DATA1, and DATA2. The third program, PAAP (Practical Advanced Analysis Program), generates the output files P.OUT1 and P.OUT2. The schematic diagram in Figure 5.3 shows the program operation procedure. These three programs are executed by typing the command RUN.

This program has been tested in two computer processors: (1) an IBM 486 or equivalent personal computer system using Microsoft's FORTRAN 77 compiler v1.00 and Lahey's FORTRAN 77 compiler v5.01, and (2) a Sun 5 using a Sun FORTRAN 77 compiler. The total size of the three programs is as small as 111 KB (= 0.111 MB), and so the 3.5" high density diskette (1.44MB) can accommodate the three programs and several example problems.

The program can generate the output files P.OUT1 and P.OUT2 in a reasonable time period described in the following. The run time on an IBM 486 PC with memory of 640K to get the output files for the eight-story frame shown in Figure 6.38 is taken as 4 min 10 s and 2 min 30 s in real time rather than CPU time by using Microsoft FORTRAN and Lahey FORTRAN, respectively. In the Sun 5, the run time varies approximately 2 to 3 min depending on the degree of occupancy by users.

The input data format of the proposed program is the same as that of a usual linear elastic analysis program except that geometric imperfection and incremental loads are considered. The input data consists of 13 data sets including 5 control data, 3 section property data, 3 element data, 1 boundary data, and 1 load data set. The coordinate systems both in the input and output are based on the global coordinates system. Units are not specified by the program and must be input consistently. The input format is summarized in Table 5.1 [5]. The proposed program may be used for practical design by taking advantage of the small program size, short run time, and simple input format.

The proposed analysis program can be directly used or obtained by simple modifications of usual in-house programs. The modifications are based on the concept described in Chapter 2. To capture second-order effects, stability functions are recommended because they lead to large savings in modeling and solution effects by the use of one or two elements per member. The simplified stability functions reported by Chen and Lui [2] are suggested here. The AISC-LRFD cross-section plastic strength curves expressed as Equation (5.3) are for both strong- and weak-axis bending. The CRC E_t may be incorporated to account for the stiffness degradation associated with residual stresses. The parameter η may be used to present a gradual stiffness reduction associated with flexure at sections. The η function is assumed to vary according to the parabolic expression in Equation (5.6), and is a function of the force-state parameter in Equation (5.7). For geometric imperfections, the explicit imperfection modeling method, the equivalent notional load method, or the further reduced tangent method may be used. The reduced tangent modulus method further reduces the tangent modulus E_t to account for stiffness degradation due to geometrical imperfections. The resistance factors ϕ_b and ϕ_c are built in the analysis program by incorporating the resistance factors in the LRFD cross-section strength.

References

[1] AISC (1994) *Load and Resistance Factor Design Specification*, American Institute of Steel Construction, 2nd ed., Chicago, IL.

[2] Chen, W.F. and Lui, E.M. (1992) *Stability Design of Steel Frames*, CRC Press, Boca Raton, FL, 380 pp.

[3] Cook, R.D., Malkus, D.S., and Plesha, M.E. (1989) *Concepts and Applications of Finite Element Analysis*, 3rd ed., John Wiley & Sons, New York, 630 pp.

[4] Hughes, T. J. R. (1987) *The Finite Element Method: Linear Static and Dynamic Finite Element Analysis*, Prentice-Hall, Englewood Cliffs, NJ, 803 pp.

[5] Kim, S.E. (1996) Practical Advanced Analysis for Steel Frame Design, Ph.D. dissertation, School of Civil Engineering, Purdue University, West Lafayette, IN, 271 pp.

[6] Kim, S.E. and Chen, W.F. (1996) Practical advanced analysis for steel frame design, The ASCE Structural Congress XIV Special Proceedings Volume on Analysis and Computation, Chicago, IL, April, 19-30.

[7] Liew, J.Y.R. (1992) Advanced Analysis for Frame Design, Ph.D. dissertation, School of Civil Engineering, Purdue University, West Lafayette, IN, 393 pp.

chapter six

Design Examples

6.1 *Introduction*

In previous chapters, the concept, verifications, and computer program of the practical advanced analysis method for steel frame design have been presented. The present advanced analysis method has been developed and refined to achieve both simplicity in use and, as far as possible, a realistic representation of the behavior and strength. The advanced analysis method captures the limit state strength and stability of a structural system and its individual members. As a result, the method can be used for practical frame design without the tedious separate member capacity checks including the calculation of K-factor.

The aim of this chapter is to provide further confirmation of the validity of the LRFD-based advanced analysis methods for practical frame design. The comparative design examples in this chapter show the detailed design procedure for advanced and LRFD design procedures [4]. The design procedures are similar to those described in Chapter 4, and may be grouped into four basic steps: (1) load condition, (2) structural modeling, (3) analysis, and (4) limit state check. The design examples cover simple structures, truss structures, braced frames, unbraced frames, and semi-rigid frames. The three practical models including an explicit imperfection, an equivalent notional load, and a further reduced tangent modulus modeling are used for the design examples. Member sizes determined by advanced procedures are compared with those determined by the LRFD, and good agreement is generally observed. The advanced analysis method achieves the LRFD design requirements without the calculation of K-factor. The methods are practical and therefore recommended for general use.

The design examples are limited to two-dimensional steel frames so that the spatial behavior is not considered. Lateral torsional buckling is assumed to be prevented by adequate lateral braces. Compact W-sections are assumed so that sections can develop their full plastic moment capacity without buckling locally. All loads are statically applied.

6.2 Simple Structures

6.2.1 Three-Span Continuous Beam

Figure 6.1 shows a three-span continuous beam subject to a uniformly distributed load of 7 k/ft. All members are continuously braced laterally and A36 steel is used. Herein, both the design procedures by the advanced analysis and the LRFD method will be presented in what follows.

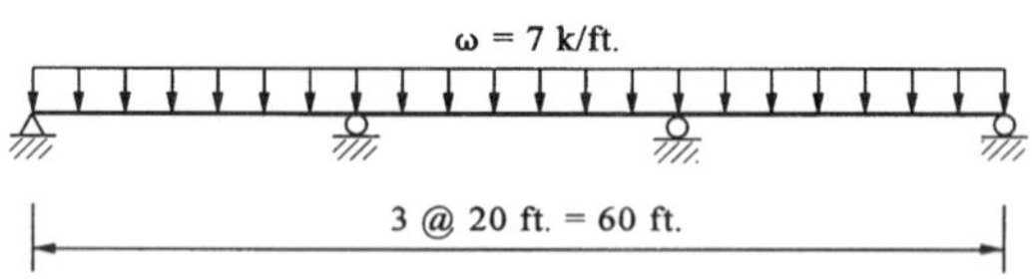

Figure 6.1 Configuration and load condition of the three span continuous beam subject to uniform distributed load.

Design by the Advanced Analysis

Step 1: Load Condition and Preliminary Member Sizing

The critical factored load combination is shown in Figure 6.1. Approximate analysis may be conducted by assuming the locations of inflection points as discussed in Section 4.8.1. Using the approximate method described in Section 4.8.1 or Beam Diagrams and Formulas in the AISC-LRFD Specification, the maximum moment is $0.1wL^2 = 280$ kip-ft. A preliminary member size could be W21x50 ($\phi_b M_p = 297$ kip-ft) referring to the Beam Selection Table in the LRFD Specification.

Step 2: Structural Modeling

Each beam may be modeled with two elements (Section 4.9.1), resulting in a total of six elements. The uniformly distributed loads can be converted to concentrated loads as shown in Figure 6.2. Geometric imperfections are not considered in this example because they do not influence the ultimate strength of beams. Incremental loads may be determined by dividing the factored loads by a scaling factor. When a scaling factor of 50 is used, the incremental load condition is as shown in Figure 6.3.

Step 3: Analysis

Referring to the input instructions in Section 5.3.2, the input data may be easily generated as listed in Table 6.1. Programs DATAGEN, INPUT, and PAAP are executed in sequence by entering the batch file command RUN. The program will continue to increment the load until the beam has reached its ultimate state.

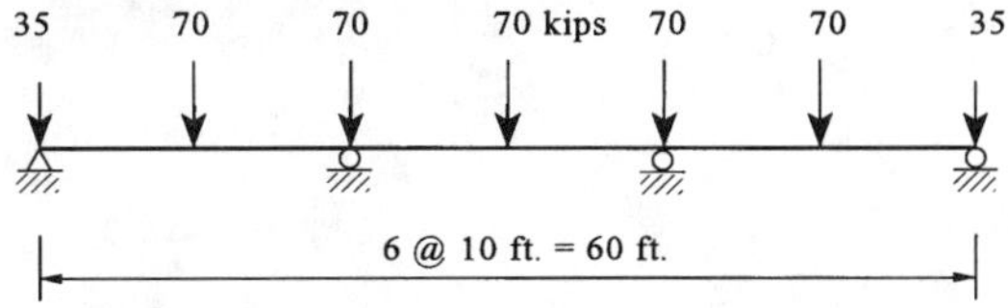

Figure 6.2 Concentrated load condition converted from the distributed load for the three span continuous beam.

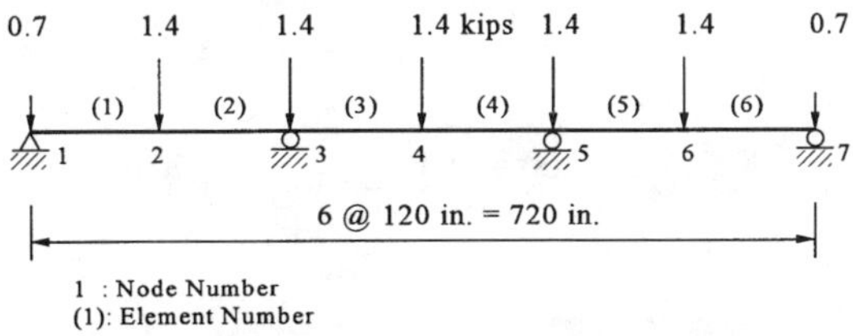

Figure 6.3 Modeling of the three span continuous beam.

Step 4: Check of Load-Carrying Capacity

The incremental displacements and member forces are listed in P.OUT2 (Table 6.2). The first plastic hinge forms at Load Step 62, and the corresponding load is 86.8 kips at the interior concentrated load point. The last plastic hinge forms at Load Step 69, corresponding to a load of 88.8 kips at the same location. The deflection increases drastically at Load Step 70 at Node 6, and thus indicates system failure.

One may choose the load-carrying capacity corresponding to the first plastic hinge forming. The load-carrying capacity is greater than the factored load of 70 kips on interior nodal points. When the member size is reduced to W21x44, the load-carrying capacity results in 75.4 kips at the formation of the first plastic hinge and thus the member size is adequate.

Table 6.1 Input Data P.DAT for the Three-Span Continuous Beam

```
Three span continuous beam
    0    1
    7    4  100
    0    1    0
    0    6    0
    1      14.7       984.0       110.0    29000.0       36.0
    1     120.0         0.0    1      1      2      6
    1    1    1
    3         1           3    2
    1                  -0.7                  2      6
    2                  -1.4                  5
```

Table 6.2 Output P.OUT2 for the Three-Span Continuous Beam

```
============================
= LOAD-DEFLECTION BEHAVIOR =
============================


   -----------------
   - LOAD STEP   1 -
   -----------------
       LOAD SEQUENCE  1
       DETERMINANT OF STIFFNESS MATRIX = 0.124 TIMES 10 TO THE POWER  72
       TRACE OF STIFFNESS MATRIX       = 0.115 TIMES 10 TO THE POWER   8

            NODE      ITEM                X              Y              R

               1   NODAL FORCE     0.0000D+00     0.0000D+00     0.0000D+00
                   DISPLACEMENT    0.0000D+00     0.0000D+00    -0.1060D-03

               2   NODAL FORCE     0.0000D+00    -0.1400D+01     0.0000D+00
                   DISPLACEMENT    0.0000D+00    -0.7771D-02     0.1766D-04

               3   NODAL FORCE     0.0000D+00     0.0000D+00     0.0000D+00
                   DISPLACEMENT    0.0000D+00     0.0000D+00     0.3532D-04

               4   NODAL FORCE     0.0000D+00    -0.1400D+01     0.0000D+00
                   DISPLACEMENT    0.0000D+00    -0.1413D-02     0.5082D-20

               5   NODAL FORCE     0.0000D+00     0.0000D+00     0.0000D+00
                   DISPLACEMENT    0.0000D+00     0.0000D+00    -0.3532D-04

               6   NODAL FORCE     0.0000D+00    -0.1400D+01     0.0000D+00
                   DISPLACEMENT    0.0000D+00    -0.7771D-02    -0.1766D-04

               7   NODAL FORCE     0.0000D+00     0.0000D+00     0.0000D+00
                   DISPLACEMENT    0.0000D+00     0.0000D+00     0.1060D-03

       ELEMENT FORCES IN GLOBAL COORDINATE
       -----------------------------------

         FRAME ELEMENT :                  ELEMENT FORCES
         ELEMENT NUMBER        Fx                Fy                Mz

                 1   i   0.6654152D-13   0.4900000D+00  -0.7105427D-14
                     j  -0.6654152D-13  -0.4900000D+00   0.5880000D+02
                 2   i   0.1235771D-12  -0.9100000D+00  -0.5880000D+02
                     j  -0.1235771D-12   0.9100000D+00  -0.5040000D+02
                 3   i   0.5713576D-15   0.7000000D+00   0.5040000D+02
                     j  -0.5713576D-15  -0.7000000D+00   0.3360000D+02
                 4   i   0.5713559D-15  -0.7000000D+00  -0.3360000D+02
                     j  -0.5713559D-15   0.7000000D+00  -0.5040000D+02
                 5   i   0.1235771D-12   0.9100000D+00   0.5040000D+02
                     j  -0.1235771D-12  -0.9100000D+00   0.5880000D+02
                 6   i   0.6654153D-13  -0.4900000D+00  -0.5880000D+02
                     j  -0.6654153D-13   0.4900000D+00  -0.1421085D-13
```

Table 6.2 Output P.OUT2 for the Three-Span Continuous Beam (cont.)

```
-----------------
- LOAD STEP  62 -
-----------------
    LOAD SEQUENCE  1
    DETERMINANT OF STIFFNESS MATRIX = 0.163 TIMES 10 TO THE POWER  60
    TRACE OF STIFFNESS MATRIX       = 0.391 TIMES 10 TO THE POWER   7

         NODE      ITEM                 X              Y              R

           1   NODAL FORCE     0.0000D+00     0.0000D+00     0.0000D+00
               DISPLACEMENT    0.0000D+00     0.0000D+00    -0.9107D-02

           2   NODAL FORCE     0.0000D+00    -0.8680D+02     0.0000D+00
               DISPLACEMENT   -0.2478D-02    -0.7939D+00     0.9859D-03

           3   NODAL FORCE     0.0000D+00     0.0000D+00     0.0000D+00
               DISPLACEMENT   -0.4906D-02     0.0000D+00     0.3640D-02

           4   NODAL FORCE     0.0000D+00    -0.8680D+02     0.0000D+00
               DISPLACEMENT   -0.4907D-02    -0.4170D-01    -0.1321D-18

           5   NODAL FORCE     0.0000D+00     0.0000D+00     0.0000D+00
               DISPLACEMENT   -0.4909D-02     0.0000D+00    -0.3640D-02

           6   NODAL FORCE     0.0000D+00    -0.8680D+02     0.0000D+00
               DISPLACEMENT   -0.7337D-02    -0.7939D+00    -0.9859D-03

           7   NODAL FORCE     0.0000D+00     0.0000D+00     0.0000D+00
               DISPLACEMENT   -0.9815D-02     0.0000D+00     0.9107D-02


 * PLASTIC HINGE FORMS AT THE B-TH END OF FRAME  ELEMENT  1 *

 * PLASTIC HINGE FORMS AT THE B-TH END OF FRAME  ELEMENT  5 *

    ELEMENT FORCES IN GLOBAL COORDINATE
    -----------------------------------

      FRAME ELEMENT :                 ELEMENT FORCES
      ELEMENT NUMBER          Fx              Fy               Mz

             1  i   -0.1316785D+00   0.2969069D+02  -0.2891909D-11
                j    0.1316785D+00  -0.2969069D+02   0.3562700D+04
             2  i   -0.2483277D+00  -0.5718211D+02  -0.3562700D+04
                j    0.2483277D+00   0.5718211D+02  -0.3298806D+04
             3  i   -0.2611214D-03   0.4340000D+02   0.3306705D+04
                j    0.2611214D-03  -0.4340000D+02   0.1901295D+04
             4  i   -0.2611214D-03  -0.4340000D+02  -0.1901295D+04
                j    0.2611214D-03   0.4340000D+02  -0.3306705D+04
             5  i   -0.2479574D+00   0.5723808D+02   0.3306705D+04
                j    0.2479574D+00  -0.5723808D+02   0.3561518D+04
             6  i   -0.1326554D+00  -0.2954301D+02  -0.3561518D+04
                j    0.1326554D+00   0.2954301D+02   0.1653878D+02
```

Table 6.2 Output P.OUT2 for the Three-Span Continuous Beam (cont.)

```
-----------------
- LOAD STEP  69 -
-----------------
    LOAD SEQUENCE  1
    DETERMINANT OF STIFFNESS MATRIX = 0.184 TIMES 10 TO THE POWER  54
    TRACE OF STIFFNESS MATRIX       = 0.297 TIMES 10 TO THE POWER   7

         NODE      ITEM               X              Y              R

            1   NODAL FORCE      0.0000D+00     0.0000D+00     0.0000D+00
                DISPLACEMENT     0.0000D+00     0.0000D+00    -0.1391D-01

            2   NODAL FORCE      0.0000D+00    -0.8876D+02     0.0000D+00
                DISPLACEMENT    -0.7386D-02    -0.1370D+01     0.1676D-01

            3   NODAL FORCE      0.0000D+00     0.0000D+00     0.0000D+00
                DISPLACEMENT    -0.1468D-01     0.0000D+00     0.6028D-02

            4   NODAL FORCE      0.0000D+00    -0.8876D+02     0.0000D+00
                DISPLACEMENT    -0.1468D-01    -0.1665D-02     0.6007D-06

            5   NODAL FORCE      0.0000D+00     0.0000D+00     0.0000D+00
                DISPLACEMENT    -0.1467D-01     0.0000D+00    -0.6065D-02

            6   NODAL FORCE      0.0000D+00    -0.8876D+02     0.0000D+00
                DISPLACEMENT    -0.2234D-01    -0.1406D+01     0.5074D-02

            7   NODAL FORCE      0.0000D+00     0.0000D+00     0.0000D+00
                DISPLACEMENT    -0.3009D-01     0.0000D+00     0.1420D-01


 * PLASTIC HINGE FORMS AT THE A-TH END OF FRAME  ELEMENT  5 *

    ELEMENT FORCES IN GLOBAL COORDINATE
    -----------------------------------

      FRAME ELEMENT :                  ELEMENT FORCES
      ELEMENT NUMBER          Fx                Fy                Mz

            1   i   -0.2741717D+00    0.2968347D+02   -0.1629732D+00
                j    0.2741717D+00   -0.2968347D+02    0.3561572D+04
            2   i   -0.5279389D+00   -0.5914729D+02   -0.3561572D+04
                j    0.5279389D+00    0.5914729D+02   -0.3534918D+04
            3   i    0.1435809D-01    0.4437286D+02    0.3542755D+04
                j   -0.1435809D-01   -0.4437286D+02    0.1781988D+04
            4   i    0.1435943D-01   -0.4438714D+02   -0.1781988D+04
                j   -0.1435943D-01    0.4438714D+02   -0.3544468D+04
            5   i   -0.5440442D+00    0.5932864D+02    0.3559092D+04
                j    0.5440442D+00   -0.5932864D+02    0.3559092D+04
            6   i   -0.2831634D+00   -0.2953980D+02   -0.3559829D+04
                j    0.2831634D+00    0.2953980D+02    0.1569449D+02


-----------------
- LOAD STEP  70 -
-----------------
    LOAD SEQUENCE  1
    DETERMINANT OF STIFFNESS MATRIX = 0.685 TIMES 10 TO THE POWER  53
    TRACE OF STIFFNESS MATRIX       = 0.296 TIMES 10 TO THE POWER   7

         NODE      ITEM               X              Y              R

            1   NODAL FORCE      0.0000D+00     0.0000D+00     0.0000D+00
                DISPLACEMENT     0.0000D+00     0.0000D+00    -0.1715D-01
```

Table 6.2 Output P.OUT2 for the Three-Span Continuous Beam (cont.)

```
     2    NODAL FORCE      0.0000D+00    -0.8904D+02     0.0000D+00
          DISPLACEMENT    -0.1181D-01    -0.1760D+01     0.2457D-01

     3    NODAL FORCE      0.0000D+00     0.0000D+00     0.0000D+00
          DISPLACEMENT    -0.2350D-01     0.0000D+00     0.7543D-02

     4    NODAL FORCE      0.0000D+00    -0.8904D+02     0.0000D+00
          DISPLACEMENT    -0.2350D-01    -0.2586D-03    -0.1115D-04

     5    NODAL FORCE      0.0000D+00     0.0000D+00     0.0000D+00
          DISPLACEMENT    -0.2350D-01     0.0000D+00    -0.6077D-02

     6    NODAL FORCE      0.0000D+00    -0.8904D+02     0.0000D+00
          DISPLACEMENT    -0.2462D+00    -0.1998D+02     0.1601D+00

     7    NODAL FORCE      0.0000D+00     0.0000D+00     0.0000D+00
          DISPLACEMENT    -0.4703D+00     0.0000D+00     0.1690D+00


* PLASTIC HINGE FORMS AT THE B-TH END OF FRAME  ELEMENT  2 *

  ELEMENT FORCES IN GLOBAL COORDINATE
  ----------------------------------

    FRAME ELEMENT :                      ELEMENT FORCES
    ELEMENT NUMBER          Fx                 Fy                 Mz

         1   i    -0.3707086D+00    0.2967884D+02   -0.3821170D+00
             j     0.3707086D+00   -0.2967884D+02    0.3560807D+04
         2   i    -0.7214817D+00   -0.5933781D+02   -0.3560807D+04
             j     0.7214817D+00    0.5933781D+02   -0.3557695D+04
         3   i     0.1487738D-01    0.4435781D+02    0.3557695D+04
             j    -0.1487738D-01   -0.4435781D+02    0.1765243D+04
         4   i     0.1487979D-01   -0.4450923D+02   -0.1782016D+04
             j    -0.1487979D-01    0.4450923D+02   -0.3559092D+04
         5   i    -0.9671251D+01    0.6052913D+02    0.3486749D+04
             j     0.9671251D+01   -0.6052913D+02    0.3486749D+04
         6   i    -0.4827963D+01   -0.3013265D+02   -0.3486749D+04
             j     0.4827963D+01    0.3013265D+02    0.1546503D+02
```

Step 5: Check of Serviceability

Referring to P.OUT2 (Table 6.2), the deflection at Node 2 and 6 corresponding to Load Step 1 is 0.00777 in. This deflection may be considered elastic because the behavior of the beam is linear and elastic under a small load. The total deflection of 0.389 in. can be obtained by multiplying the deflection of 0.00777 in. by the scaling factor of 50. The deflection ratio over the beam length is 1/618 which satisfies the limitation of 1/360. Since the factored load is used for the calculation of deflection above, the deflection corresponding to the service load will be smaller than that above.

Design by the AISC-LRFD Method

The beam members carry only bending moments, not axial forces. The bending moment diagram for this beam is shown in Figure 6.4. The member size is governed by the maximum negative moment of 280 kip-ft at the interior supports. The resulting member size may be selected as W21x50 ($\phi_b M_p = 297$ kip-ft) by the use of the Beam Selection Table in the LRFD Specification.

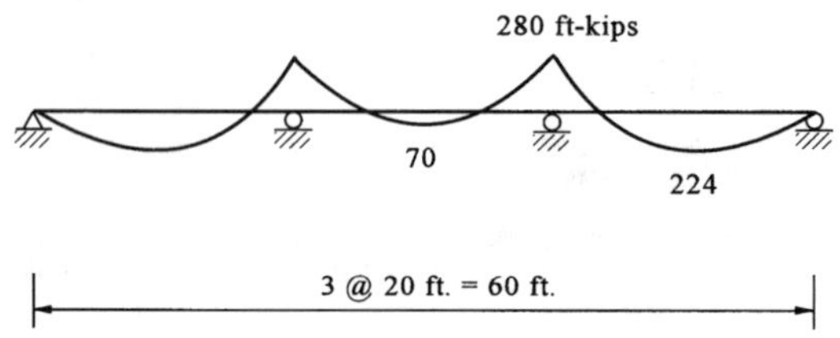

Figure 6.4 Bending moment diagram of the three span continuous beam.

Comparison of Results

Advanced analysis predicts one size smaller member than does the LRFD (Figure 6.5) because it allows for inelastic moment redistribution. The LRFD method looks easier than advanced analysis for the design of this continuous beam because the K-factor calculation and interaction equation checks are unnecessary. However, in general, advanced analysis is a much easier method for designing the frame as a whole.

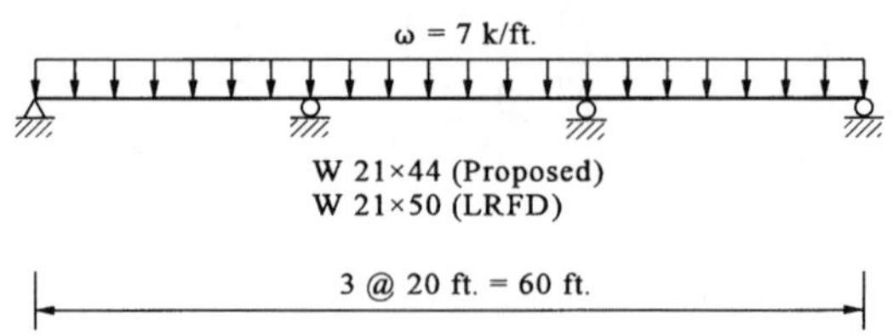

Figure 6.5 Comparison of member sizes of the three span continuous beam.

6.2.2 *Two-Story Column*

Figure 6.6 shows a continuous column having a lower column of 15 ft and an upper column of 10 ft subject to a factored axial load of 200 kips. The strength of weak-axis will control the member size since additional supports are not provided in weak-axis direction. A wide flange section of A36 steel is used.

Design by the Advanced Analysis

Step 1: Load Condition and Preliminary Member Sizing

The critical factored load combination is shown in Figure 6.6. Since the shorter upper column helps the strength of the longer lower column, the

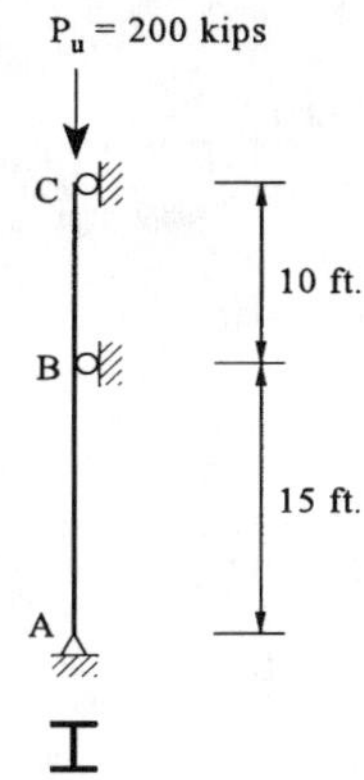

Figure 6.6 Configuration and load condition of the two-story column subject to axial load.

upper column has the K-factor of 1.294 although the column is braced (Section 3.8.1). The effective length results in 12.94 ft (= 10 ft $\times$ 1.294). The preliminary member size is selected as W8x31 (ϕP_n = 204 kips) using the Column Design Table in the LRFD Specification.

Step 2: Structural Modeling

Each column member may be modeled with two elements as recommended in Section 4.9.1. All three imperfection models are used. In the explicit imperfection model, the two-element columns are assumed to have an initial geometric imperfection equal to 0.01 ft (0.12 in) and 0.015 ft (0.18 in) at the mid-height of the upper and lower column shown in Figure 6.7(a), respectively. In the equivalent notional load method, the notional load equal to 0.004 times the gravity load is applied at mid-height of the upper and lower column shown in Figure 6.7(b). In the further reduced tangent modulus method, the reduced tangent modulus factor equal to 0.85 is used by the program, and is shown in Figure 6.7(c). An incremental load of 4 kips is determined by dividing the factored load of 200 kips by a scaling factor of 50 (Figure 6.7).

Step 3: Analysis

Referring to the input instructions described in Section 5.3.2, the input data may be easily generated as listed in Table 6.3. Program DATAGEN, INPUT, and PAAP are executed in sequence by entering the batch file command RUN on the screen.

Step 4: Check of Load-Carrying Capacity

The incremental displacements and member forces are listed in P.OUT2 (Tables 6.4a through 6.4c). The load-carrying capacity predicted by the explicit imperfection model and the equivalent notional load

Table 6.3a Input Data P.DAT of the Explicit Imperfection Modeling for the Two-Story Column

```
Two-story column
     1    1
     5    3   100
     0    1     0
     0    4     0
     1       9.13        37.1          14.1    29000.0         36.0
     1       0.18        90.0      1     1     2
     2      -0.18        90.0      1     2     3
     3       0.12        60.0      1     3     4
     4      -0.12        60.0      1     4     5
     1    1     1
     3    1
     5    1
     5                   -4.0
```

Table 6.3b Input Data P.DAT of the Equivalent Notional Load Modeling for the Two-Story Column

```
Two-story column
     2    1
     5    3   100
     0    1     0
     0    4     0
     1       9.13        37.1          14.1    29000.0         36.0
     1                   90.0      1     1     2
     2                   90.0      1     2     3
     3                   60.0      1     3     4
     4                   60.0      1     4     5
     1    1     1
     3    1
     5    1
     2      0.016
     4      0.016
     5                   -4.0
```

Table 6.3c Input Data P.DAT of the Further Reduced Tangent Modulus for the Two-Story Column

```
Two-story column
     3    1
     5    3   100
     0    1     0
     0    4     0
     1       9.13        37.1          14.1    29000.0         36.0     1
     1                   90.0      1     1     2
     2                   90.0      1     2     3
     3                   60.0      1     3     4
     4                   60.0      1     4     5
     1    1     1
     3    1
     5    1
     5                   -4.0
```

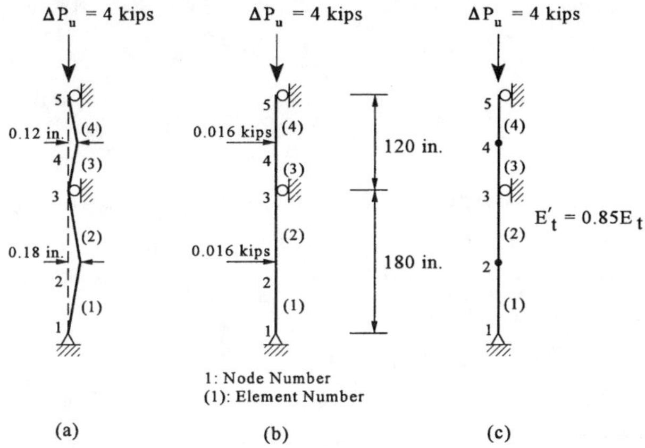

Figure 6.7 Modeling of the two-story column.

model is 209.6 kips. The further reduced tangent modulus model predicts 213.2 kips. Since the load-carrying capacities are slightly greater than the applied load, the column member size of W8x31 is adequate.

Design by the AISC-LRFD Method

Since the column member has an axial force without bending moments, interaction equation checks are not required. For the upper column, the slenderness parameter is

$$\lambda_{cy} = \frac{KL}{\pi r_y}\sqrt{\frac{F_y}{E}} = \frac{(1.294)(10)(12)}{(\pi)(2.02)}\sqrt{\frac{36}{29,000}} = 0.862$$

where $K = 1.294$ (exact value obtained in Section 3.8.1) is used instead of $K = 1.0$ (from the alignment chart) for a comparison purpose with the proposed method.

The column strength is:

$$\phi_c P_n = \phi_c A_g \left(0.658^{\lambda_c^2}\right) F_y = (0.85)(9.13)(0.658)^{(1.016)^2}(36) = 204.7 \text{ kips}$$

The lower column has the same strength as that of the upper column. Since the column strength is greater than the applied load, the column size of W8x31 is adequate.

Comparison of Results

The advanced analysis method and the LRFD method predict the same member size of W8x31. The LRFD method, however, has a difficulty estimating K for an unequal length two-story column because no specific procedures are provided. Since the alignment chart was developed

Table 6.4a Output P.OUT2 of the Explicit Imperfection Modeling for the Two-Story Column

```
============================
= LOAD-DEFLECTION BEHAVIOR =
============================


 -----------------
 - LOAD STEP   1 -
 -----------------
     LOAD SEQUENCE  1
     DETERMINANT OF STIFFNESS MATRIX = 0.370 TIMES 10 TO THE POWER  42
     TRACE OF STIFFNESS MATRIX       = 0.504 TIMES 10 TO THE POWER   6

          NODE      ITEM                 X              Y              R

             1   NODAL FORCE       0.0000D+00     0.0000D+00     0.0000D+00
                 DISPLACEMENT      0.0000D+00     0.0000D+00    -0.1706D-04

             2   NODAL FORCE       0.0000D+00     0.0000D+00     0.0000D+00
                 DISPLACEMENT      0.9233D-03    -0.1362D-02     0.3262D-05

             3   NODAL FORCE       0.0000D+00     0.0000D+00     0.0000D+00
                 DISPLACEMENT      0.0000D+00    -0.2723D-02     0.4015D-05

             4   NODAL FORCE       0.0000D+00     0.0000D+00     0.0000D+00
                 DISPLACEMENT      0.1421D-03    -0.3630D-02    -0.2175D-05

             5   NODAL FORCE       0.0000D+00    -0.4000D+01     0.0000D+00
                 DISPLACEMENT      0.0000D+00    -0.4537D-02     0.4684D-05

     ELEMENT FORCES IN GLOBAL COORDINATE
     -----------------------------------

       FRAME ELEMENT :                     ELEMENT FORCES
       ELEMENT NUMBER          Fx                Fy                Mz

               1  i     0.2641157D-02     0.4000000D+01     0.0000000D+00
                  j    -0.2641157D-02    -0.4000000D+01     0.4860000D+00
               2  i     0.2558843D-02     0.4000000D+01    -0.4860000D+00
                  j    -0.2558843D-02    -0.4000000D+01    -0.4680000D+00
               3  i    -0.3890408D-02     0.4000000D+01     0.4680000D+00
                  j     0.3890408D-02    -0.4000000D+01     0.2460000D+00
               4  i    -0.3909592D-02     0.4000000D+01    -0.2460000D+00
                  j     0.3909592D-02    -0.4000000D+01    -0.2775558D-16


 -----------------
 - LOAD STEP  55 -
 -----------------
     LOAD SEQUENCE  1
     DETERMINANT OF STIFFNESS MATRIX = 0.096 TIMES 10 TO THE POWER  36
```

Table 6.4a Output P.OUT2 of the Explicit Imperfection Modeling for the Two-Story Column (cont.)

```
TRACE OF STIFFNESS MATRIX        = 0.222 TIMES 10 TO THE POWER   6

     NODE       ITEM               X             Y             R

       1    NODAL FORCE       0.0000D+00    0.0000D+00    0.0000D+00
            DISPLACEMENT      0.0000D+00    0.0000D+00   -0.2490D-02

       2    NODAL FORCE       0.0000D+00    0.0000D+00    0.0000D+00
            DISPLACEMENT      0.1450D+00   -0.7380D-01    0.3253D-03

       3    NODAL FORCE       0.0000D+00    0.0000D+00    0.0000D+00
            DISPLACEMENT      0.0000D+00   -0.1476D+00    0.1080D-02

       4    NODAL FORCE       0.0000D+00    0.0000D+00    0.0000D+00
            DISPLACEMENT     -0.1216D-01   -0.1965D+00   -0.2038D-03

       5    NODAL FORCE       0.0000D+00   -0.2096D+03    0.0000D+00
            DISPLACEMENT      0.0000D+00   -0.2454D+00   -0.1861D-03

ELEMENT FORCES IN GLOBAL COORDINATE
-----------------------------------

  FRAME ELEMENT :                    ELEMENT FORCES
  ELEMENT NUMBER          Fx               Fy                Mz

        1   i    0.2170939D+00    0.2096315D+03   -0.1904032D-13
            j   -0.2170939D+00   -0.2096315D+03    0.4865267D+02
        2   i    0.2056631D+00    0.2096310D+03   -0.4865267D+02
            j   -0.2056631D+00   -0.2096310D+03   -0.3804781D+02
        3   i   -0.3173482D+00    0.2096322D+03    0.3804781D+02
            j    0.3173482D+00   -0.2096322D+03    0.3618901D+01
        4   i   -0.3167815D+00    0.2096320D+03   -0.3618901D+01
            j    0.3167815D+00   -0.2096320D+03   -0.1049508D-14

> DETERMINANT OF THE TANGENT STIFFNESS MATRIX IS NEGATIVE AT LOAD
> STEP  56 WITH THE PRESENT LOAD INCREMENT.  THUS, LOAD INCREMENT
> IS REDUCED TO 1/5 OF PREVIOUS VALUE FOR THIS LOAD STEP.

> DETERMINANT OF THE TANGENT STIFFNESS MATRIX IS NEGATIVE AT LOAD
> STEP  56 WITH THE PRESENT LOAD INCREMENT.  THUS, LOAD INCREMENT
> IS REDUCED TO 1/5 OF PREVIOUS VALUE FOR THIS LOAD STEP.

>> SOLUTION DOES NOT CONVERGE, A LIMIT STATE MAY HAVE BEEN REACHED <<
```

Table 6.4b Output P.OUT2 of the Equivalent Notional Load Modeling for the Two-Story Column

```
============================
= LOAD-DEFLECTION BEHAVIOR =
============================


    -----------------
    - LOAD STEP   1 -
    -----------------
        LOAD SEQUENCE  1
        DETERMINANT OF STIFFNESS MATRIX = 0.370 TIMES 10 TO THE POWER  42
        TRACE OF STIFFNESS MATRIX       = 0.504 TIMES 10 TO THE POWER   6

             NODE      ITEM                 X              Y              R

               1    NODAL FORCE      0.0000D+00     0.0000D+00     0.0000D+00
                    DISPLACEMENT     0.0000D+00     0.0000D+00    -0.1706D-04

               2    NODAL FORCE      0.1600D-01     0.0000D+00     0.0000D+00
                    DISPLACEMENT     0.9260D-03    -0.1360D-02     0.3262D-05

               3    NODAL FORCE      0.0000D+00     0.0000D+00     0.0000D+00
                    DISPLACEMENT     0.0000D+00    -0.2719D-02     0.4015D-05

               4    NODAL FORCE      0.1600D-01     0.0000D+00     0.0000D+00
                    DISPLACEMENT     0.1439D-03    -0.3626D-02    -0.2175D-05

               5    NODAL FORCE      0.0000D+00    -0.4000D+01     0.0000D+00
                    DISPLACEMENT     0.0000D+00    -0.4532D-02     0.4684D-05

        ELEMENT FORCES IN GLOBAL COORDINATE
        -----------------------------------

          FRAME ELEMENT :                  ELEMENT FORCES
          ELEMENT NUMBER        Fx                Fy               Mz

                 1   i   -0.5358843D-02   0.4000000D+01   0.5551115D-16
                     j    0.5358843D-02  -0.4000000D+01   0.4860000D+00
                 2   i    0.1055884D-01   0.4000000D+01  -0.4860000D+00
                     j   -0.1055884D-01  -0.4000000D+01  -0.4680000D+00
                 3   i   -0.1189041D-01   0.4000000D+01   0.4680000D+00
                     j    0.1189041D-01  -0.4000000D+01   0.2460000D+00
                 4   i    0.4090408D-02   0.4000000D+01  -0.2460000D+00
                     j   -0.4090408D-02  -0.4000000D+01   0.5551115D-16


    -----------------
    - LOAD STEP  55 -
    -----------------
        LOAD SEQUENCE  1
        DETERMINANT OF STIFFNESS MATRIX = 0.088 TIMES 10 TO THE POWER  36
```

Table 6.4b Output P.OUT2 of the Equivalent Notional Load Modeling for the Two-Story Column (cont.)

```
TRACE OF STIFFNESS MATRIX          = 0.222 TIMES 10 TO THE POWER   6

   NODE       ITEM                X             Y              R

     1    NODAL FORCE       0.0000D+00     0.0000D+00     0.0000D+00
          DISPLACEMENT      0.0000D+00     0.0000D+00    -0.2490D-02

     2    NODAL FORCE       0.8385D+00     0.0000D+00     0.0000D+00
          DISPLACEMENT      0.1452D+00    -0.7351D-01     0.3253D-03

     3    NODAL FORCE       0.0000D+00     0.0000D+00     0.0000D+00
          DISPLACEMENT      0.0000D+00    -0.1470D+00     0.1080D-02

     4    NODAL FORCE       0.8385D+00     0.0000D+00     0.0000D+00
          DISPLACEMENT     -0.1206D-01    -0.1960D+00    -0.2038D-03

     5    NODAL FORCE       0.0000D+00    -0.2096D+03     0.0000D+00
          DISPLACEMENT      0.0000D+00    -0.2449D+00    -0.1862D-03

ELEMENT FORCES IN GLOBAL COORDINATE
-----------------------------------

  FRAME ELEMENT :                  ELEMENT FORCES
  ELEMENT NUMBER        Fx               Fy               Mz

        1    i   -0.2021600D+00   0.2096315D+03   -0.9270362D-14
             j    0.2021600D+00  -0.2096315D+03    0.4865594D+02
        2    i    0.6249339D+00   0.2096310D+03   -0.4865594D+02
             j   -0.6249339D+00  -0.2096310D+03   -0.3804954D+02
        3    i   -0.7366271D+00   0.2096322D+03    0.3804954D+02
             j    0.7366271D+00  -0.2096322D+03    0.3617083D+01
        4    i    0.1024681D+00   0.2096320D+03   -0.3617083D+01
             j   -0.1024681D+00  -0.2096320D+03    0.3186687D-14

> DETERMINANT OF THE TANGENT STIFFNESS MATRIX IS NEGATIVE AT LOAD
> STEP  56 WITH THE PRESENT LOAD INCREMENT.  THUS, LOAD INCREMENT
> IS REDUCED TO 1/5 OF PREVIOUS VALUE FOR THIS LOAD STEP.


> DETERMINANT OF THE TANGENT STIFFNESS MATRIX IS NEGATIVE AT LOAD
> STEP  56 WITH THE PRESENT LOAD INCREMENT.  THUS, LOAD INCREMENT
> IS REDUCED TO 1/5 OF PREVIOUS VALUE FOR THIS LOAD STEP.

>> SOLUTION DOES NOT CONVERGE, A LIMIT STATE MAY HAVE BEEN REACHED <<
```

Table 6.4c Output P.OUT2 of the Further Reduced Tangent Modulus Modeling for the Two-Story Column

```
============================
= LOAD-DEFLECTION BEHAVIOR =
============================


    -----------------
    - LOAD STEP   1 -
    -----------------
        LOAD SEQUENCE  1
        DETERMINANT OF STIFFNESS MATRIX = 0.617 TIMES 10 TO THE POWER  41
        TRACE OF STIFFNESS MATRIX       = 0.429 TIMES 10 TO THE POWER   6

             NODE      ITEM               X              Y              R

                1   NODAL FORCE      0.0000D+00     0.0000D+00     0.0000D+00
                    DISPLACEMENT     0.0000D+00     0.0000D+00    -0.1088D-20

                2   NODAL FORCE      0.0000D+00     0.0000D+00     0.0000D+00
                    DISPLACEMENT    -0.8378D-32    -0.1600D-02    -0.1088D-20

                3   NODAL FORCE      0.0000D+00     0.0000D+00     0.0000D+00
                    DISPLACEMENT     0.0000D+00    -0.3199D-02    -0.1088D-20

                4   NODAL FORCE      0.0000D+00     0.0000D+00     0.0000D+00
                    DISPLACEMENT     0.3545D-32    -0.4266D-02    -0.1088D-20

                5   NODAL FORCE      0.0000D+00    -0.4000D+01     0.0000D+00
                    DISPLACEMENT     0.0000D+00    -0.5332D-02    -0.1088D-20

        ELEMENT FORCES IN GLOBAL COORDINATE
        -----------------------------------

          FRAME ELEMENT :                    ELEMENT FORCES
          ELEMENT NUMBER         Fx                Fy               Mz

                 1   i   -0.6432491D-15    0.4000000D+01   -0.3822248D-32
                     j    0.6432491D-15   -0.4000000D+01   -0.3115132D-29
                 2   i    0.2449294D-15    0.4000000D+01    0.3096021D-29
                     j   -0.2449294D-15   -0.4000000D+01    0.1054940D-29
                 3   i    0.2449294D-15    0.4000000D+01   -0.1037740D-29
                     j   -0.2449294D-15   -0.4000000D+01    0.2465350D-29
                 4   i   -0.6432491D-15    0.4000000D+01   -0.2459617D-29
                     j    0.6432491D-15   -0.4000000D+01   -0.3440023D-31

    -----------------
    - LOAD STEP  60 -
    -----------------
        LOAD SEQUENCE  1
        DETERMINANT OF STIFFNESS MATRIX = 0.084 TIMES 10 TO THE POWER  35
        TRACE OF STIFFNESS MATRIX       = 0.199 TIMES 10 TO THE POWER   6
```

Table 6.4c Output P.OUT2 of the Further Reduced Tangent Modulus Modeling for the Two-Story Column (cont.)

```
NODE      ITEM                  X              Y              R

  1     NODAL FORCE      0.0000D+00     0.0000D+00     0.0000D+00
        DISPLACEMENT     0.0000D+00     0.0000D+00     0.2928D-15

  2     NODAL FORCE      0.0000D+00     0.0000D+00     0.0000D+00
        DISPLACEMENT    -0.1709D-13    -0.8828D-01    -0.2067D-16

  3     NODAL FORCE      0.0000D+00     0.0000D+00     0.0000D+00
        DISPLACEMENT     0.0000D+00    -0.1766D+00    -0.2070D-15

  4     NODAL FORCE      0.0000D+00     0.0000D+00     0.0000D+00
        DISPLACEMENT     0.6558D-14    -0.2354D+00     0.1189D-16

  5     NODAL FORCE      0.0000D+00    -0.2132D+03     0.0000D+00
        DISPLACEMENT     0.0000D+00    -0.2943D+00     0.1547D-15

ELEMENT FORCES IN GLOBAL COORDINATE
-----------------------------------

  FRAME ELEMENT :                 ELEMENT FORCES
  ELEMENT NUMBER        Fx               Fy               Mz

        1  i   -0.3502340D-13   0.2132480D+03  -0.1154896D-26
           j    0.3502340D-13  -0.2132480D+03  -0.4195805D-11
        2  i   -0.4244499D-14   0.2132480D+03   0.4195805D-11
           j    0.4244499D-14  -0.2132480D+03   0.1622941D-11
        3  i   -0.1453399D-13   0.2132480D+03  -0.1622941D-11
           j    0.1453399D-13  -0.2132480D+03   0.3278441D-11
        4  i    0.2034779D-13   0.2132480D+03  -0.3278441D-11
           j   -0.2034779D-13  -0.2132480D+03  -0.8267606D-27

> DETERMINANT OF THE TANGENT STIFFNESS MATRIX IS NEGATIVE AT LOAD
> STEP  61 WITH THE PRESENT LOAD INCREMENT.  THUS, LOAD INCREMENT
> IS REDUCED TO 1/5 OF PREVIOUS VALUE FOR THIS LOAD STEP.

> DETERMINANT OF THE TANGENT STIFFNESS MATRIX IS NEGATIVE AT LOAD
> STEP  61 WITH THE PRESENT LOAD INCREMENT.  THUS, LOAD INCREMENT
> IS REDUCED TO 1/5 OF PREVIOUS VALUE FOR THIS LOAD STEP.

>> SOLUTION DOES NOT CONVERGE, A LIMIT STATE MAY HAVE BEEN REACHED <<
```

under the assumptions that column lengths are equal, the chart may not be used for the unequal column case (Section 3.8.1). When the strength of this column is calculated by AISC-LRFD procedures, formal differential equation or slope-deflection equation methods should be used to determine K, but this is not practical. The advanced analysis method eliminates this difficulty because the calculation of K is unnecessary.

6.3 Truss Structures

6.3.1 Roof Truss

Figure 6.8 shows a hinged-jointed roof truss subject to gravity loads of 20 kips at the joints. A36 steel pipe is used. All member sizes are assumed identical.

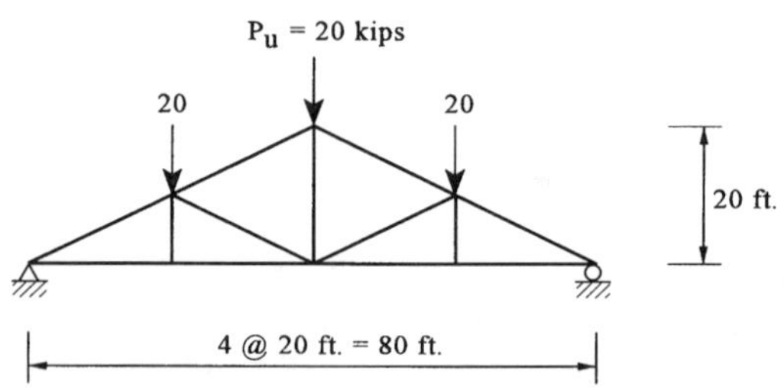

Figure 6.8 Configuration and load condition of the hinged-jointed roof truss.

Design by the Advanced Analysis

Step 1: Load Condition and Preliminary Member Sizing

The critical factored load condition is shown in Figure 6.8. The member forces of the truss may be obtained (Figure 6.9) by using equilibrium conditions. The maximum compressive force is 67.1 kips. The effective length is the same as the actual length (22.4 ft) because K is 1.0. The preliminary member size of steel pipe is a 6 in. diameter with 0.28 in. thickness ($\phi P_n = 81$ kips) by using the Column Design Table in the LRFD Specification.

Step 2: Structural Modeling

Each member is modeled with one truss element without geometric imperfection because the program computes the axial strength of the truss member with the LRFD column strength equations which indirectly account for geometric imperfections. An incremental load of 0.5 kips is determined by dividing the factored load of 20 kips by a scaling factor of 40 as shown in Figure 6.10.

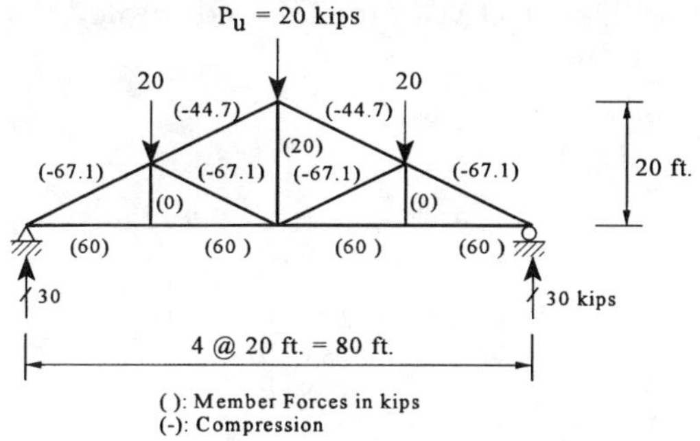

Figure 6.9 Member forces of the hinged-jointed roof truss.

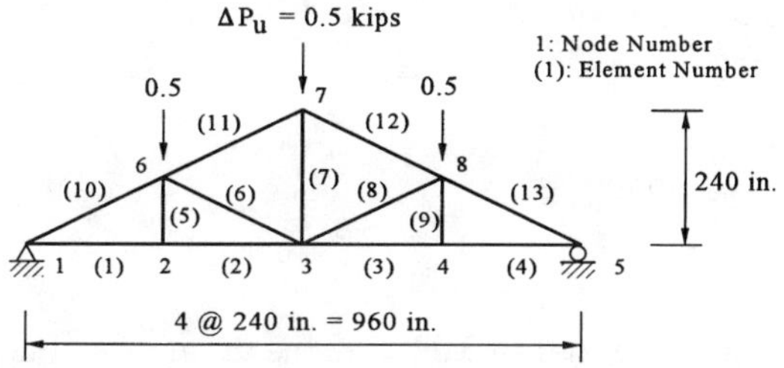

Figure 6.10 Modeling of the hinged-jointed roof truss.

Step 3: Analysis

Referring to the input instructions described in Section 5.3.2, the input data may be easily generated as listed in Table 6.5. Note that the total number of supports (NBOUND) in the hinged-jointed truss must be equal to the total number of nodal points because the nodes of a truss element are restrained against rotation. Program DATAGEN, INPUT, and PAAP are executed in sequence by entering the batch file command RUN on the screen.

Step 4: Check of Load-Carrying Capacity

The incremental displacements and member forces are listed in P.OUT2 (Table 6.6). Truss Elements (10) and (13) fail at Load Step 48, with loads at Nodes 6, 7, and 8 being 24 kips. Since this truss is statically determinant, failure of one member leads to failure of the whole system. Load Step 49 shows a sharp increase in displacement and indicates a system failure. The member force of Element (10) is 80.4 kips (F_x = 72.0 kip, F_y = 35.7 kips). Since the load-carrying capacity of 24 kips at Node 6, 7, and 8 is greater than the applied load of 20 kips, the member size is adequate.

Step 5: Check of Serviceability

Referring to P.OUT2 (Table 6.6), the deflection at Node 3 correspond-

Table 6.5 Input Data P.DAT of the Hinged-Jointed Roof Truss

```
Roof truss
    0     1
    8     8   100
    0     0     1
    0     0    13
    1        5.58        28.1    29000.0          36.0
    1       240.0         0.0      1    1     2      4
    5         0.0       120.0      1    2     6
    6      -240.0       120.0      1    3     6
    7         0.0       240.0      1    3     7
    8       240.0       120.0      1    3     8
    9         0.0       120.0      1    4     8
   10       240.0       120.0      1    1     6
   11       240.0       120.0      1    6     7
   12       240.0      -120.0      1    7     8
   13       240.0      -120.0      1    8     5
    1     1     1    1
    2                1    3
    5           1    1
    6                1    3
    6                  -0.5                   3
```

ing to Load Step 1 is equal to 0.02 in. This deflection may be considered elastic because the behavior of the beam is linear and elastic under small loads. The total deflection of 0.8 in. is obtained by multiplying the deflection of 0.02 in. by the scaling factor of 40. The deflection ratio over the span length is 1/1200 which meets the limitation 1/360. The deflection at the service load will be smaller than that above because the factored load is used for the calculation of deflection above.

Design by the AISC-LRFD Method

Since the truss member has the axial force but not bending moment, the interaction equation checks are not required. The slenderness parameter of Element (10) is

$$\lambda_c = \frac{KL}{\pi r}\sqrt{\frac{F_y}{E}} = \frac{(22.4)(12)}{(\pi)(2.25)}\sqrt{\frac{36}{29,000}} = 1.34$$

The axial strength is

$$\phi_c P_n = \phi_c A_g \left(0.658^{\lambda_c^2}\right) F_y = (0.85)(5.58)\left(0.658^{(1.34)^2}\right)(36) = 80.5$$

Comparing this axial strength of 80.5 kips with the member force of 67.1 kips, the member size appears adequate.

Comparison of Results

The advanced analysis and LRFD method predict the same member size of steel pipe with 6 in. diameter and 0.28 in. thickness. The load-

Table 6.6 Output P.OUT2 of the Hinged-Jointed Roof Truss

```
=============================
= LOAD-DEFLECTION BEHAVIOR =
=============================


  -----------------
  - LOAD STEP   1 -
  -----------------
      LOAD SEQUENCE  1
      DETERMINANT OF STIFFNESS MATRIX = 0.320 TIMES 10 TO THE POWER  37
      TRACE OF STIFFNESS MATRIX       = 0.180 TIMES 10 TO THE POWER   5

           NODE      ITEM              X             Y              R

              1   NODAL FORCE     0.0000D+00     0.0000D+00     0.0000D+00
                  DISPLACEMENT    0.0000D+00     0.0000D+00     0.0000D+00

              2   NODAL FORCE     0.0000D+00     0.0000D+00     0.0000D+00
                  DISPLACEMENT    0.2225D-02    -0.1860D-01     0.0000D+00

              3   NODAL FORCE     0.0000D+00     0.0000D+00     0.0000D+00
                  DISPLACEMENT    0.4449D-02    -0.2000D-01     0.0000D+00

              4   NODAL FORCE     0.0000D+00     0.0000D+00     0.0000D+00
                  DISPLACEMENT    0.6674D-02    -0.1860D-01     0.0000D+00

              5   NODAL FORCE     0.0000D+00     0.0000D+00     0.0000D+00
                  DISPLACEMENT    0.8899D-02     0.0000D+00     0.0000D+00

              6   NODAL FORCE     0.0000D+00    -0.5000D+00     0.0000D+00
                  DISPLACEMENT    0.6189D-02    -0.1860D-01     0.0000D+00

              7   NODAL FORCE     0.0000D+00    -0.5000D+00     0.0000D+00
                  DISPLACEMENT    0.4449D-02    -0.1926D-01     0.0000D+00

              8   NODAL FORCE     0.0000D+00    -0.5000D+00     0.0000D+00
                  DISPLACEMENT    0.2709D-02    -0.1860D-01     0.0000D+00

      ELEMENT FORCES IN GLOBAL COORDINATE
      -----------------------------------


        TRUSS ELEMENT :               ELEMENT FORCES
        ELEMENT NO.      Fxi           Fyi           Fxj            Fyj

               1     -0.150D+01     0.116D-03     0.150D+01    -0.116D-03
               2     -0.150D+01     0.879D-05     0.150D+01    -0.879D-05
               3     -0.150D+01    -0.879D-05     0.150D+01     0.879D-05
               4     -0.150D+01    -0.116D-03     0.150D+01     0.116D-03
               5      0.000D+00     0.000D+00     0.000D+00     0.000D+00
               6     -0.500D+00     0.250D+00     0.500D+00    -0.250D+00
               7      0.804D-16    -0.500D+00    -0.804D-16     0.500D+00
               8      0.500D+00     0.250D+00    -0.500D+00    -0.250D+00
               9      0.000D+00     0.000D+00     0.000D+00     0.000D+00
              10      0.150D+01     0.750D+00    -0.150D+01    -0.750D+00
              11      0.100D+01     0.500D+00    -0.100D+01    -0.500D+00
              12      0.100D+01    -0.500D+00    -0.100D+01     0.500D+00
              13      0.150D+01    -0.750D+00    -0.150D+01     0.750D+00
```

Table 6.6 Output P.OUT2 of the Hinged-Jointed Roof Truss (cont.)

```
-----------------
- LOAD STEP  48 -
-----------------
    LOAD SEQUENCE  1
    DETERMINANT OF STIFFNESS MATRIX = 0.045 TIMES 10 TO THE POWER  32
    TRACE OF STIFFNESS MATRIX       = 0.163 TIMES 10 TO THE POWER   5

         NODE      ITEM                X              Y              R

           1    NODAL FORCE       0.0000D+00     0.0000D+00     0.0000D+00
                DISPLACEMENT      0.0000D+00     0.0000D+00     0.0000D+00

           2    NODAL FORCE       0.0000D+00     0.0000D+00     0.0000D+00
                DISPLACEMENT      0.1053D+00    -0.8932D+00     0.0000D+00

           3    NODAL FORCE       0.0000D+00     0.0000D+00     0.0000D+00
                DISPLACEMENT      0.2122D+00    -0.9610D+00     0.0000D+00

           4    NODAL FORCE       0.0000D+00     0.0000D+00     0.0000D+00
                DISPLACEMENT      0.3191D+00    -0.8932D+00     0.0000D+00

           5    NODAL FORCE       0.0000D+00     0.0000D+00     0.0000D+00
                DISPLACEMENT      0.4244D+00     0.0000D+00     0.0000D+00

           6    NODAL FORCE       0.0000D+00    -0.2400D+02     0.0000D+00
                DISPLACEMENT      0.2958D+00    -0.8935D+00     0.0000D+00

           7    NODAL FORCE       0.0000D+00    -0.2400D+02     0.0000D+00
                DISPLACEMENT      0.2122D+00    -0.9254D+00     0.0000D+00

           8    NODAL FORCE       0.0000D+00    -0.2400D+02     0.0000D+00
                DISPLACEMENT      0.1286D+00    -0.8935D+00     0.0000D+00

 * TRUSS ELEMENT 10 HAS FAILED BY BUCKLING *

 * TRUSS ELEMENT 13 HAS FAILED BY BUCKLING *

    ELEMENT FORCES IN GLOBAL COORDINATE
    -----------------------------------

      TRUSS ELEMENT :                  ELEMENT FORCES
      ELEMENT NO.      Fxi            Fyi             Fxj             Fyj

            1      -0.721D+02     0.268D+00      0.721D+02     -0.268D+00
            2      -0.721D+02     0.204D-01      0.721D+02     -0.204D-01
            3      -0.721D+02    -0.204D-01      0.721D+02      0.204D-01
            4      -0.721D+02    -0.268D+00      0.721D+02      0.268D+00
            5       0.385D-03     0.243D+00     -0.385D-03     -0.243D+00
            6      -0.241D+02     0.120D+02      0.241D+02     -0.120D+02
            7       0.387D-14    -0.240D+02     -0.387D-14      0.240D+02
            8       0.241D+02     0.120D+02     -0.241D+02     -0.120D+02
            9      -0.385D-03     0.243D+00      0.385D-03     -0.243D+00
           10       0.720D+02     0.357D+02     -0.720D+02     -0.357D+02
```

Table 6.6 Output P.OUT2 of the Hinged-Jointed Roof Truss (cont.)

11	0.480D+02	0.240D+02	-0.480D+02	-0.240D+02
12	0.480D+02	-0.240D+02	-0.480D+02	0.240D+02
13	0.720D+02	-0.357D+02	-0.720D+02	0.357D+02

```
-----------------
- LOAD STEP  49 -
-----------------
    LOAD SEQUENCE  1
    DETERMINANT OF STIFFNESS MATRIX = 0.045 TIMES 10 TO THE POWER  32
    TRACE OF STIFFNESS MATRIX       = 0.163 TIMES 10 TO THE POWER   5
```

NODE	ITEM	X	Y	R
1	NODAL FORCE	0.0000D+00	0.0000D+00	0.0000D+00
	DISPLACEMENT	0.0000D+00	0.0000D+00	0.0000D+00
2	NODAL FORCE	0.0000D+00	0.0000D+00	0.0000D+00
	DISPLACEMENT	0.9600D-01	-0.3390D+01	0.0000D+00
3	NODAL FORCE	0.0000D+00	0.0000D+00	0.0000D+00
	DISPLACEMENT	0.2029D+00	-0.3459D+01	0.0000D+00
4	NODAL FORCE	0.0000D+00	0.0000D+00	0.0000D+00
	DISPLACEMENT	0.3098D+00	-0.3390D+01	0.0000D+00
5	NODAL FORCE	0.0000D+00	0.0000D+00	0.0000D+00
	DISPLACEMENT	0.4058D+00	0.0000D+00	0.0000D+00
6	NODAL FORCE	0.0000D+00	-0.2450D+02	0.0000D+00
	DISPLACEMENT	0.2865D+00	-0.3391D+01	0.0000D+00
7	NODAL FORCE	0.0000D+00	-0.2450D+02	0.0000D+00
	DISPLACEMENT	0.2029D+00	-0.3424D+01	0.0000D+00
8	NODAL FORCE	0.0000D+00	-0.2450D+02	0.0000D+00
	DISPLACEMENT	0.1194D+00	-0.3391D+01	0.0000D+00

ELEMENT FORCES IN GLOBAL COORDINATE

TRUSS ELEMENT : ELEMENT FORCES

ELEMENT NO.	Fxi	Fyi	Fxj	Fyj
1	-0.721D+02	0.102D+01	0.721D+02	-0.102D+01
2	-0.721D+02	0.208D-01	0.721D+02	-0.208D-01
3	-0.721D+02	-0.208D-01	0.721D+02	0.208D-01
4	-0.721D+02	-0.102D+01	0.721D+02	0.102D+01
5	0.157D-02	0.992D+00	-0.157D-02	-0.992D+00
6	-0.238D+02	0.119D+02	0.238D+02	-0.119D+02
7	0.566D-13	-0.238D+02	-0.566D-13	0.238D+02
8	0.238D+02	0.119D+02	-0.238D+02	-0.119D+02
9	-0.157D-02	0.992D+00	0.157D-02	-0.992D+00
10	0.723D+02	0.351D+02	-0.723D+02	-0.351D+02
11	0.483D+02	0.241D+02	-0.483D+02	-0.241D+02
12	0.483D+02	-0.241D+02	-0.483D+02	0.241D+02
13	0.723D+02	-0.351D+02	-0.723D+02	0.351D+02

carrying capacities of Element (10) predicted by these two methods are the same as 80.5 kips. This is because the truss system is statically determinant, rendering inelastic moment redistribution of little or no benefit.

6.3.2 *Pratt Truss*

Figure 6.11 shows a Pratt truss used for medium spans. The Pratt truss has an advantage in that the diagonals except end posts are all in tension for uniformly distributed loads. A square structural tube will be used for the upper chords and vertical members, and a W-section will be used for the lower chords and diagonals. A46 steel is used for compression members and A36 steel for tension members.

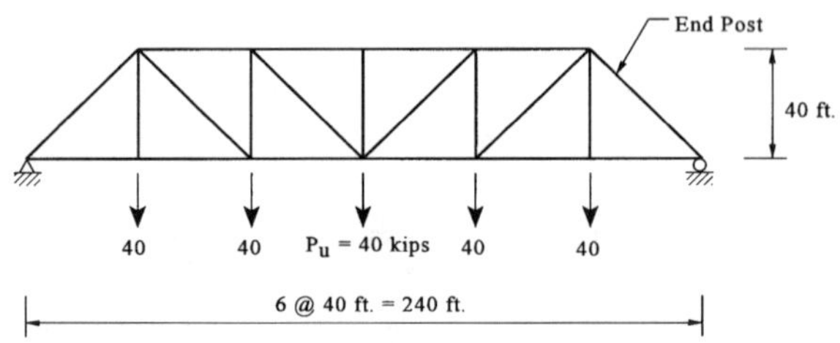

Figure 6.11 Configuration and load condition of the hinged-jointed pratt truss.

Design by the Advanced Analysis

Step 1: Load Condition and Preliminary Member Sizing

The critical factored load condition is shown in Figure 6.11. The member forces of the truss may be obtained in Figure 6.12 by enforcing equilibrium conditions. The end post is the most critical compression member since the effective length KL is largest at 56.6 ft ($= 40$ ft $\times \sqrt{2} \times 1.0$). The preliminary size of the end post is a structural tube 10x10x5/8 by using column strength equations in the AISC-LRFD Specification.

The preliminary size for the lower chords is governed by the maximum slenderness ratio of 300 specified in the AISC-LRFD Specification. The slenderness limitation prevents undesirable vibrations or deflections due to wind loads. The required radius of gyration for diagonals is calculated as 2.26 in. ($r_y \leq L/300 = 40$ ft $\times 12 \times \sqrt{2}/300$). The preliminary size for the tension members is W10x49 ($r_y = 2.54$ in.).

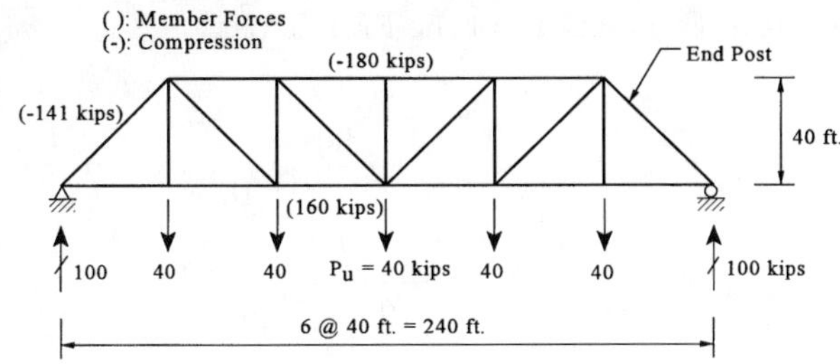

Figure 6.12 Member forces of the hinged-jointed pratt truss.

Step 2: Structural Modeling

Each member is modeled with one truss element without geometric imperfection as described in Section 6.3.1. An incremental load of 1 kip is determined by dividing the factored load of 40 kips and by a scaling factor of 40 as shown in Figure 6.13.

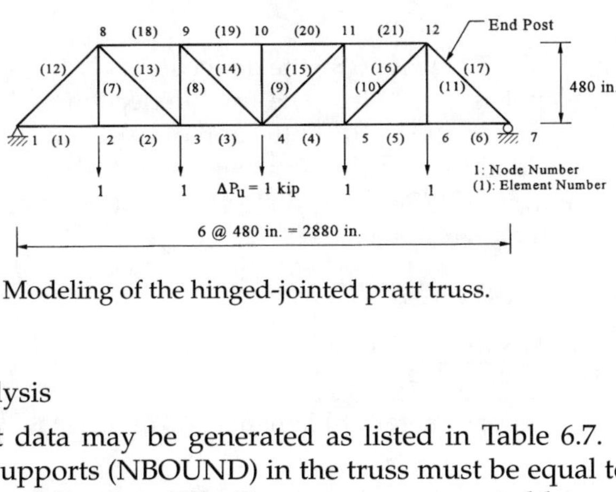

Figure 6.13 Modeling of the hinged-jointed pratt truss.

Step 3: Analysis

The input data may be generated as listed in Table 6.7. The total number of supports (NBOUND) in the truss must be equal to the total number of nodal points (12). Programs are executed by entering the batch file command RUN.

Step 4: Check of Load-Carrying Capacity

The incremental displacement and member forces are listed in P.OUT2 (Table 6.8). Truss Elements (12) and (17) fail at Load Step 43, with loads at Nodes 2 through 6 being 43 kips. Since the Pratt truss is determinant, failure of one member leads to failure of the structural system. Load Step 44 increases displacement abruptly and it indicates a system failure. The member force of Element (12) and (17) is 148 kips ($F_x = F_y = 105$ kips). Since the load-carrying capacity of 43 kips is greater than the applied load of 40 kips, the member size is adequate.

Table 6.7 Input Data P.DAT of the Pratt Truss

```
Pratt Truss
    0    1
   12   12  100
    0    0    2
    0    0   21
    1    22.40     321.0  29000.0      46.0
    2    14.40      93.4  29000.0      36.0
    1    480.0       0.0    2    1    2    6
    7      0.0     480.0    1    2    8    5
   12    480.0     480.0    1    1    8
   13   -480.0     480.0    2    3    8    2
   15    480.0     480.0    2    4   11    2
   17   -480.0     480.0    1    7   12
   18    480.0       0.0    1    8    9    4
    1    1    1    1
    2                1    5
    7            1   1
    8                1    5
    2                 -1.0               5
```

Step 5: Check of Serviceability

Referring to P.OUT2 (Table 6.8), the deflection at Node 4 corresponding to Load Step 1 is 0.0429 in. The total deflection of 1.716 in. can be computed by multiplying the deflection of 0.0429 in. by the scaling factor of 40. The deflection ratio over the span length is 1/1678, which meets the limitation of 1/360. The deflection at the service load will be smaller than that above.

Design by the AISC-LRFD Method

Member Capacity Check for End Post
(Tube 10x10x5/8: $A = 22.4$, $r = 3.78$)

The slenderness parameter of Elements (12) and (17) is

$$\lambda_c = \frac{KL}{\pi r}\sqrt{\frac{F_y}{E}} = \frac{(1.0)(56.6)(12)}{(\pi)(3.78)}\sqrt{\frac{46}{29,000}} = 2.28 > 1.5$$

The axial strength:

$$\phi_c P_n = \phi_c A_g \left(\frac{0.877}{\lambda_c^2}\right) F_y = (0.85)(22.4)\left(\frac{0.877}{2.28^2}\right)(46) = 148 \text{ kips}$$

Comparing this axial strength of 148 kips with the member force of 141 kips, the member size appears adequate.

Table 6.8 Output P.OUT2 of the Pratt Truss

```
============================
= LOAD-DEFLECTION BEHAVIOR =
============================


  -----------------
  - LOAD STEP   1 -
  -----------------
      LOAD SEQUENCE  1
      DETERMINANT OF STIFFNESS MATRIX = 0.487 TIMES 10 TO THE POWER  63
      TRACE OF STIFFNESS MATRIX       = 0.412 TIMES 10 TO THE POWER   5

           NODE      ITEM               X              Y              R

             1    NODAL FORCE      0.0000D+00     0.0000D+00     0.0000D+00
                  DISPLACEMENT     0.0000D+00     0.0000D+00     0.0000D+00

             2    NODAL FORCE      0.0000D+00    -0.1000D+01     0.0000D+00
                  DISPLACEMENT     0.2874D-02    -0.2259D-01     0.0000D+00

             3    NODAL FORCE      0.0000D+00    -0.1000D+01     0.0000D+00
                  DISPLACEMENT     0.5747D-02    -0.3761D-01     0.0000D+00

             4    NODAL FORCE      0.0000D+00    -0.1000D+01     0.0000D+00
                  DISPLACEMENT     0.1034D-01    -0.4293D-01     0.0000D+00

             5    NODAL FORCE      0.0000D+00    -0.1000D+01     0.0000D+00
                  DISPLACEMENT     0.1494D-01    -0.3761D-01     0.0000D+00

             6    NODAL FORCE      0.0000D+00    -0.1000D+01     0.0000D+00
                  DISPLACEMENT     0.1782D-01    -0.2259D-01     0.0000D+00

             7    NODAL FORCE      0.0000D+00     0.0000D+00     0.0000D+00
                  DISPLACEMENT     0.2069D-01     0.0000D+00     0.0000D+00

             8    NODAL FORCE      0.0000D+00     0.0000D+00     0.0000D+00
                  DISPLACEMENT     0.1663D-01    -0.2185D-01     0.0000D+00

             9    NODAL FORCE      0.0000D+00     0.0000D+00     0.0000D+00
                  DISPLACEMENT     0.1367D-01    -0.3798D-01     0.0000D+00

            10    NODAL FORCE      0.0000D+00     0.0000D+00     0.0000D+00
                  DISPLACEMENT     0.1034D-01    -0.4293D-01     0.0000D+00

            11    NODAL FORCE      0.0000D+00     0.0000D+00     0.0000D+00
                  DISPLACEMENT     0.7020D-02    -0.3798D-01     0.0000D+00

            12    NODAL FORCE      0.0000D+00     0.0000D+00     0.0000D+00
                  DISPLACEMENT     0.4064D-02    -0.2185D-01     0.0000D+00

      ELEMENT FORCES IN GLOBAL COORDINATE
      -----------------------------------


        TRUSS ELEMENT :               ELEMENT FORCES
        ELEMENT NO.     Fxi           Fyi           Fxj           Fyj
```

Table 6.8 Output P.OUT2 of the Pratt Truss (cont.)

1	-0.250D+01	0.118D-03	0.250D+01	-0.118D-03
2	-0.250D+01	0.782D-04	0.250D+01	-0.782D-04
3	-0.400D+01	0.443D-04	0.400D+01	-0.443D-04
4	-0.400D+01	-0.443D-04	0.400D+01	0.443D-04
5	-0.250D+01	-0.782D-04	0.250D+01	0.782D-04
6	-0.250D+01	-0.118D-03	0.250D+01	0.118D-03
7	-0.287D-04	-0.100D+01	0.287D-04	0.100D+01
8	0.825D-05	0.500D+00	-0.825D-05	-0.500D+00
9	0.000D+00	0.000D+00	0.000D+00	0.000D+00
10	-0.825D-05	0.500D+00	0.825D-05	-0.500D+00
11	0.287D-04	-0.100D+01	-0.287D-04	0.100D+01
12	0.250D+01	0.250D+01	-0.250D+01	-0.250D+01
13	0.150D+01	-0.150D+01	-0.150D+01	0.150D+01
14	0.500D+00	-0.500D+00	-0.500D+00	0.500D+00
15	-0.500D+00	-0.500D+00	0.500D+00	0.500D+00
16	-0.150D+01	-0.150D+01	0.150D+01	0.150D+01
17	-0.250D+01	0.250D+01	0.250D+01	-0.250D+01
18	0.400D+01	-0.134D-03	-0.400D+01	0.134D-03
19	0.450D+01	-0.464D-04	-0.450D+01	0.464D-04
20	0.450D+01	0.464D-04	-0.450D+01	-0.464D-04
21	0.400D+01	0.134D-03	-0.400D+01	-0.134D-03

```
-----------------
- LOAD STEP  43 -
-----------------
    LOAD SEQUENCE  1
    DETERMINANT OF STIFFNESS MATRIX = 0.059 TIMES 10 TO THE POWER  57
    TRACE OF STIFFNESS MATRIX       = 0.389 TIMES 10 TO THE POWER   5
```

NODE	ITEM	X	Y	R
1	NODAL FORCE	0.0000D+00	0.0000D+00	0.0000D+00
	DISPLACEMENT	0.0000D+00	0.0000D+00	0.0000D+00
2	NODAL FORCE	0.0000D+00	-0.4300D+02	0.0000D+00
	DISPLACEMENT	0.1228D+00	-0.9718D+00	0.0000D+00
3	NODAL FORCE	0.0000D+00	-0.4300D+02	0.0000D+00
	DISPLACEMENT	0.2460D+00	-0.1618D+01	0.0000D+00
4	NODAL FORCE	0.0000D+00	-0.4300D+02	0.0000D+00
	DISPLACEMENT	0.4437D+00	-0.1847D+01	0.0000D+00
5	NODAL FORCE	0.0000D+00	-0.4300D+02	0.0000D+00
	DISPLACEMENT	0.6415D+00	-0.1618D+01	0.0000D+00
6	NODAL FORCE	0.0000D+00	-0.4300D+02	0.0000D+00
	DISPLACEMENT	0.7647D+00	-0.9718D+00	0.0000D+00
7	NODAL FORCE	0.0000D+00	0.0000D+00	0.0000D+00
	DISPLACEMENT	0.8875D+00	0.0000D+00	0.0000D+00
8	NODAL FORCE	0.0000D+00	0.0000D+00	0.0000D+00
	DISPLACEMENT	0.7144D+00	-0.9404D+00	0.0000D+00
9	NODAL FORCE	0.0000D+00	0.0000D+00	0.0000D+00
	DISPLACEMENT	0.5868D+00	-0.1634D+01	0.0000D+00
10	NODAL FORCE	0.0000D+00	0.0000D+00	0.0000D+00
	DISPLACEMENT	0.4437D+00	-0.1847D+01	0.0000D+00

Table 6.8 Output P.OUT2 of the Pratt Truss (cont.)

```
  11   NODAL FORCE     0.0000D+00    0.0000D+00    0.0000D+00
       DISPLACEMENT    0.3007D+00   -0.1634D+01    0.0000D+00

  12   NODAL FORCE     0.0000D+00    0.0000D+00    0.0000D+00
       DISPLACEMENT    0.1731D+00   -0.9404D+00    0.0000D+00

* TRUSS ELEMENT 12 HAS FAILED BY BUCKLING *

* TRUSS ELEMENT 17 HAS FAILED BY BUCKLING *

  ELEMENT FORCES IN GLOBAL COORDINATE
  -----------------------------------

    TRUSS ELEMENT :               ELEMENT FORCES
    ELEMENT NO.     Fxi          Fyi          Fxj          Fyj

         1     -0.108D+03    0.218D+00    0.108D+03   -0.218D+00
         2     -0.108D+03    0.145D+00    0.108D+03   -0.145D+00
         3     -0.172D+03    0.821D-01    0.172D+03   -0.821D-01
         4     -0.172D+03   -0.821D-01    0.172D+03    0.821D-01
         5     -0.108D+03   -0.145D+00    0.108D+03    0.145D+00
         6     -0.108D+03   -0.218D+00    0.108D+03    0.218D+00
         7     -0.529D-01   -0.429D+02    0.529D-01    0.429D+02
         8      0.154D-01    0.217D+02   -0.154D-01   -0.217D+02
         9      0.103D-16    0.168D+00   -0.103D-16   -0.168D+00
        10     -0.154D-01    0.217D+02    0.154D-01   -0.217D+02
        11      0.529D-01   -0.429D+02   -0.529D-01    0.429D+02
        12      0.105D+03    0.105D+03   -0.105D+03   -0.105D+03
        13      0.644D+02   -0.646D+02   -0.644D+02    0.646D+02
        14      0.215D+02   -0.215D+02   -0.215D+02    0.215D+02
        15     -0.215D+02   -0.215D+02    0.215D+02    0.215D+02
        16     -0.644D+02   -0.646D+02    0.644D+02    0.646D+02
        17     -0.105D+03    0.105D+03    0.105D+03   -0.105D+03
        18      0.172D+03   -0.249D+00   -0.172D+03    0.249D+00
        19      0.194D+03   -0.859D-01   -0.194D+03    0.859D-01
        20      0.194D+03    0.859D-01   -0.194D+03   -0.859D-01
        21      0.172D+03    0.249D+00   -0.172D+03   -0.249D+00
```

Member Capacity Check for Top Chord
(Tubing 10x10x5/8: $A = 22.4, r = 3.78$)

The slenderness parameter of Elements (19) and (20) is

$$\lambda_c = \frac{KL}{\pi r}\sqrt{\frac{F_y}{E}} = \frac{(1.0)(40)(12)}{(\pi)(3.78)}\sqrt{\frac{46}{29,000}} = 1.61 > 1.5$$

The axial strength is

$$\phi_c P_n = \phi_c A_g \left(\frac{0.877}{\lambda_c^2}\right) F_y = (0.85)(22.4)\left(\frac{0.877}{1.61^2}\right)(46) = 296 \text{ kips}$$

Comparing this axial strength of 296 kips with the member force of 180 kips, the member size is adequate.

Comparison of Results

The advanced analysis and LRFD method predict the same size structural tube: 10x10x5/8. The load-carrying capacities of Element (12) and (17) predicted by these two methods are identical as 148 kips. This is because the truss system is statically determinant as described in Section 6.3.1.

6.4 Braced Frames

6.4.1 Simple Braced Frame

Figure 6.14 shows a simple braced frame consisting of a column and a beam subject to a gravity load of 180 kips on the top of the column. All members are assumed to be continuously braced laterally, and the same wide flange section of A36 steel is used for both the beam and the column.

Design by the Advanced Analysis

Step 1: Load Condition and Preliminary Member Sizing

The critical factored load combination is shown in Figure 6.14. Since the column is restrained by the beam, the effective length factor of the column is assumed to be 0.8, and the resulting effective length is equal to 9.6 ft (= 12 ft × 0.8). Referring to the column design equations in the AISC-LRFD Specification, the preliminary member size of W8x24 is selected for the column and the beam. The LRFD Column Design Table is not applicable here because the frame is laterally braced (strong-axis controls).

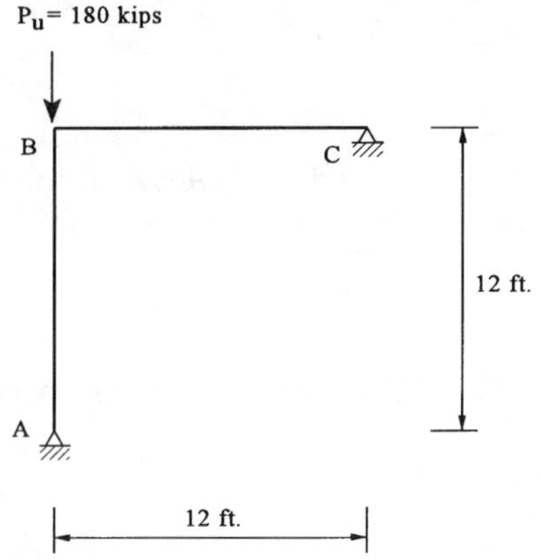

Figure 6.14 Configuration and load condition of the simple braced frame.

Step 2: Structural Modeling

Each member is modeled with two elements as recommended in Section 4.9.1. All three imperfection models are used. In the explicit imperfection model, the two-element column is assumed to have an initial out-of-straightness equal to 0.144 in. at mid-height of the column shown in Figure 6.15(a). In the equivalent notional load modeling, the notional load equal to 0.004 times the gravity load is applied at mid-height of the column shown in Figure 6.15(b). In the further reduced tangent modulus modeling, the reduced tangent modulus factor equal to 0.85 is built in the program shown in Figure 6.15(c). An incremental load of 3.6 kips is computed by dividing the factored load of 180 kips by a scaling factor of 50 (Figure 6.15).

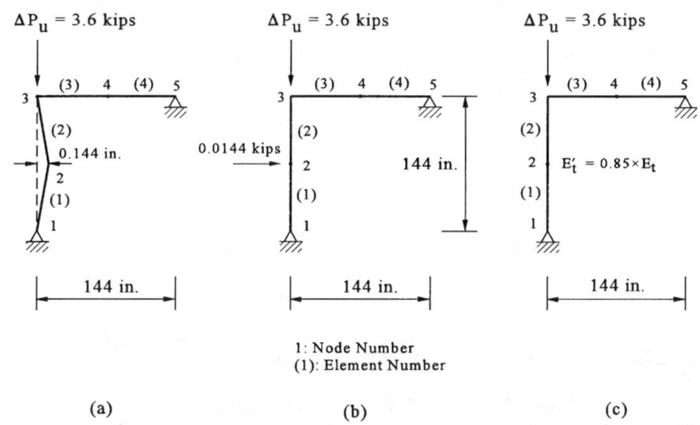

Figure 6.15 Modeling of the simple braced frame.

Step 3: Analysis

The input data may be listed in Tables 6.9a through 6.9c. Programs are executed by entering the batch file command RUN.

Table 6.9a Input Data P.DAT of the Explicit Imperfection Modeling for the Simple Braced Frame

```
Simple braced frame
   1      1
   5      2   100
   0      1     0
   0      4     0
   1         7.08          82.8           23.2     29000.0          36.0
   1        0.144          72.0      1     1      2
   2       -0.144          72.0      1     2      3
   3         72.0           0.0      1     3      4
   4         72.0           0.0      1     4      5
   1      1     1             2      4
   3                       -3.6
```

Table 6.9b Input Data P.DAT of the Equivalent Notional Load Modeling for the Simple Braced Frame

Frame

```
Simple braced frame
   2      1
   5      2   100
   0      1     0
   0      4     0
   1         7.08          82.8           23.2     29000.0          36.0
   1          0.0          72.0      1     1      2
   2          0.0          72.0      1     2      3
   3         72.0           0.0      1     3      4
   4         72.0           0.0      1     4      5
   1      1     1             2      4
   3                       -3.6
   2      0.0144
```

Step 4: Check of Load-Carrying Capacity

The incremental displacements and member forces are listed in P.OUT2 (Tables 6.10a through 6.10c). The load-carrying capacity pre-

Table 6.9c Input Data P.DAT of the Further Reduced Tangent Modulus Modeling for the Simple Braced Frame
Braced Frame

```
Simple braced frame
    3    1
    5    2   100
    0    2     0
    0    4     0
    1       7.08         82.8        23.2    29000.0         36.0     1
    2       7.08         82.8        23.2    29000.0         36.0
    1        0.0         72.0     1     1      2
    2        0.0         72.0     1     2      3
    3       72.0          0.0     2     3      4
    4       72.0          0.0     2     4      5
    1    1     1            2     4
    3                    -3.6
```

dicted by the explicit imperfection model and the equivalent notional load model is 194.4 kips. The further reduced tangent modulus model predicts 194.8 kips. Since the load-carrying capacities are greater than the applied load, the member size of W8x24 appears adequate.

Design by the AISC-LRFD Method

The column strength may be calculated with:

$$G_A = \infty \text{ (pinned-end) and } G_B = 1$$

Note that the theoretical instead of the recommended G value is used for comparison with advanced analysis. To obtain accurate K-factor, the inelastic reduction factor τ [Equations (1.7a and 1.7b)] is calculated as:

$$P_u = 180 \text{ kips}$$

$$P_y = (7.08)(36) = 255 \text{ kips}$$

$$\tfrac{P_u}{P_y} = \tfrac{180}{255} = 0.706 > \tfrac{1}{3}$$

$$\tau = -3.205\left(\tfrac{P_u}{P_y}\right)\ln\left(\tfrac{P_u/P_y}{0.85}\right) = -3.205(0.706)\ln\left(\tfrac{0.706}{0.85}\right) = 0.42 > \tfrac{1}{3}$$

To account for the far end of the beam being hinged [Equation (1.8b)]:

$$(G_B)_{\text{adjust}} = \frac{\tau(G_B)}{1.5} = \frac{(0.42)(1)}{1.5} = 0.28$$

Table 6.10a Output P.OUT2 of the Explicit Imperfection Modeling for the Simple Braced Frame

```
============================
= LOAD-DEFLECTION BEHAVIOR =
============================


  -----------------
  - LOAD STEP   1 -
  -----------------
      LOAD SEQUENCE  1
      DETERMINANT OF STIFFNESS MATRIX = 0.498 TIMES 10 TO THE POWER  44
      TRACE OF STIFFNESS MATRIX       = 0.108 TIMES 10 TO THE POWER   7

           NODE      ITEM               X              Y              R

              1   NODAL FORCE     0.0000D+00     0.0000D+00     0.0000D+00
                  DISPLACEMENT    0.0000D+00     0.0000D+00    -0.1020D-04

              2   NODAL FORCE     0.0000D+00     0.0000D+00     0.0000D+00
                  DISPLACEMENT    0.5016D-03    -0.1263D-02    -0.6019D-06

              3   NODAL FORCE     0.0000D+00    -0.3600D+01     0.0000D+00
                  DISPLACEMENT   -0.1189D-05    -0.2526D-02     0.1266D-04

              4   NODAL FORCE     0.0000D+00     0.0000D+00     0.0000D+00
                  DISPLACEMENT   -0.5946D-06    -0.1395D-02     0.1815D-04

              5   NODAL FORCE     0.0000D+00     0.0000D+00     0.0000D+00
                  DISPLACEMENT    0.0000D+00     0.0000D+00     0.1998D-04

      ELEMENT FORCES IN GLOBAL COORDINATE
      -----------------------------------

        FRAME ELEMENT :                  ELEMENT FORCES
        ELEMENT NUMBER          Fx                Fy                Mz

                1   i   -0.1670549D-02   0.3598304D+01   -0.1110223D-15
                    j    0.1670549D-02  -0.3598304D+01    0.6402492D+00
                2   i   -0.1720994D-02   0.3598304D+01   -0.6402492D+00
                    j    0.1720994D-02  -0.3598304D+01    0.2441869D+00
                3   i   -0.1695742D-02  -0.1695769D-02   -0.2441869D+00
                    j    0.1695742D-02   0.1695769D-02    0.1220934D+00
                4   i   -0.1695742D-02  -0.1695775D-02   -0.1220934D+00
                    j    0.1695742D-02   0.1695775D-02    0.4579670D-15


  -----------------
  - LOAD STEP  55 -
  -----------------
      LOAD SEQUENCE  1
      DETERMINANT OF STIFFNESS MATRIX = 0.557 TIMES 10 TO THE POWER  37
      TRACE OF STIFFNESS MATRIX       = 0.588 TIMES 10 TO THE POWER   6

           NODE      ITEM               X              Y              R
```

Table 6.10a Output P.OUT2 of the Explicit Imperfection Modeling for the Simple Braced Frame (cont.)

```
 1   NODAL FORCE      0.0000D+00    0.0000D+00    0.0000D+00
     DISPLACEMENT     0.0000D+00    0.0000D+00   -0.1118D-02

 2   NODAL FORCE      0.0000D+00    0.0000D+00    0.0000D+00
     DISPLACEMENT     0.6015D-01   -0.7826D-01    0.3705D-05

 3   NODAL FORCE      0.0000D+00   -0.1944D+03    0.0000D+00
     DISPLACEMENT     0.5959D-04   -0.1565D+00    0.9890D-03

 4   NODAL FORCE      0.0000D+00    0.0000D+00    0.0000D+00
     DISPLACEMENT     0.3265D-04   -0.8090D-01    0.1099D-02

 5   NODAL FORCE      0.0000D+00    0.0000D+00    0.0000D+00
     DISPLACEMENT     0.0000D+00    0.0000D+00    0.1136D-02

ELEMENT FORCES IN GLOBAL COORDINATE
-----------------------------------

  FRAME ELEMENT :                  ELEMENT FORCES
  ELEMENT NUMBER        Fx               Fy               Mz

        1   i   -0.3073341D-01    0.1943944D+03    0.1505740D-13
            j    0.3073341D-01   -0.1943944D+03    0.4194077D+02
        2   i   -0.3713530D-01    0.1943945D+03   -0.4194077D+02
            j    0.3713530D-01   -0.1943945D+03    0.4898117D+01
        3   i   -0.3402290D-01   -0.3405139D-01   -0.4898117D+01
            j    0.3402290D-01    0.3405139D-01    0.2448991D+01
        4   i   -0.3402384D-01   -0.3405201D-01   -0.2448991D+01
            j    0.3402384D-01    0.3405201D-01    0.1896400D-13

> DETERMINANT OF THE TANGENT STIFFNESS MATRIX IS NEGATIVE AT LOAD
> STEP  56 WITH THE PRESENT LOAD INCREMENT.  THUS, LOAD INCREMENT
> IS REDUCED TO 1/5 OF PREVIOUS VALUE FOR THIS LOAD STEP.


> DETERMINANT OF THE TANGENT STIFFNESS MATRIX IS NEGATIVE AT LOAD
> STEP  56 WITH THE PRESENT LOAD INCREMENT.  THUS, LOAD INCREMENT
> IS REDUCED TO 1/5 OF PREVIOUS VALUE FOR THIS LOAD STEP.

>> SOLUTION DOES NOT CONVERGE, A LIMIT STATE MAY HAVE BEEN REACHED <<
```

Table 6.10b Output P.OUT2 of the Equivalent Notional Load Modeling for the Simple Braced Frame

Frame

```
============================
= LOAD-DEFLECTION BEHAVIOR =
============================


  -----------------
  - LOAD STEP   1 -
  -----------------
      LOAD SEQUENCE  1
      DETERMINANT OF STIFFNESS MATRIX = 0.499 TIMES 10 TO THE POWER  44
      TRACE OF STIFFNESS MATRIX       = 0.108 TIMES 10 TO THE POWER   7

           NODE     ITEM                X              Y              R

             1    NODAL FORCE      0.0000D+00     0.0000D+00     0.0000D+00
                  DISPLACEMENT     0.0000D+00     0.0000D+00    -0.1024D-04

             2    NODAL FORCE      0.1440D-01     0.0000D+00     0.0000D+00
                  DISPLACEMENT     0.5070D-03    -0.1262D-02    -0.6380D-06

             3    NODAL FORCE      0.0000D+00    -0.3600D+01     0.0000D+00
                  DISPLACEMENT     0.3858D-05    -0.2524D-02     0.1264D-04

             4    NODAL FORCE      0.0000D+00     0.0000D+00     0.0000D+00
                  DISPLACEMENT     0.1929D-05    -0.1394D-02     0.1814D-04

             5    NODAL FORCE      0.0000D+00     0.0000D+00     0.0000D+00
                  DISPLACEMENT     0.0000D+00     0.0000D+00     0.1997D-04

      ELEMENT FORCES IN GLOBAL COORDINATE
      -----------------------------------

        FRAME ELEMENT :                 ELEMENT FORCES
        ELEMENT NUMBER          Fx              Fy              Mz

               1   i    -0.8873428D-02   0.3598301D+01   0.5551115D-16
                   j     0.8873428D-02  -0.3598301D+01   0.6407113D+00
               2   i     0.5476086D-02   0.3598301D+01  -0.6407113D+00
                   j    -0.5476086D-02  -0.3598301D+01   0.2446225D+00
               3   i     0.5501232D-02  -0.1698681D-02  -0.2446225D+00
                   j    -0.5501232D-02   0.1698681D-02   0.1223113D+00
               4   i     0.5501232D-02  -0.1698661D-02  -0.1223113D+00
                   j    -0.5501232D-02   0.1698661D-02   0.4579670D-15


  -----------------
  - LOAD STEP  56 -
  -----------------
      LOAD SEQUENCE  1
      DETERMINANT OF STIFFNESS MATRIX = 0.449 TIMES 10 TO THE POWER  37
      TRACE OF STIFFNESS MATRIX       = 0.587 TIMES 10 TO THE POWER   6

           NODE     ITEM                X              Y              R
```

Table 6.10b Output P.OUT2 of the Equivalent Notional Load Modeling for the Simple Braced Frame (cont.)

```
  1    NODAL FORCE      0.0000D+00    0.0000D+00    0.0000D+00
       DISPLACEMENT     0.0000D+00    0.0000D+00   -0.1139D-02

  2    NODAL FORCE      0.7777D+00    0.0000D+00    0.0000D+00
       DISPLACEMENT     0.6164D-01   -0.7814D-01    0.3486D-05

  3    NODAL FORCE      0.0000D+00   -0.1944D+03    0.0000D+00
       DISPLACEMENT     0.3332D-03   -0.1563D+00    0.9910D-03

  4    NODAL FORCE      0.0000D+00    0.0000D+00    0.0000D+00
       DISPLACEMENT     0.1693D-03   -0.8069D-01    0.1097D-02

  5    NODAL FORCE      0.0000D+00    0.0000D+00    0.0000D+00
       DISPLACEMENT     0.0000D+00    0.0000D+00    0.1133D-02

ELEMENT FORCES IN GLOBAL COORDINATE
-----------------------------------

  FRAME ELEMENT :              ELEMENT FORCES
  ELEMENT NUMBER        Fx                Fy                Mz

       1   i   -0.4179972D+00    0.1944018D+03   -0.1942890D-13
           j    0.4179972D+00   -0.1944018D+03    0.4209223D+02
       2   i    0.3533330D+00    0.1944019D+03   -0.4209223D+02
           j   -0.3533330D+00   -0.1944019D+03    0.4720662D+01
       3   i    0.3564413D+00   -0.3239552D-01   -0.4720662D+01
           j   -0.3564413D+00    0.3239552D-01    0.2361240D+01
       4   i    0.3564404D+00   -0.3239555D-01   -0.2361240D+01
           j   -0.3564404D+00    0.3239555D-01    0.7404667D-13

> DETERMINANT OF THE TANGENT STIFFNESS MATRIX IS NEGATIVE AT LOAD
> STEP  57 WITH THE PRESENT LOAD INCREMENT.  THUS, LOAD INCREMENT
> IS REDUCED TO 1/5 OF PREVIOUS VALUE FOR THIS LOAD STEP.

>> SOLUTION DOES NOT CONVERGE, A LIMIT STATE MAY HAVE BEEN REACHED <<
```

Table 6.10c Output P.OUT2 of the Further Reduced Tangent Modulus Modeling for the Simple Braced Frame

Braced Frame

```
=============================
= LOAD-DEFLECTION BEHAVIOR =
=============================


  -----------------
  - LOAD STEP   1 -
  -----------------
      LOAD SEQUENCE  1
      DETERMINANT OF STIFFNESS MATRIX = 0.205 TIMES 10 TO THE POWER  44
      TRACE OF STIFFNESS MATRIX       = 0.100 TIMES 10 TO THE POWER   7

           NODE      ITEM                X              Y              R

             1    NODAL FORCE       0.0000D+00     0.0000D+00     0.0000D+00
                  DISPLACEMENT      0.0000D+00     0.0000D+00    -0.5550D-05

             2    NODAL FORCE       0.0000D+00     0.0000D+00     0.0000D+00
                  DISPLACEMENT      0.2994D-03    -0.1484D-02    -0.1375D-05

             3    NODAL FORCE       0.0000D+00    -0.3600D+01     0.0000D+00
                  DISPLACEMENT     -0.2305D-05    -0.2968D-02     0.1115D-04

             4    NODAL FORCE       0.0000D+00     0.0000D+00     0.0000D+00
                  DISPLACEMENT     -0.1153D-05    -0.1739D-02     0.2179D-04

             5    NODAL FORCE       0.0000D+00     0.0000D+00     0.0000D+00
                  DISPLACEMENT      0.0000D+00     0.0000D+00     0.2534D-04

      ELEMENT FORCES IN GLOBAL COORDINATE
      -----------------------------------

        FRAME ELEMENT :                    ELEMENT FORCES
        ELEMENT NUMBER          Fx               Fy               Mz

              1   i   -0.3271970D-02   0.3596713D+01  -0.2775558D-16
                  j    0.3271970D-02  -0.3596713D+01   0.2366587D+00
              2   i   -0.3301997D-02   0.3596713D+01  -0.2366587D+00
                  j    0.3301997D-02  -0.3596713D+01   0.4733173D+00
              3   i   -0.3286926D-02  -0.3286982D-02  -0.4733173D+00
                  j    0.3286926D-02   0.3286982D-02   0.2366587D+00
              4   i   -0.3286926D-02  -0.3287005D-02  -0.2366587D+00
                  j    0.3286926D-02   0.3287005D-02  -0.2220446D-15


  -----------------
  - LOAD STEP  61 -
  -----------------
      LOAD SEQUENCE  1
      DETERMINANT OF STIFFNESS MATRIX = 0.111 TIMES 10 TO THE POWER  37
      TRACE OF STIFFNESS MATRIX       = 0.583 TIMES 10 TO THE POWER   6

           NODE      ITEM                X              Y              R
```

Table 6.10c Output P.OUT2 of the Further Reduced Tangent Modulus Modeling for the Simple Braced Frame (cont.)

```
 1   NODAL FORCE     0.0000D+00     0.0000D+00     0.0000D+00
     DISPLACEMENT    0.0000D+00     0.0000D+00    -0.7154D-03

 2   NODAL FORCE     0.0000D+00     0.0000D+00     0.0000D+00
     DISPLACEMENT    0.3990D-01    -0.9246D-01    -0.5979D-04

 3   NODAL FORCE     0.0000D+00    -0.1950D+03     0.0000D+00
     DISPLACEMENT    0.1734D-04    -0.1849D+00     0.8714D-03

 4   NODAL FORCE     0.0000D+00     0.0000D+00     0.0000D+00
     DISPLACEMENT    0.2271D-04    -0.1036D+00     0.1336D-02

 5   NODAL FORCE     0.0000D+00     0.0000D+00     0.0000D+00
     DISPLACEMENT    0.0000D+00     0.0000D+00     0.1490D-02

ELEMENT FORCES IN GLOBAL COORDINATE
-----------------------------------

FRAME ELEMENT :                    ELEMENT FORCES
ELEMENT NUMBER          Fx               Fy                Mz

    1   i   -0.1417610D+00   0.1948727D+03  -0.3150258D-14
        j    0.1417610D+00  -0.1948727D+03   0.1799153D+02
    2   i   -0.1449591D+00   0.1948728D+03  -0.1799153D+02
        j    0.1449591D+00  -0.1948728D+03   0.2064723D+02
    3   i   -0.1435839D+00  -0.1435675D+00  -0.2064723D+02
        j    0.1435839D+00   0.1435675D+00   0.1032205D+02
    4   i   -0.1436055D+00  -0.1435686D+00  -0.1032205D+02
        j    0.1436055D+00   0.1435686D+00  -0.2919887D-13

> DETERMINANT OF THE TANGENT STIFFNESS MATRIX IS NEGATIVE AT LOAD
> STEP  62 WITH THE PRESENT LOAD INCREMENT.  THUS, LOAD INCREMENT
> IS REDUCED TO 1/5 OF PREVIOUS VALUE FOR THIS LOAD STEP.

>> SOLUTION DOES NOT CONVERGE, A LIMIT STATE MAY HAVE BEEN REACHED <<
```

From the nomograph shown in Figure 1.12(a)

$$K_x = 0.78$$

The slenderness parameter is

$$\lambda_{cx} = \frac{(K_x)(L)}{\pi r_x}\sqrt{\frac{F_y}{E}} = \frac{(0.78)(144)}{(\pi)(3.42)}\sqrt{\frac{36}{29,000}} = 0.368$$

Therefore, the column strength is

$$\phi_c P_n = \phi_c A_g \left(0.658^{\lambda_c^2}\right) F_y = (0.85)(7.08)\left(0.658^{0.368^2}\right)(36) = 186 \text{ kips}$$

Since the column strength is greater than the applied load, the column size of W8x24 is adequate.

Comparison of Results

The advanced analysis method and the LRFD method predict the same member size of W8x24. This is because the frame is failed by the column buckling rather than the formation of mechanism with plastic hinges, and so does not render inelastic moment redistribution.

6.4.2 *Braced Eight-Story Frame*

Figure 6.16 shows a braced eight-story, one-bay frame with hinge supports. All beams and braces are pin-connected to the columns. The column, beam, and brace sizes are assumed to be identical, respectively. All beams are continuously braced in weak-axis direction. The member is arranged such that the weak-axis of columns is subjected to major bending action. A36 steel is used for all members.

Design by the Advanced Analysis

Step 1: Load Condition

The load conditions are:

Roof loads:

Dead load = 45 psf
Snow load = 30 psf

Typical floor loads:

Dead load = 55 psf
Live load = 50 psf

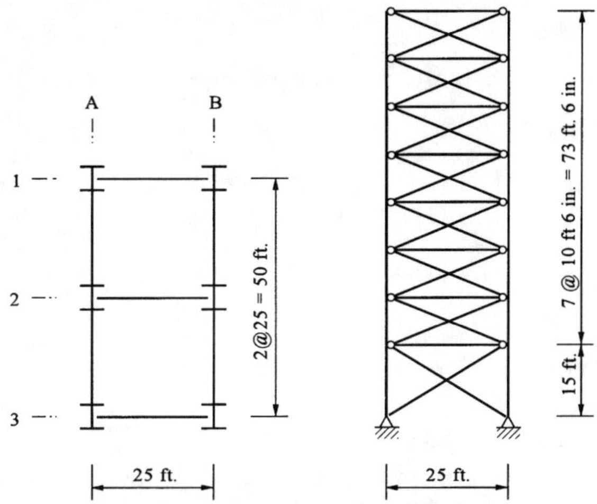

Figure 6.16 Configuration of the braced eight-story frame.

Partition load = 20 psf

Wind load:

Wind pressure = 30 psf

The load calculations are:

Dead load:

$$D = (0.045)(25)(25/2) = 14.1 \text{ kips for roof}$$
$$D = (0.055)(25)(25/2) = 17.2 \text{ kips for each floor}$$

Live load

$$L = (0.05 + 0.02)(25)(25/2) = 21.9 \text{ kips}$$
$$A_I = (25)(25 \times 2)(7) = 8750 \text{ ft}^2$$
$$\text{Reduced } L = 21.9\left(0.25 + \frac{15}{\sqrt{8750}}\right) = 8.99 \text{ kips}$$
$$\text{Minimum } L = (0.4)(21.9) = 8.76 \text{ kips}$$

The live load reduction is for the design of the first floor columns which are the most critical because the same size columns will be used in this design. The same live load reduction is applied to all other stories for simplicity. If designing column sizes differently for other stories, the live load reduction procedures proposed by Ziemian and McGuire [8] as described in Section 4.3.2 may be used. Note that the different column sizes should be used for actual design.

Snow load

$$S = (0.03)(25)(25/2) = 9.375 \text{ kips}$$

Wind Load

$W = (0.03)(15 + 10.5)(1/2)(25) = 9.56$ kips	for 2nd floor level
$W = (0.03)(10.5)(25) = 7.88$ kips	for 3rd through 8th floor level
$W = (0.03)(10.5)(1/2)(25) = 3.94$ kips	for roof level

The load combinations are:

$1.2D + 1.6L + 0.5S$ (Load case 1)

$$1.2(14.1) + 1.6(0) + 0.5(9.375) = 21.6 \text{ kips at roof}$$

$$1.2(17.2) + 1.6(8.99) + 0.5(0) = 35.0 \text{ kips at each floor}$$

$1.2D + 1.3W + 0.5L + 0.5S$ (Load case 2)

Vertical:	$1.2(14.1) + 0.5(9.375) = 21.6$ kips	at roof
Vertical:	$1.2(17.2) + 0.5(8.99) = 25.1$ kips	at each floor
Horizontal:	$1.3(9.56) = 12.4$ kips	at floor level
	$1.3(7.88) = 10.2$ kips	at 3rd through 8th floor level
	$1.3(3.94) = 5.12$ kips	at roof level

Step 2: Preliminary Member Sizing

From the load combination in Figure 6.17, the critical forces on the first floor can be simply calculated. The sums of the gravity loads are 267 kips in load combination 1 and 197 kips in combination 2. The total wind load with respect to the second floor level is 78.7 kips, and the coupling reactions at supports caused by the wind load are

$$P'_u = \frac{(5.12)(88.5) + (10.2)(78 + 67.5 + 57 + 46.5 + 36 + 25.5) + (12.4)(15)}{25}$$

$$= 152 \text{ kips}$$

The preliminary member sizes are determined as follows.

Column

$$P_{u.\max} = 197 + 152 = 349 \text{ kips in load case 2}$$

$$(KL)_y = (1.0)(15) = 15 \text{ ft}$$

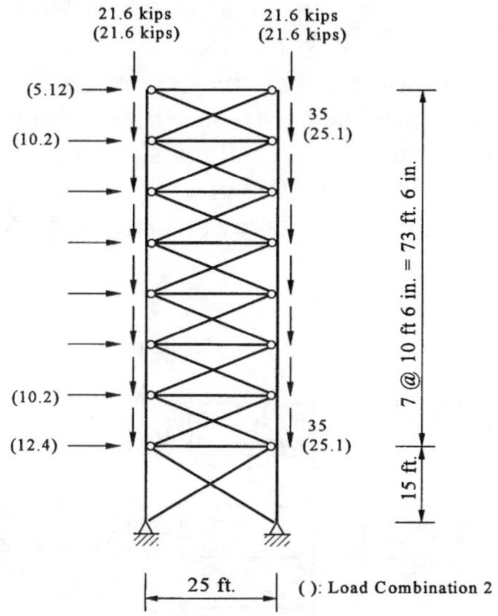

Figure 6.17 Load condition of the braced eight-story frame.

Try W14x61 ($\phi P_n = 412$ kips)

Brace

$(T)(25/29) = 78.7$ kips (Figure 6.19)

$T = 91$ kips

$A_{\text{req'd}} = \frac{T}{\phi F_y} = \frac{91}{(0.9)(36)} = 2.8 \text{ in.}^2$

Try L4x4x7/16 (A = 3.31 in.^{2})

Beam

Influence area, $A_I = (25)(25x2) = 1250 \text{ ft}^2$

$D = (0.055)(25) = 1.375$ k/ft

$L = (0.05 + 0.02)(25) = 1.75$ k/ft

Reduced load, $L = (1.75)\left(0.25 + \dfrac{15}{\sqrt{1250}}\right) = 1.18$ k/ft

Minimum load, $L = (0.4)(1.75) = 0.7$ k/ft

$1.2D + 1.6L = 1.2(1.375) + 1.6(1.18) = 3.54$ k/ft

$M_u = \left(\frac{1}{8}\right) wL^2 = \left(\frac{1}{8}\right)(3.54)(25)^2 = 276$ kip-ft

The lightest section is W21x50 ($\phi M_p = 297$ kip-ft)

Step 3: Structural Modeling

Each column is modeled with two elements because the frame is braced. Each beam is modeled with one truss element because the beam is simply supported and can be easily designed by a simple static independent with global analysis. The brace members are modeled as one truss element. Figure 6.18 shows the structural modeling.

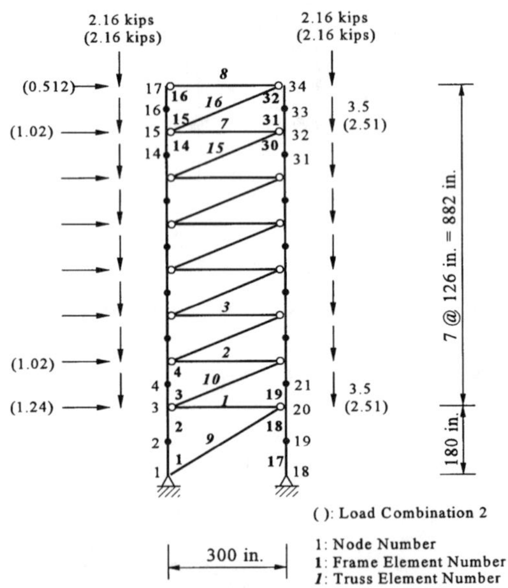

Figure 6.18 Modeling of the braced eight-story frame.

Only the further reduced tangent modulus method is used for this example because the explicit imperfection model and the equivalent notional load model invoke tedious efforts for large braced frames. This is because geometric imperfection should be modeled at the middle of a column, or the notional load should be applied at the middle of a column. In the further reduced tangent modulus model, the program automatically accounts for geometric imperfection effects.

Herein, the scaling factor of 10 is selected because this frame is highly indeterminate and the load incremental size is automatically reduced if changes of the element stiffness parameter η exceed the predefined value 0.1.

Step 4: Analysis

The input data may be generated as shown in Table 6.11. Programs

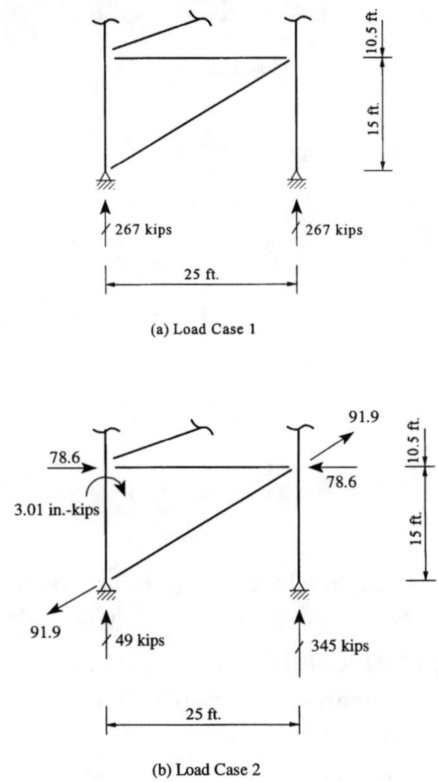

Figure 6.19 Member forces by the first-order analysis for the braced eight-story frame.

DATAGEN, INPUT, and PAAP are executed in sequence by entering the batch file command RUN.

Step 5: Check of Load-Carrying Capacity

According to the output P.OUT2, the column size can be reduced to W14x53 from the preliminary member size W14x61. The ultimate load-carrying capacity is equal to 22.4 kips with respect to the gravity load at roof in load combination 2. Since this load-carrying capacity is 3.7% greater than the applied load of 21.6 kips, the reduced member sizes are adequate.

Step 6: Serviceability Check

The lateral displacement at the roof level by the service wind load (1.0W) without geometric imperfection is 1.74 in. and the drift ratio is 1/610 which satisfies the drift limit of 1/400. The beam deflection is calculated as 0.363 in. by the simple equation of $(5wL^4)/(384EI)$, where

Table 6.11 Input Data P.DAT of the Braced Eight-Story Frame

```
Braced eight-story frame
  3    1
 34    2  100
  0    1    2
  0   32   16
  1     15.60       57.7        22.0   29000.0          36.0     1
  1     14.70      984.0     29000.0      36.0
  2      3.31       4.97     29000.0      36.0
  1       0.0       90.0     1    1    2    2
  3       0.0       63.0     1    3    4   14
 17       0.0       90.0     1   18   19    2
 19       0.0       63.0     1   20   21   14
  1     300.0        0.0     1    3   20    8    2
  9     300.0      180.0     2    1   20
 10     300.0      126.0     2    3   22    7    2
  1    1    1          2    17
  3      1.24      -2.51
  5      1.02      -2.51               6    2
 17     0.512      -2.16
 20                -2.51               7    2
 34                -2.16
```

w includes only service live load 1.18 k/ft, and the resulting deflection ratio is equal to 1/825 meeting the allowable deflection ratio of 1/360.

Design by the AISC-LRFD Method

The first-order elastic analysis is shown in Figure 6.19. Member capacity checks are carried out in what follows.

Member Capacity Check for Columns
(W14x61: $A = 17.9$, $I_y = 107$, $r_y = 2.45$, $Z_y = 32.8$)

- For $1.2D + 1.6L + 0.5S$

  Calculation of P_n

  $$K_x = K_y = 1.0$$

  $$\lambda_{cy} = \frac{(1.0)(180)}{(\pi)(2.45)}\sqrt{\frac{36}{29,000}} = 0.824 < 1.50$$

  $$P_n = (0.658)^{(0.824)^2}(36)(17.9) = 485 \text{ kips}$$

  Resulting P_u

  $$P_u = 267 \text{ kips}$$

  Interaction equation check

  $$\frac{P_u}{\phi P_n} = \frac{267}{(0.85)(485)} = 0.65 < 1.0 \text{ O.K.}$$

- For $1.2D + 1.3W + 0.5L + 0.5S$

  Calculation of P_n

$$\lambda_{cy} = \frac{(1.0)(180)}{(\pi)(2.45)}\sqrt{\frac{36}{29,000}} = 0.824 < 1.5$$

$$P_n = (0.658)^{(0.824)^2}(36)(17.9) = 485 \text{ kips}$$

Calculation of M_n

$$\phi M_n = \phi M_p = (0.9)(32.8)(36) = 1,063 \text{ kip-in}$$

Resulting P_u

$$P_u = 345 \text{ kips}$$

Calculation of M_u

$$C_m = 0.6 - 0.4\left(\frac{0.0}{9.6}\right) = 0.6$$

$$P_{e1} = \frac{(\pi^2)(29,000)(107)}{[(1.0)(15)(12)]^2} = 945 \text{ kips}$$

$$B_1 = \frac{0.6}{\left(1 - \frac{345}{945}\right)} = 0.945$$

Use $B_1 = 1.0$

$$M_u = (1.0)(9.6) = 9.6$$

Interaction equation check

$$\frac{P_u}{\phi P_n} + \frac{8M_u}{9\phi M_n} = \frac{345}{(0.85)(485)} + \frac{(8)(9.6)}{(9)(1,063)} = 0.84 < 1.0 \text{ O.K.}$$

Use W14x61

Member Capacity Check for Beams
(W21x50; $A = 14.7$, $I_x = 984$, $r_x = 8.18$, $Z_x = 110$)

- For $1.2D + 1.6L + 0.5S$

  From preliminary member design

$$\frac{M_u}{\phi M_p} = \frac{276}{297} = 0.93 < 1.0$$

- For $1.2D + 1.3W + 0.5L + 0.5S$

  Calculation of P_n

  The P_n is controlled by the strong-axis buckling because the beam is laterally braced.

$$K_x = 1.0$$

$$\lambda_{cx} = \frac{(1.0)(300)}{(\pi)(8.18)}\sqrt{\frac{36}{29,000}} = 0.411 < 1.50$$

$$P_n = (0.658)^{(0.411)^2}(36)(14.7) = 493 \text{ kips}$$

  Calculation of M_n

$$\phi M_n = (0.9)(110)(36) = 3,564 \text{ kip-in}$$

Resulting P_u

$$P_u = 78.6 \text{ kips}$$

Resulting M_u

$$M_u = \left(\frac{1}{8}\right)[(1.2)(1.375) + (0.5)(1.18)](25)^2 = 175 \text{ k-ft}$$

Interaction equation check

$$\frac{P_u}{\phi P_n} = \frac{78.6}{(0.85)(493)} = 0.188 < 0.2$$

$$\frac{P_u}{2\phi P_n} + \frac{M_u}{\phi M_n} = \frac{0.188}{2} + \frac{(175)(12)}{3,564} = 0.683 < 1.0 \text{ O.K.}$$

Use W21x50

Member Capacity Check for Braces
(L4x4x7/16: $A = 3.31$)

- For $1.2D + 1.3W + 0.5L + 0.5S$

$$\phi P_n = (0.9)(3.31)(36) = 107 \text{ kips}$$

$$P_u = 92 \text{ kips}$$

$$\frac{P_u}{\phi P_n} = \frac{92}{107} = 0.86 < 1.0 \text{ O.K.}$$

Use L4x4x7/16

Comparison of Results

The member sizes determined by the two methods are compared in Table 6.12. The two methods predict the same size for the beams and the braces, but not the columns. The LRFD method does not provide a procedure for determining an accurate K-factor for unequal story columns. Since the shorter upper columns restrain the longer first story columns, the first story columns should have K-factor less than 1.0. In the member capacity checks in the LRFD method, this effect was neglected by the use of K-factor equal to 1.0 and it results in a larger column in the LRFD method.

Table 6.12 Comparison of Member Sizes of the Braced Eight-Story Frame

Member	LRFD	Proposed	Remark
Column	W14x61	W14x53	One size smaller
Beam	W21x50	W21x50	Same
Brace	L4x4x7/16	L4x4x7/16	Same

In this example, the same member sizes for the columns, beams, and braces were assumed for simplicity. In practical design, the members should be optimized with different sizes. The advanced analysis method can optimize member sizes in a way that the conventional LRFD method cannot because the adequacy of its members can be checked simply by comparing the applied load with the load-carrying capacity read directly from the computer output P.OUT2. On the contrary, the LRFD method requires a repeat of separate member capacity checks for every member corresponding to each step in the optimization process.

6.5 Unbraced Frames

6.5.1 One-Story Two-Bay Frame

Figure 6.20 shows a fixed-supported two-bay portal frame subject to gravity and lateral loads. Detailed procedures for selecting appropriate member sizes are presented in the following. All members are assumed to be laterally braced. W16 of A36 steel is used for beams and W14 for columns.

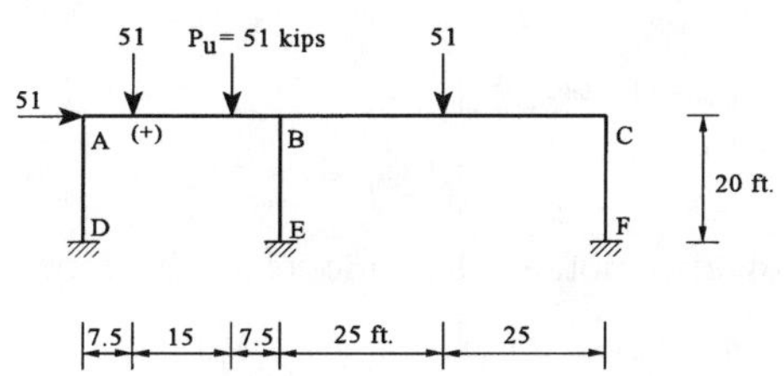

Figure 6.20 Configuration and load condition of the unbraced two-bay frame.

Design by the Advanced Analysis

Step 1: Load Condition and Preliminary Member Sizing

The critical load combination is shown in Figure 6.20. Preliminary sizes are W16x89 beams and W14x68 columns.

Step 2: Structural Modeling

Each column is modeled with one element, and the beams have three in the left and two in the right. In the explicit imperfection model, the geometric imperfection of 0.04 ft at the top of each column is calculated by 0.2% times the column height of 20 ft as shown in Figure 6.21. In the equivalent notional load model, the notional load of 0.306 kips results from 0.2% times the total gravity load of 153 kips and is added to the lat-

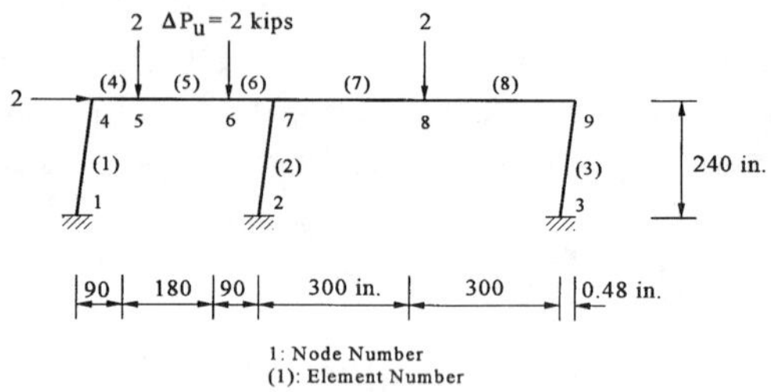

Figure 6.21 Explicit imperfection modeling for the unbraced two-bay frame.

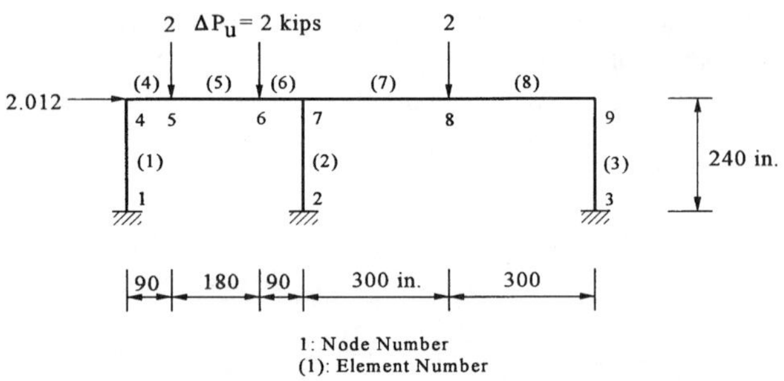

Figure 6.22 Equivalent notional load model for the unbraced two-bay frame.

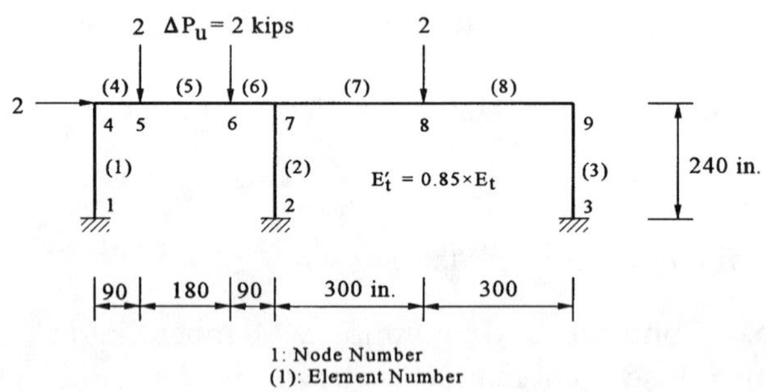

Figure 6.23 Further reduced tangent modulus model for the unbraced two-bay frame.

eral load shown in Figure 6.22. In the further reduced tangent modulus model, the program automatically accounts for geometric imperfection effects (Figure 6.23). Users can choose from one of the three methods to generate their input data. Herein, the incremental concentrated load is 2 kips resulting from a 25.5 scaling factor.

Step 3: Analysis

Referring to the input instructions in Section 5.3.2, the input data may be easily generated as listed in Tables 6.13a through 6.13c. Programs DATAGEN, INPUT, and PAAP are executed in sequence by entering the batch file command RUN. The program will continue to increment the load until the frame has reached its ultimate state.

Table 6.13a Input Data P.DAT of the Explicit Imperfection Modeling for the Two-Bay Frame

```
Two-bay frame
      1    1
      9    3   100
      0    2     0
      0    8     0
      1      20.00        723.0        115.0    29000.0        36.0
      2      26.20       1300.0        175.0    29000.0        36.0
      1       0.48        240.0     1     1     4
      2       0.48        240.0     1     2     7
      3       0.48        240.0     1     3     9
      4       90.0          0.0     2     4     5
      5      180.0          0.0     2     5     6
      6       90.0          0.0     2     6     7
      7      300.0          0.0     2     7     8     2
      1    1     1     1    3
      4       2.00
      5                   -2.00                 2
      8                   -2.00
```

Step 4: Check of Load-Carrying Capacity

Referring to the output file P.OUT2, the load-carrying capacity of the structural system is summarized in Table 6.14. The explicit imperfection model and the equivalent notional load model predict a load-carrying capacity of 51.96 kips when the first plastic hinge forms and of 63.68 kips at the ultimate state. The further reduced tangent modulus model predicts load-carrying capacities of 50.35 kips and 63.60 kips. If the load-carrying capacity at the first plastic hinge forming is adopted, the preliminary member sizes are adequate. When the load-carrying capacity at the ultimate state is selected, the member sizes may be reduced.

Table 6.13b Input Data P.DAT of the Equivalent Notional Load Modeling for the Two-Bay Frame

```
Two-bay frame
    2    1
    9    3   100
    0    2     0
    0    8     0
    1      20.00       723.0       115.0    29000.0        36.0
    2      26.20      1300.0       175.0    29000.0        36.0
    1       0.00       240.0    1    1     4
    2       0.00       240.0    1    2     7
    3       0.00       240.0    1    3     9
    4       90.0         0.0    2    4     5
    5      180.0         0.0    2    5     6
    6       90.0         0.0    2    6     7
    7      300.0         0.0    2    7     8      2
    1    1     1    1    3
    4      2.012
    5                  -2.00               2
    8                  -2.00
```

Step 5: Check of Serviceability

Member sizes may be governed by the serviceability condition. When a total lateral load of 51 kips is regarded as 1.3W, the lateral displacement corresponding to 1.0W (without geometric imperfection) is computed as 0.92 in. from the computer output P.OUT2. The drift ratio is 1/261, which does not satisfy the limitation 1/400. This indicates that the initial member sizes should be increased.

Step 6: Check of Ductility

Even though the ductility check should be performed in every case, it is given for this case only because its procedure is straightforward. The compactness and lateral unbraced length of Beam BC and Column BE are checked in what follows.

Check for Compactness of Beam BC (W16x89)

$$\frac{b_f}{2t_f} = 5.9$$

$$\frac{65}{\sqrt{F_y}} = \frac{65}{\sqrt{36}} = 10.8 > 5.9$$

Table 6.13c Input Data P.DAT of the Further Reduced Tangent Modulus Modeling for the Two-Bay Frame

Frame

```
Two-bay frame
     3    1
     9    3   100
     0    2     0
     0    8     0
     1     20.00       723.0       115.0   29000.0       36.0     1
     2     26.20      1300.0       175.0   29000.0       36.0
     1      0.00       240.0    1    1     4
     2      0.00       240.0    1    2     7
     3      0.00       240.0    1    3     9
     4      90.0         0.0    2    4     5
     5     180.0         0.0    2    5     6
     6      90.0         0.0    2    6     7
     7     300.0         0.0    2    7     8    2
     1    1     1    1     3
     4      2.00
     5                 -2.00               2
     8                 -2.00
```

Table 6.14 Load-Carrying Capacity of Two-Bay Frame

Modeling method	First plastic hinge forming	Ultimate state
Explicit imperfection modeling Equivalent notional load modeling	51.96	63.68
Further reduced tangent modulus model	50.35	63.60

Note: unit: kips.

Therefore, the flange is compact.

$$\frac{h}{t_w} = 27.0$$

$$\frac{P_u}{\phi_b P_y} = \frac{21.8}{(0.9)(22.6)(36)} = 0.0298 \leq 0.125$$

$$\frac{640}{\sqrt{F_y}}\left(1 - \frac{2.75 P_u}{\phi_b P_y}\right) = \frac{640}{\sqrt{36}}[1 - (2.75)\ (0.0298)] = 98 > 27.0$$

where $P_u = 21.8$ kips is obtained from the output file P.OUT2. Thus, the web is compact.

Check for Unbraced Length of Beam BC (W16x89)

$$L_{pd} = \frac{[3,600 + 2,200(M_1/M_2)]r_y}{F_y}$$

$$= \frac{[3,600 - 2,200(2,752/3,648)](2.49)}{36}$$

$$= 134 \text{ in. } (= 11 \text{ ft })$$

where $M_1 = 2,752$ kip-in and $M_2 = 3,648$ kip-in are obtained from the output file P.OUT2. As a result, Beam BC should be laterally braced at the mid-span and less than an 11-ft interval.

Check for Compactness of Column EB (W14x68)

$$\frac{b_f}{2t_f} = 7.0$$

$$\frac{65}{\sqrt{F_y}} = \frac{65}{\sqrt{36}} = 10.8 > 7.0$$

Therefore, the flange is compact.

$$\frac{h}{t_w} = 27.5$$

$$\frac{P_u}{\phi_b P_y} = \frac{96.6}{(0.9)(15.6)(36)} = 0.191 \geq 0.125$$

$$\frac{191}{\sqrt{F_y}}\left(2.33 - \frac{P_u}{\phi_b P_y}\right) = \frac{191}{\sqrt{36}}(2.33 - 0.191) = 68 > 27.5$$

where $P_u = 96.6$ kips is obtained from the output P.OUT2. Thus, the web is compact.

Check for Unbraced Length of Column EB (W14x68)

$$L_{pd} = \frac{[3,600 + 2,200(M_1/M_2)]r_y}{F_y}$$

$$= \frac{[3,600 + 2,200(1,111/2,511)](2.46)}{36}$$

$$= 313 \text{ in. } (= 26 \text{ ft })$$

where $M_1 = 1,111$ kip-in and $M_2 = 2,511$ kip-in are obtained from the output P.OUT2. Since the column height is equal to 20 ft, no lateral intermediate brace is required.

Design by the AISC-LRFD Method

The first order elastic analysis is shown in Figure 6.24. Member capacity checks corresponding to the interaction equations are carried out in what follows.

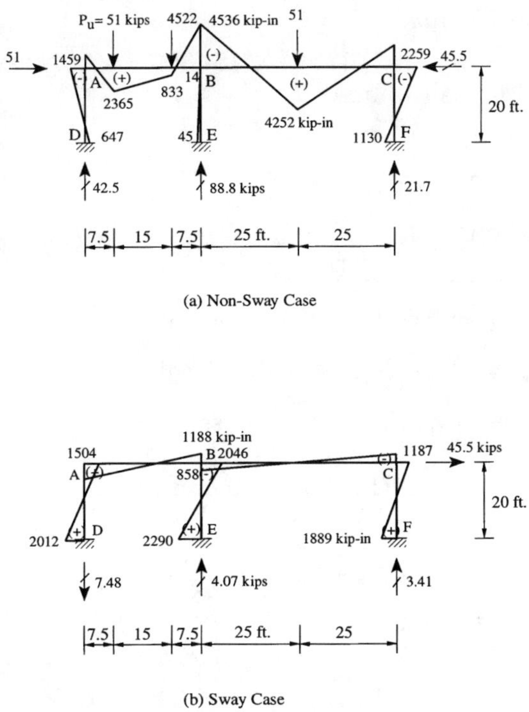

Figure 6.24 Member forces by the first-order analysis for the unbraced two-bay frame.

Member Capacity Check for Columns FC
($W14x68$: $A = 20$, $I_x = 723$, $r_x = 6.01$, $Z_x = 115$)

- Calculation of P_n

$$P_u = 21.7 + 3.41 = 25.1 \text{ kips}$$
$$P_y = (20)(36) = 720 \text{ kips}$$
$$\frac{P_u}{P_y} = \frac{25.1}{720} = 0.035 < \frac{1}{3}$$
$$\tau = 1.0$$
$$G_C = \frac{723/20}{1300/50} = 1.39, \quad G_F = 0$$
$$K_{\text{unbraced}} = 1.2 \text{ From alignment chart}$$
$$\lambda_c = \frac{(1.2)(20)(12)}{(\pi)(6.01)}\sqrt{\frac{36}{29,000}} = 0.537$$
$$P_n = (0.658)^{(0.537)^2}(36)(20.0) = 638 \text{ kips}$$

- Calculation of M_n

$$M_n = M_p = (Z_x)(F_y) = (115)(36) = 4,140 \text{ kip-in}$$
for laterally braced

- Resulting P_u

$$P_u = 25.1 \text{ kips}$$

- Calculation of M_u

$$C_m = 0.6 - 0.4\frac{1130}{2259} = 0.40$$
$$K_{\text{brace}} = 0.64 \quad \text{From alignment chart}$$
$$P_{e1} = \frac{(\pi^2)(29,000)(723)}{[(0.64)(20)(12)]^2} = 8,771 \text{ kips}$$
$$B_1 = \frac{0.4}{1 - \frac{21.7}{8.771}} = 0.4 < 1.0 \quad \text{Use } B_1 = 1.0$$
$$\sum P_u = (30)(1.7)(3) = 153 \text{ kips}$$
$$K_{FC} = 1.2$$
$$K_{DA} = 1.13(G_D = 0.834, G_A = 0)$$
$$K_{EB} = 1.10(G_E = 0.521, G_B = 0)$$
$$\sum P_{e2} = \frac{(\pi^2)(29,000)(723)}{[(20)(12)]^2}\left(\frac{1}{1.2^2} + \frac{1}{1.13^2} + \frac{1}{1.02^2}\right)$$
$$= 8,762 \text{ kips}$$
$$B_2 = \frac{1}{1 - \frac{153}{8,762}} = 1.02$$
$$M_{nt} = 2,259 \text{ kip-in}, M_{lt} = 1,187 \text{ kip-in at top of column } FC$$

$M_{nt} = 1,130$ kip-in, $M_{lt} = 1,889$ kip-in at bottom of column FC

$$M_u = (1.0)(2,259) + (1.02)(1,187) = 3,470 \text{ kip-in}$$

$$M_u = (1.0)(1,130) + (1.02)(1,889) = 3,057 \text{ kip-in}$$

- Interaction equation check

$$\frac{P_u}{\phi P_n} = \frac{(25.11)}{(0.85)(638)} = 0.046 < 0.2$$

$$\frac{P_u}{2\phi P_n} + \frac{M_u}{\phi M_n} = \frac{0.046}{2} + \frac{3{,}470}{(0.9)(4{,}140)} = 0.95 < 1.0 \text{ O.K.}$$

- Use W14x68

Member Capacity Check for Beam AB
(W16x89: $A = 26.2$, $I_x = 1{,}300$, $r_x = 7.05$, $Z_x = 175$)

- Neglect axial force
- Flexure strength check

$M_{nt} = 4,522$ kip-in, $M_{lt} = 1,188$ kip-in

$$M_u = (1.0)(4,522) + (1.02)(1,188) = 5,734 \text{ kip-in}$$

$M_n = (Z_x)(F_y) = (175)(36) = 6,300$ kip-in For laterally braced

$$\frac{M_u}{\phi M_n} = \frac{5{,}734}{(0.9)(6{,}300)} = 1.01 \text{ O.K.}$$

- Use W16x89

Comparison of Results

When the load-carrying capacity corresponding to the formation of the first plastic hinge is taken to signify failure, the advanced and the conventional LRFD method predict the same member sizes (column: W14x68; beam: W16x89). This example is a good illustration of how the two methods predict the same member sizes if inelastic moment redistribution is not considered.

6.5.2 *Leaning Column Frame*

Figure 6.25 shows a frame with leaning columns used by LeMessurier [6]. The exterior columns lean on the central column which supplies lateral rigidity of the frame. Lateral braces are assumed to fully support the beams and the columns. The frame is subjected to a distributed gravity load and a concentrated horizontal load. A36 steel is used.

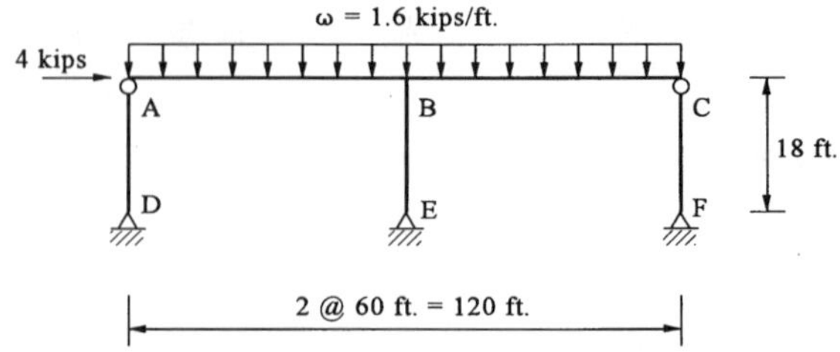

Figure 6.25 Configuration and load condition of the leaning column frame.

Design by the Advanced Analysis

Step 1: Load Condition and Preliminary Member Sizing

The critical load combination is assumed in Figure 6.25. Preliminary member sizes are selected as W27x94 for the beams, W14x43 for the central column, and W6x9 for the exterior leaning columns.

Step 2: Structural Modeling

Each column is modeled with one element and each beam is modeled four. One may use two elements instead of four per beam as discussed in Section 4.9. In the explicit imperfection model, the geometric imperfection of 0.036 ft at the top of each column is obtained by 0.2% times the column height of 18 ft as shown in Figure 6.26. In the equivalent notional load model, the notional load of 0.384 kips is calculated by 0.2% times the total gravity load of 192 kips, and is added to the lateral load shown in Figure 6.27. In the further reduced tangent modulus model, the program automatically accounts for geometric imperfections as shown in Figure 6.28. Users can choose one of the three methods for preparing their input data. Herein, the uniform gravity load of 1.6 kip/ft in Figure 6.25 is converted to four concentrated loads of 24 kips. The incremental gravity loads are 1.2 kips calculated by dividing the concentrated load by a scaling factor of 20. The horizontal load increment is 0.2 kips.

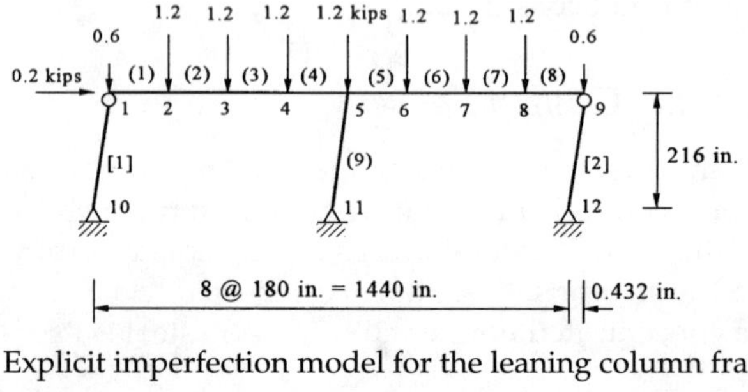

Figure 6.26 Explicit imperfection model for the leaning column frame.

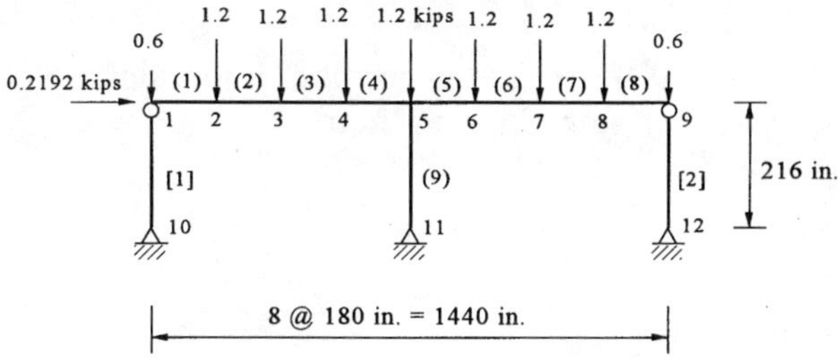

Figure 6.27 Equivalent notional load model for the leaning column frame.

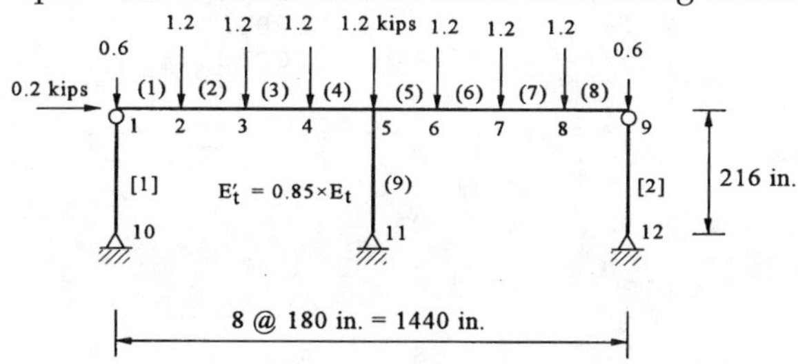

Figure 6.28 Further reduced tangent modulus model for the leaning column frame.

Step 3: Analysis

The input data is listed in Tables 6.15a through 6.15c. Programs are executed in sequence by entering the batch file command RUN.

Table 6.15a Input Data P.DAT of the Explicit Imperfection Modeling for the Leaning Column Frame

```
Two-bay leaning column frame
   1    1
  12    3   200
   0    2     1
   0    9     2
   1      12.6        428.0         69.6    29000.0         36.0
   2      27.7       3270.0        278.0    29000.0         36.0
   1      2.68         16.4      29000.0       36.0
   1     180.0          0.0      2     1     2     8
   9     0.432        216.0      1    11     5
   1     0.432        216.0      1    10     1
   2     0.432        216.0      1    12     9
  10    1     1     1     2      2
  11    1     1     0
   1       0.2         -0.6
   2                   -1.2                  7
   9                   -0.6
```

Table 6.15b Input Data P.DAT of the Equivalent Notional Load Modeling for the Leaning Column Frame
Frame

```
Two-bay leaning column frame
     2     1
    12     3   200
     0     2     1
     0     9     2
     1       12.6        428.0         69.6    29000.0          36.0
     2       27.7       3270.0        278.0    29000.0          36.0
     1       2.68         16.4      29000.0         36.0
     1      180.0          0.0      2     1      2     8
     9        0.0        216.0      1    11      5
     1        0.0        216.0      1    10      1
     2        0.0        216.0      1    12      9
    10     1     1     1     2      2
    11     1     1     0
     1     0.2192         -0.6
     2                    -1.2                   7
     9                    -0.6
```

Table 6.15c Input Data P.DAT of the Further Reduced Tangent Modulus Modeling for the Leaning Column Frame
Column Frame

```
Two-bay leaning column frame
     3     1
    12     3   200
     0     2     1
     0     9     2
     1       12.6        428.0         69.6    29000.0          36.0    1
     2       27.7       3270.0        278.0    29000.0          36.0
     1       2.68         16.4      29000.0         36.0
     1      180.0          0.0      2     1      2     8
     9        0.0        216.0      1    11      5
     1        0.0        216.0      1    10      1
     2        0.0        216.0      1    12      9
    10     1     1     1     2      2
    11     1     1     0
     1        0.2         -0.6
     2                    -1.2                   7
     9                    -0.6
```

Step 4: Check of Load-Carrying Capacity

According to the output file P.OUT2, the explicit and notional load models predict the same ultimate load-carrying capacity (4.18 kips), and the reduced modulus method predicts 4.17 kips at the horizontal load point. Since they are greater than the applied factored load of 4 kips, the preliminary selection of member sizes is adequate.

Step 5: Check of Serviceability

Member sizes in this example are governed by the serviceability condition. When the total lateral load of 4 kips in Figure 6.25 is regarded as 1.3W, the lateral displacement corresponding to 1.0W is computed as 1.015 in. from the output P.OUT2. The drift ratio is 1/213, which does not satisfy the 1/400 limit. The initial choice of members should thus be increased.

Design by the AISC-LRFD Method

The preliminary member sizes are the same as those of the advanced analysis except the beam size is W33x118 instead of W27x94 due to no inelastic redistribution allowed in LRFD method. The first-order elastic analysis is shown in Figure 6.29. Member capacity checks using interaction equations follow.

Member Capacity Check of the Central Column
(W14x43: $A = 12.6$, $I_x = 428$, $r_x = 5.82$, $Z_x = 69.6$)

- Calculation of P_n

$$P_u = 118.6 \text{ kips}$$

$$P_y = (12.6)(36) = 453.6 \text{ kips}$$

$$\frac{P_u}{P_y} = 0.26 < \frac{1}{3}$$

$$\tau = 1.0$$

$$G_B = \frac{428/18}{(2)(5,900)/(60)} = 0.121$$

$$G_{B,\text{adjust}} = (0.121)(2) = 0.242 \text{ For far end hinge}$$

$$G_E = \infty$$

$$K_{\text{unbraced}} = 2.05 \text{ From alignment chart}$$

To account for the leaning column effect

$$K_{\text{story}} = K_{\text{unbraced}}\sqrt{\frac{\sum P_u}{P_{ui}}} = 2.05\sqrt{\frac{118.6 + 36.8 + 36.8}{118.6}} = 2.61$$

$$K_{\text{story,min}} = \sqrt{\frac{5}{8}}(2.05) = 1.62 < 2.61$$

$$\lambda_c = \frac{(2.61)(18)(12)}{(\pi)(5.82)}\sqrt{\frac{36}{29,000}} = 1.09 < 1.5$$

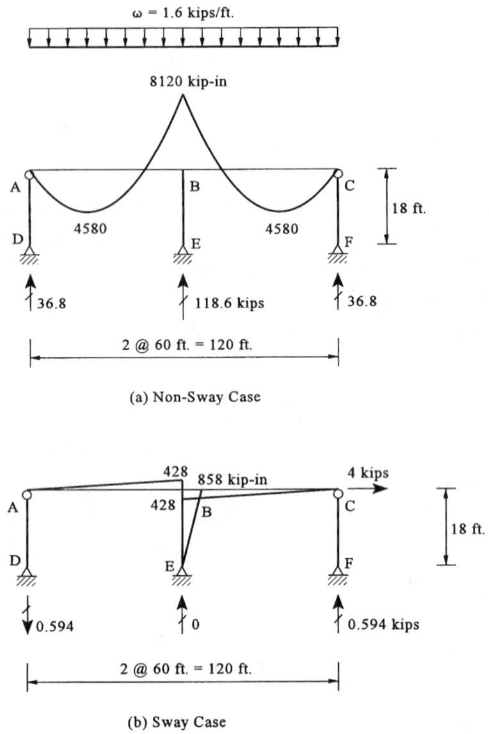

Figure 6.29 Member forces by the first-order analysis for the leaning column frame.

$$P_n = (0.658)^{1.09^2}(36)(12.6) = 277 \text{ kips}$$

- Calculation of M_n

  $M_n = M_p = (Z_x)(F_y) = (69.6)(36) = 2,506$ kip-in For laterally braced

- Resulting P_u

  $$P_u = 118.6 \text{ kips}$$

- Calculation of M_u

  $$C_m = 0.60 - 0.4\frac{0}{6.22} = 0.6$$

  $$G_B = \frac{0.121}{1.5} = 0.081 \text{ For far end hinge in braced case}$$

  $$K_{\text{brace}} = 0.73 \text{ From alignment chart}$$

  $$P_{e1} = \frac{(\pi^2)(29,000)(428)}{[(0.73)(18)(12)]^2} = 4,927 \text{ kips}$$

  $$B_1 = \frac{0.6}{1-\frac{118.6}{4,927}} = 0.61 < 1.0 \quad \text{Use } B_1 = 1.0$$

$$\sum P_u = 36.8 + 118.6 + 36.8 = 192 \text{ kips}$$

$$\sum P_{e2} = \frac{(\pi^2)(29,000)(428)}{[(2.05)(18)(12)]^2} = 625 \text{ kips}$$

$$B_2 = \frac{1}{1-\frac{192}{625}} = 1.44$$

$$M_{nt} = 0, \quad M_{lt} = 858 \text{ kip-in at top of column}$$

$$M_u = (1.0)(0.0) + (1.44)(858) = 1,236 \text{ kip-in}$$

- Interaction equation check

$$\frac{P_u}{\phi P_n} = \frac{118.6}{(0.85)(277)} = 0.504 > 0.2$$

$$\frac{P_u}{\phi P_n} + \frac{8M_u}{9\phi M_n} = 0.504 + \left(\frac{8}{9}\right)\left(\frac{1,236}{(0.9)(2,506)}\right) = 0.99 < 1.0 \text{ O.K.}$$

- Use W14x43

Leaning Column Check (W6x9: $A = 2.68$, $I_x = 16.4$, $r_x = 2.47$, $Z_x = 6.23$)

$$K = 1.0$$

$$\lambda_c = \frac{(1.0)(18)(12)}{(\pi)(2.47)}\sqrt{\frac{36}{29,000}} = 0.98$$

$$P_n = (0.658)^{(0.98^2)}(36)(2.68) = 64.5 \text{ kips}$$

$$P_u = 36.8 \text{ kips}$$

$$\frac{P_u}{\phi P_n} = \frac{(36.8)}{(0.85)(64.5)} = 0.67 \text{ O.K.}$$

Use W6x9

Member Capacity Check for Beams AB (W33x118: $Z_x = 415$)

- Neglect axial force
- Flexure strength check

$$M_{nt} = 8,120 \text{ kip-in}, \quad M_{lt} = 428 \text{ kip-in}$$

$$M_u = (1.0)(8,120) + (1.44)(428) = 8,736 \text{ kip-in}$$

$M_n = (Z_x)(F_y) = (415)(36) = 14,940$ kip-in For laterally braced

$\frac{M_u}{\phi M_n} = \frac{8,736}{(0.9)(14,940)} = 0.65 \ll 1.0$ W33x118 is too large

Try W27x94 ($Z_x = 278$)

$$M_p = (278)(36) = 10,008 \text{ kip-in}$$

$$\frac{M_u}{\phi M_n} = \frac{8,736}{(0.9)(10,008)} = 0.97 < 1.0 \text{ O.K.}$$

Use W27x94

Comparison of Results

The three models lead to selection of the same members, and these also match those the LRFD method predicts (see Figure 6.30). This is because the leaning column frame is not highly redundant and thus does not benefit greatly from inelastic moment redistribution being considered by the advanced method. Note that the leaning column size is W6x9 under the wind load combination, but it should be increased in actual design. This is because the design of the leaning column is governed by gravity load combination omitted in this example.

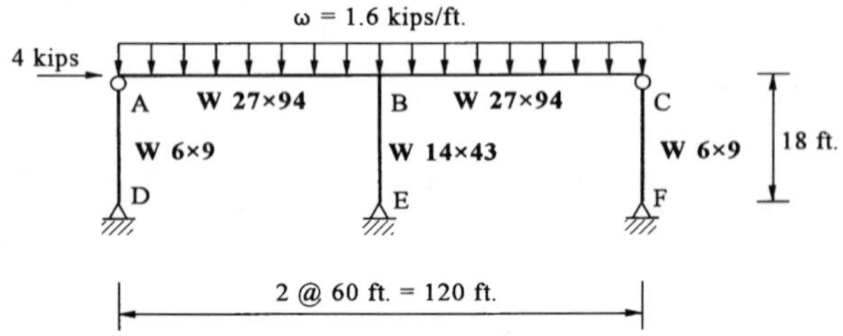

Figure 6.30 Comparison of member sizes of the leaning column frame.

6.5.3 Two-Story Frame

Figure 6.31 shows a fixed-supported two-story frame subject to gravity and lateral loads. All members are assumed to be laterally braced, and wide flange sections of A36 steel are used for beams and columns.

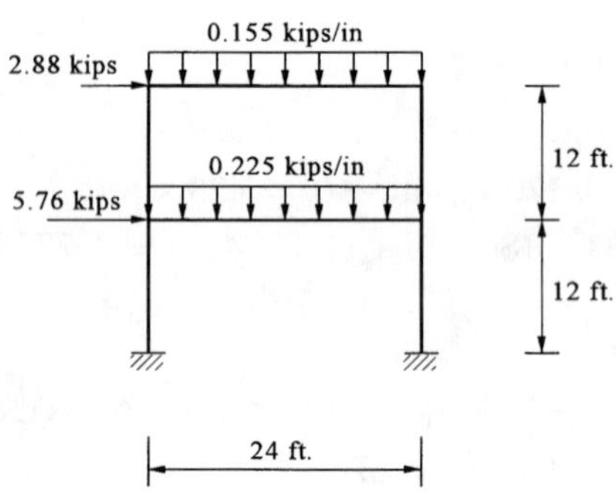

Figure 6.31 Configuration and load condition of the unbraced two-story frame.

Design by the Advanced Analysis

Step 1: Load Condition and Preliminary Member Sizing

The critical load combination is shown in Figure 6.31. Preliminary member sizes are selected as W14x22 for the roof beam, W16x26 for the floor beam, and W10x26 for the columns.

Step 2: Structural Modeling

Each column is modeled with one element and the beams are modeled with two. The uniform gravity loads in Figure 6.31 are converted to concentrated loads (Figure 6.32). In the explicit imperfection model, the geometric imperfections of 0.576 in. at the roof level and 0.288 in. at the second floor level in Figure 6.33 are calculated by 0.2% times column height of 24 ft and 12 ft, respectively. In the equivalent notional load modeling, a notional load of 0.0893 kips at the roof level and 0.1296 kips at second floor level in Figure 6.34 results from 0.2% times the total gravity load of 44.64 kips and 64.8 kips, respectively, and they are added to the lateral loads at each level. In the further reduced tangent modulus method, the program automatically accounts for geometric imperfection effects (Figure 6.35). Herein, the incremental horizontal load is determined by dividing the total loads by a scaling factor of 20. An incremental load of 0.144 kips is imposed at the roof level calculated by dividing the factored concentrated load of 2.88 kips by 20.

Step 3: Analysis

The input data is shown in Tables 6.16a through 6.16c. Programs are executed by entering the batch file command RUN.

Step 4: Check of Load-Carrying Capacity

From the output file P.OUT2, the explicit imperfection model and the equivalent notional load model predict the same load-carrying capacity of 3.086 kips with respect to the horizontal load at the roof level when the first hinge forms. The further reduced tangent modulus model predicts 3.117 kips. Since the deflection at limit state is too large, the load corresponding to the first plastic hinge forming is taken for the load-carrying capacity of the structural system.

Step 5: Check of Serviceability

When the lateral loads in Figure 6.31 are regarded as 1.3W, the lateral displacement at roof level corresponding to 1.0W is computed as 0.528 in. from the computer output P.OUT2. The drift ratio is 1/549, which meets the limitation 1/400. The initial member sizes are adequate.

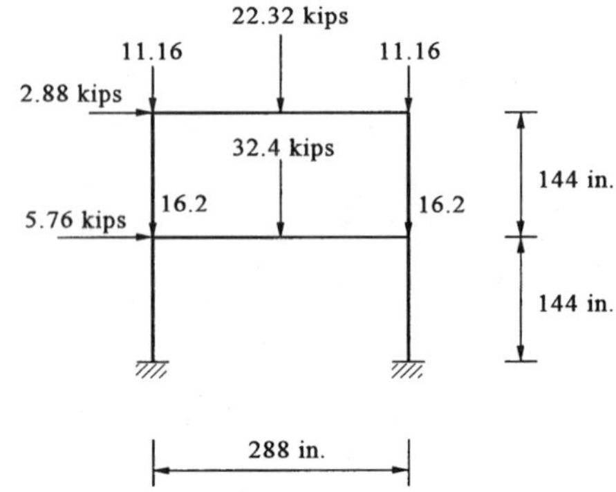

Figure 6.32 Concentrated load condition converted from the distributed load for the two-story frame.

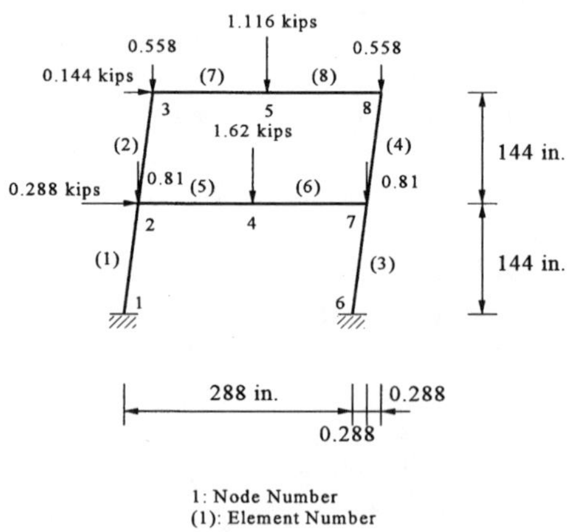

Figure 6.33 Explicit imperfection model for the two-story frame.

Design by the AISC-LRFD Method

The preliminary member sizes are the same as those of the advanced analysis except the floor beam size is W16x31. The first-order elastic analysis is shown in Figure 6.36. Member capacity checks are in what follows.

Member Capacity Check of Lower Column
(W10x26: $A = 7.61$, $I_x = 144$, $r_x = 4.35$, $Z_x = 31.3$)

- Calculation of P_n

$$P_u = 54.7 + 3.3 = 58 \text{ kips}$$

$$P_y = (7.61)(36) = 274 \text{ kips}$$

$$\frac{P_u}{P_y} = \frac{58}{274} = 0.21 < \frac{1}{3}$$

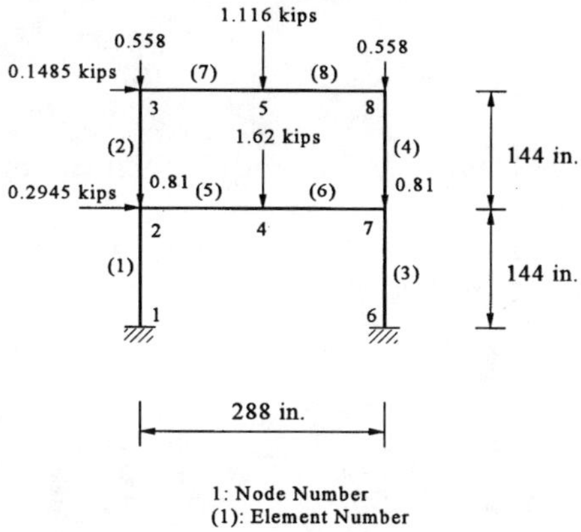

Figure 6.34 Equivalent notional load model for the two-story frame.

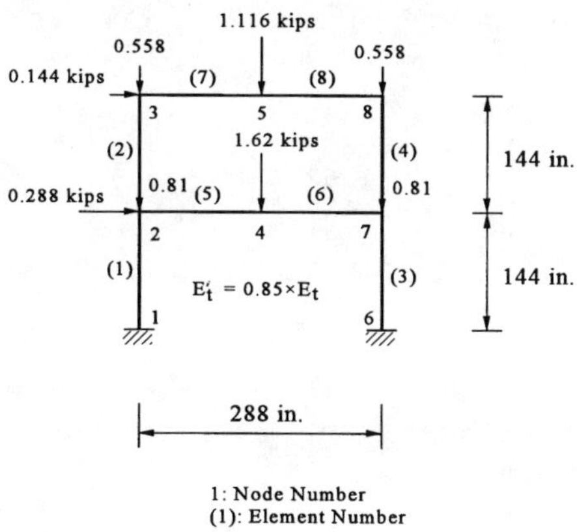

Figure 6.35 Further reduced tangent modulus model for the two-story frame.

$$\tau = 1.0$$

$$G_T = \frac{(2)(144)/(144)}{375/288} = 1.54, \quad G_B = 0$$

$$K_{\text{unbraced}} = 1.23 \text{ From alignment chart}$$

$$\lambda_c = \frac{(1.23)(144)}{(\pi)(4.35)}\sqrt{\frac{36}{29,000}} = 0.457 < 1.5$$

$$P_n = (0.658)^{(0.457^2)}(36)(7.61) = 251 \text{ kips}$$

Table 6.16a Input Data P.DAT of the Explicit Imperfection Modeling for the Two-Story Frame

```
Two-story frame
    1    1
    8    2   100
    0    3     0
    0    8     0
    1      7.61      144.0         31.3   29000.0       36.0
    2      7.68      301.0         44.2   29000.0       36.0
    3      6.49      199.0         33.2   29000.0       36.0
    1     0.288      144.0    1    1     2    2
    3     0.288      144.0    1    6     7    2
    5     144.0        0.0    2    2     4
    6     144.0        0.0    2    4     7
    7     144.0        0.0    3    3     5
    8     144.0        0.0    3    5     8
    1    1    1    1    2    5
    2     0.288      -0.81
    3     0.144     -0.558
    4                -1.62
    5               -1.116
    7                -0.81
    8               -0.558
```

Table 6.16b Input Data P.DAT of the Equivalent Notional Load Modeling for the Two-Story Frame

```
Two-story frame
    2    1
    8    2   100
    0    3     0
    0    8     0
    1      7.61      144.0         31.3   29000.0       36.0
    2      7.68      301.0         44.2   29000.0       36.0
    3      6.49      199.0         33.2   29000.0       36.0
    1     0.000      144.0    1    1     2    2
    3     0.000      144.0    1    6     7    2
    5     144.0        0.0    2    2     4
    6     144.0        0.0    2    4     7
    7     144.0        0.0    3    3     5
    8     144.0        0.0    3    5     8
    1    1    1    1    2    5
    2    0.2945      -0.81
    3    0.1485     -0.558
    4                -1.62
    5               -1.116
    7                -0.81
    8               -0.558
```

Table 6.16c Input Data P.DAT of the Further Reduced Tangent Modulus Modeling for the Two-Story Frame
Story Frame

```
Two-story frame
   3    1
   8    2   100
   0    3     0
   0    8     0
   1      7.61        144.0         31.3   29000.0       36.0    1
   2      7.68        301.0         44.2   29000.0       36.0
   3      6.49        199.0         33.2   29000.0       36.0
   1     0.000        144.0     1     1     2    2
   3     0.000        144.0     1     6     7    2
   5     144.0          0.0     2     2     4
   6     144.0          0.0     2     4     7
   7     144.0          0.0     3     3     5
   8     144.0          0.0     3     5     8
   1    1     1     1     2     5
   2    0.2880        -0.81
   3    0.1440       -0.558
   4                  -1.62
   5                 -1.116
   7                  -0.81
   8                 -0.558
```

- Calculation of M_n

  $M_n = M_p = (Z_x)(F_y) = (31.3)(36) = 1,127$ kip-in

  For laterally braced

- Resulting P_u

  $P_u = 58$ kips

- Calculation of M_u

  $C_m = 0.6 - 0.4\left(\frac{233.2}{462.4}\right) = 0.398$

  $K_{\text{brace}} = 0.65$ From alignment chart

  $P_{e1} = \frac{(\pi^2)(29{,}000)(144)}{[(0.65)(144)]^2} = 4,704$ kips

  $B_1 = \frac{0.398}{1-\frac{54.7}{4{,}704}} = 0.40$ Use $B_1 = 1.0$

  $\sum P_u = (54.7)(2) = 109$ kips

  $$\sum P_{e2} = \frac{(2)(\pi^2)(29,000)(144)}{[(1.23)(144)]^2} = 2,628 \text{ kips}$$

  $B_2 = \frac{1}{1-\frac{109.4}{2{,}628}} = 1.04$

  $M_{nt} = 462.4$ kip-in, $M_{lt} = 264.2$ kip-in at top of column

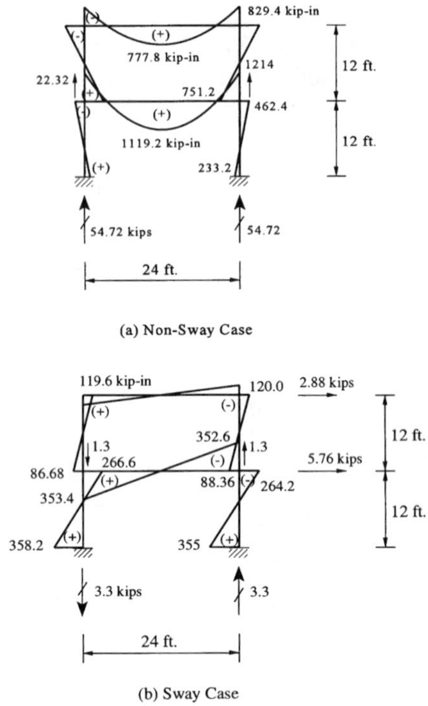

Figure 6.36 Member forces by the first-order analysis for the two-story frame.

$$M_{nt} = 233.2 \text{ kip-in}, \quad M_{lt} = 355 \text{ kip-in at bottom of column}$$

$$M_u = (1.0)(462.4) + (1.04)(264.2) = 737.2 \text{ kip-in}$$

$$M_u = (1.0)(233.2) + (1.04)(355) = 602.4 \text{ kip-in}$$

- Interaction equation check

$$\frac{P_u}{\phi P_n} = \frac{(58)}{(0.85)(251)} = 0.27 > 0.2$$

$$\frac{P_u}{\phi P_n} + \frac{8M_u}{9\phi M_n} = 0.27 + \frac{8}{9}\frac{(737.2)}{(0.9)(1,127)} = 0.92 < 1.0 \text{ O.K.}$$

- Use W10x26

Member Capacity Check of Upper Column
(W10x26: $A = 7.61$, $I_x = 144$, $r_x = 4.35$, $Z_x = 31.3$)

- Calculation of P_n

$$\tau = 1.0$$

$$G_T = \frac{144/144}{199/288} = 1.45$$

$G_B = 1.54$

$K_{\text{unbraced}} = 1.45$ From alignment chart

$$\lambda_c = \frac{(1.45)(144)}{(\pi)(4.35)}\sqrt{\frac{36}{29,000}} = 0.538 < 1.5$$

$$P_n = (0.658)^{(0.538^2)}(36)(7.61) = 243 \text{ kips}$$

- Calculation of M_n

  $M_n = M_p = (Z_x)(F_y) = (31.3)(36) = 1,127$ kip-in
  For laterally braced

- Resulting P_u

  $P_u = 22.32 + 1.3 = 23.6$ kips

- Calculation of M_u

  $C_m = 0.6 - (0.4)\frac{751.2}{829.4} = 0.238$

  $K_{\text{brace}} = 0.82$ From alignment chart

  $P_{e1} = \frac{(\pi^2)(29,000)(144)}{[(0.82)(144)]^2} = 2,956$ kips

  $B_1 = \frac{0.238}{1-\frac{22.3}{2,956}} = 0.240 \quad$ Use $B_1 = 1.0$

  $$\sum P_u = (0.155)(288) = 44.64 \text{ kips}$$

  $$\sum P_{e2} = \frac{(2)(\pi^2)(29,000)(144)}{[(1.45)(144)]^2} = 1,890 \text{ kips}$$

  $B_2 = \frac{1}{1-\frac{44.64}{1,890}} = 1.02$

  $M_{nt} = 829.4$ kip-in, $\quad M_{lt} = 120$ kip-in at top of column

  $M_{nt} = 751.2$ kip-in, $\quad M_{lt} = 88.4$ kip-in at bottom of column

  $M_u = (1.0)(829.4) + (1.02)(120) = 951.8$ kip-in

  $M_u = (1.0)(751.2) + (1.02)(88.36) = 841.3$ kip-in

- Interaction equation check

  $\frac{P_u}{\phi P_n} = \frac{(23.6)}{(0.85)(243)} = 0.114 < 0.2$

  $\frac{P_u}{2\phi P_n} + \frac{M_u}{\phi M_n} = \frac{0.114}{2} + \frac{(951.8)}{(0.9)(1,127)} = 0.995 < 1.0$ O.K.

- Use W10x26

Member Capacity Check of Roof Beam (W14x22: $Z_x = 33.2$)

- Neglect axial force

- Flexure strength check

$$M_{nt} = 829.4 \text{ kip-in}, \quad M_{lt} = 120.0 \text{ kip-in}$$

$$M_u = (1)(829.4) + (1.02)(120.0) = 951.8 \text{ kip-in}$$

$$M_n = (Z_x)(F_y) = 1,195 \text{ kip-in} \quad \text{For laterally braced}$$

$$\frac{M_u}{\phi M_n} = \frac{(951.8)}{(0.9)(1,195)} = 0.88 < 1.0 \text{ O.K.}$$

- Use W14x22

Member Capacity Check of Floor Beam (W16x31: $Z_x = 54$)

- Neglect axial force
- Flexure strength check

$$M_{nt} = 1,214 \text{ kip-in}, \quad M_{lt} = 352.6 \text{ kip-in}$$

$$M_u = (1)(1,214) + (1.04)(352.6) = 1,581 \text{ kip-in}$$

$$M_n = (Z_x)(F_y) = 1,944 \text{ kip-in} \quad \text{For laterally braced}$$

$$\frac{M_u}{\phi M_n} = \frac{(1,581)}{(0.9)(1,944)} = 0.90 < 1.0 \text{ O.K.}$$

- Use W16x31

Comparison of Results

The three advanced analysis models result in the same selection of member sizes. This choice is one size smaller than or equal to those determined by the LRFD method as shown in Figure 6.37. Advanced analysis allows inelastic redistribution throughout the formation of the first hinge, but successive hinging is not illustrated by this example.

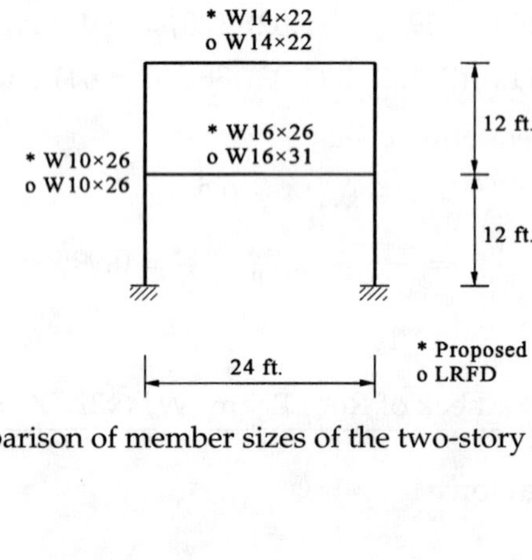

Figure 6.37 Comparison of member sizes of the two-story frame.

6.5.4 Eight-Story Frame

Figure 6.38 shows an unbraced eight-story one-bay frame with hinged supports. All beams are rigidly connected to the columns. The column and beam sizes are the same, respectively. All beams are continuously braced about their weak-axis. Bending is primarily about the strong-axis at the column. A36 steel is used for all members. The load condition is the same as that for an eight-story braced frame in Section 6.4.2.

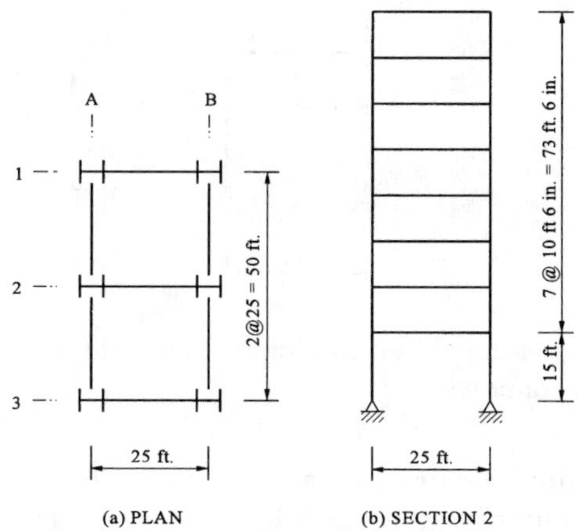

Figure 6.38 Configuration of the unbraced eight-story frame.

Design by the Advanced Analysis

Step 1: Load Condition and Preliminary Member Sizing

The critical load combination is the same as that for the eight-story braced frame in Section 6.4.2. The preliminary column and beam sizes are selected as W33x130 and W21x50.

Step 2: Structural Modeling

Each column is modeled with one element because the frame is unbraced and the maximum moment in the member occurs at the ends. Each beam is modeled with two elements. The uniform gravity loads are converted to equivalent concentrated loads in Figure 6.39.

The explicit imperfection and the further reduced tangent modulus models are used in this example because it is more difficult to prepare the input data for equivalent notional loads. Figure 6.40 shows the model for the eight-story frame. Explicit imperfection model uses an out-of-plumbness of 0.2%, and in the further reduced tangent modulus model, $0.85E_t$ is used.

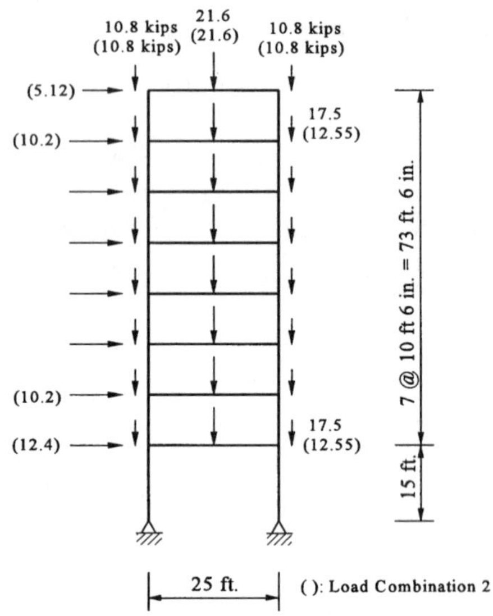

Figure 6.39 Concentrated load condition converted from the distributed load for the two-story frame.

Herein, a scaling factor of 10 is used due to the high indeterminacy. The load increment is automatically reduced if the element stiffness parameter η exceeds the predefined value 0.1. The 54 load steps required to converge on the solution are given in P.OUT2.

Step 3: Analysis

The input data may be easily generated as listed in Tables 6.17a and 6.17b. Programs are executed in sequence by typing the batch file command RUN.

Step 4: Check of Load-Carrying Capacity

From the output file P.OUT, the ultimate load-carrying capacity of the structure is obtained as 5.24 kips and 5.18 kips with respect to the lateral load at roof in load combination 2 by the imperfection method and the reduced tangent modulus method. This load-carrying capacity is 3% and 2% greater than the applied factored load of 5.12 kips. As a result, the preliminary member sizes are satisfactory.

Step 5: Check of Serviceability

The lateral drift at the roof level by the wind load (1.0W) is 5.37 in. and the drift ratio is 1/198, which does not satisfy the drift limit of 1/400. When W40x174 and W24x76 are used for column and beam members, the lateral drift is reduced to 2.64 in. and the drift ratio is 1/402, which

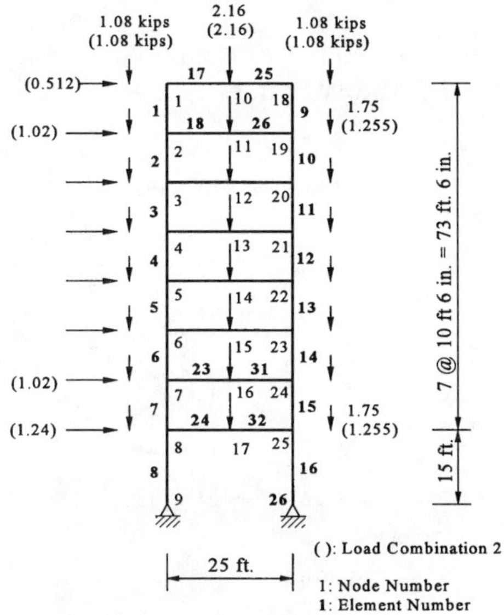

Figure 6.40 Structural modeling of the eight-story frame.

satisfies the limit 1/400. The design of this frame is thus governed by a serviceability rather than a strength.

Design by the AISC-LRFD Method

The preliminary member sizes are W40x174 columns and W24x76 beams which meet serviceability conditions described above. The first-order elastic analysis is performed for the non-sway and sway cases as shown in Figure 6.41. The lateral load combination controls the member sizes. Member capacity checks follow.

Member Capacity Check for Columns
(W40x174: $A = 51.1$, $I_x = 12{,}200$, $r_x = 15.5$, $r_y = 3.26$, $Z_x = 715$)

- For $1.2D + 1.3W + 0.5L + 0.5S$

  Calculation of P_n

  $$P_u = 193 + 152 = 345 \text{ kips}$$
  $$P_y = (51.1)(36) = 1,840 \text{ kips}$$
  $$\frac{P_u}{P_y} = 0.188 < \frac{1}{3}$$
  $$\tau = 1.0$$
  $$G_T = \frac{12{,}200/180 + 12{,}200/126}{2{,}100/300} = 23.5$$
  $$G_B = \infty$$
  $$K_x = 6.2, \quad K_y = 1.0$$

Table 6.17a Input Data P.DAT of the Explicit Imperfection Modeling for the Unbraced Eight-Story Frame
Frame

```
Unbraced eight-story frame, explicit imperfection modeling
    1      1
   26      2
    0      2      0
    0     32      0
    1      38.30     6710.00      467.00   29000.00         36.00
    2      14.70      984.00      110.00   29000.00         36.00
    1      0.252      126.00      1      2       1      7
    8      0.360      180.00      1      9       8
    9      0.252      126.00      1     19      18      7
   16      0.360      180.00      1     26      25
   17     150.00        0.00      2      1      10      8
   25     150.00        0.00      2     10      18      8
    9      1      1
   26      1      1
    1      0.512      -1.080
    2      1.020      -1.255                     6
    8      1.240      -1.255
   10                 -2.160
   11                 -2.510                     7
   18                 -1.080
   19                 -1.255                     7
```

$$\lambda_{cx} = \frac{(6.2)(180)}{(\pi)(15.5)}\sqrt{\frac{36}{29,000}} = 0.807 < 1.50$$

$$\lambda_{cy} = \frac{(1.0)(180)}{(\pi)(3.26)}\sqrt{\frac{36}{29,000}} = 0.619 < 1.50$$

$$\phi P_n = (0.85)(0.658)^{(0.807)^2}(36)(51.1) = 1,191 \text{ kips}$$

Calculation of M_n

$$\phi M_n = (0.9)(715)(36) = 23,166 \text{ kip-in}$$

Resulting P_u

$$P_u = 193 + 152 = 345 \text{ kips}$$

Calculation of M_u

$$C_m = 0.6 - 0.4\tfrac{0}{355} = 0.6$$

$$K_{\text{brace}} = 0.99$$

$$P_{e1} = \frac{(\pi^2)(29{,}000)(12{,}200)}{[(0.99)(180)]^2} = 109,962 \text{ kips}$$

$$B_1 = \frac{0.6}{1 - \frac{193}{109{,}962}} = 0.6 < 1.0$$

Use $B_1 = 1.0$

$$P_{e2} = \frac{(\pi^2)(29{,}000)(12{,}200)}{[(6.2)(180)]^2} = 2,804 \text{ kips}$$

Table 6.17b Input Data P.DAT of the Further Reduced Tangent Modulus Modeling for the Unbraced Eight-Story Frame
Eight-Story Frame

```
Unbraced eight-story frame, further reduced tangent modulus
   3    1
  26    2
   0    2    0
   0   32    0
   1     38.30    6710.00     467.00  29000.00       36.00     1
   2     14.70     984.00     110.00  29000.00       36.00
   1    0.         126.00     1    2     1     7
   8    0.         180.00     1    9     8
   9    0.         126.00     1   19    18     7
  16    0.         180.00     1   26    25
  17   150.00        0.00     2    1    10     8
  25   150.00        0.00     2   10    18     8
   9    1    1
  26    1    1
   1     0.512     -1.080
   2     1.020     -1.255              6
   8     1.240     -1.255
  10               -2.160
  11               -2.510              7
  18               -1.080
  19               -1.255              7
```

$$\sum P_u = (193)(2) = 386 \text{ kips}$$
$$\sum P_{e2} = (2,804)(2) = 5,608 \text{ kips}$$
$$B_2 = \frac{1}{1-\frac{386}{5,608}} = 1.074$$
$$M_u = (1.0)(355) + (1.074)(7,055) = 7,932 \text{ kip-in}$$

Interaction equation check

$$\frac{P_u}{\phi P_n} = \frac{345}{1,191} = 0.290 > 0.2$$
$$\frac{P_u}{\phi P_n} + \frac{8}{9}\frac{M_u}{\phi M_n} = 0.290 + \frac{8}{9}\frac{(7,932)}{(23,166)} = 0.60 < 1.0$$

Use W40x174

Member Capacity Check for Beams (W24x76: $Z_x = 200$)

$$M_u = (1.0)(920) + (6,097)(1.074) = 7,468 \text{ kip-in}$$
$$\phi M_n = (0.9)(200)(36) = 6,480 \text{ kip-in}$$
$$\frac{M_u}{\phi M_p} = \frac{7,468}{6,480} = 1.15 > 1.0 \text{ N.G.}$$

Try W27x84 ($Z_x = 244$)

$$\phi M_n = (0.9)(244)(36) = 7,906 \text{ kip-in}$$

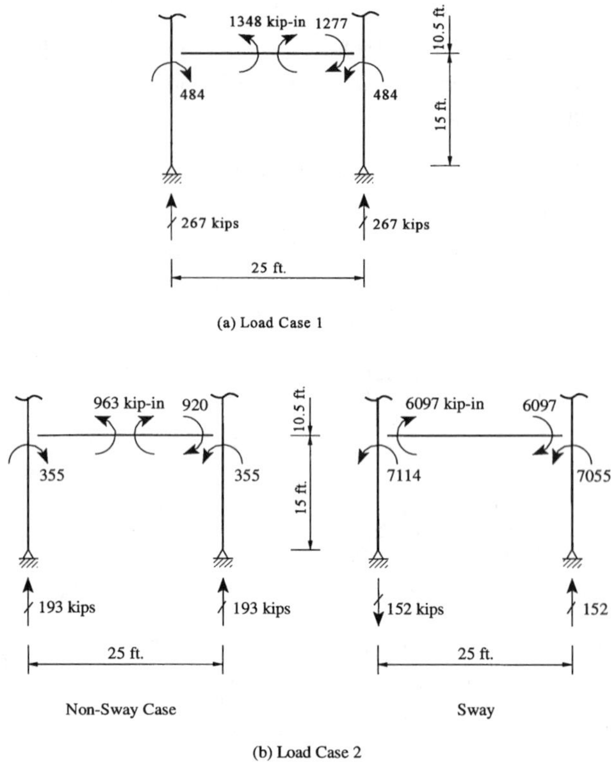

Figure 6.41 Member forces by the first-order analysis for the eight-story frame.

$$\frac{M_u}{\phi M_p} = \frac{7{,}468}{7{,}906} = 0.94 < 1.0 \text{ O.K.}$$

Use W27x84

Comparison of Results

The sizes predicted by the proposed methods are W33x130 columns and W21x50 beams. They do not, however, meet serviceability conditions, and must therefore be increased to W40x174 and W24x76 members. The LRFD method results in the same W40x174 column but a larger (W27x84) beam (Figure 6.42).

6.5.5 *Five-Bay Four-Story AISC Frame*

The AISC frame is provided in the 1991 Lecture Series [1]. The building has four stories with a penthouse centered on the top story. Each story is 12 ft high except the first which is 14 ft and the fifth which is 10 ft. There are five bays in the east-west direction spaced at 30 ft. Three bays in the

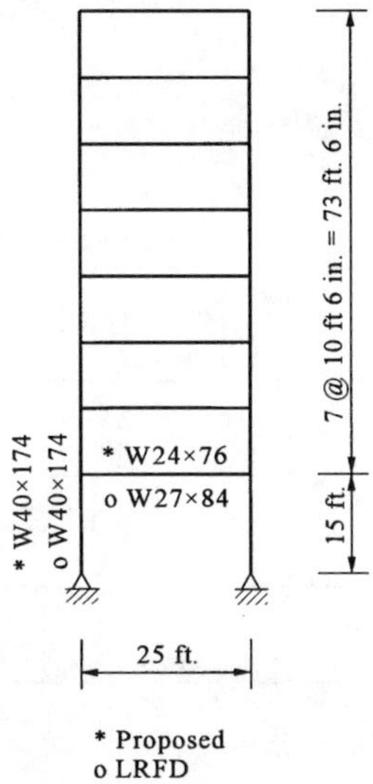

Figure 6.42 Comparison of member sizes of the eight-story frame.

north-south direction have 36 ft exterior bays and 28 ft interior bays. The frame has moment resisting exterior supports in the east-west direction and is laterally braced in the north-south direction. All columns other than these are simply supported. Plan and section views of the frame are shown in Figures 6.43 and 6.44.

Load Condition

The load condition is determined by BOCA [3] as follows:

1. Roof Loads

   Dead: 30 psf

   Live: 21 psf

2. Floor Loads

   Dead: 68 psf

   Live: Office; 75 psf

   Lobby and penthouse; 100 psf

3. Cladding: 15 psf

4. Wind Load: 80 mph, Exposure C

The following three load combinations govern the member sizes of the frame.

(1) $1.2D + 1.6L + 0.5S$

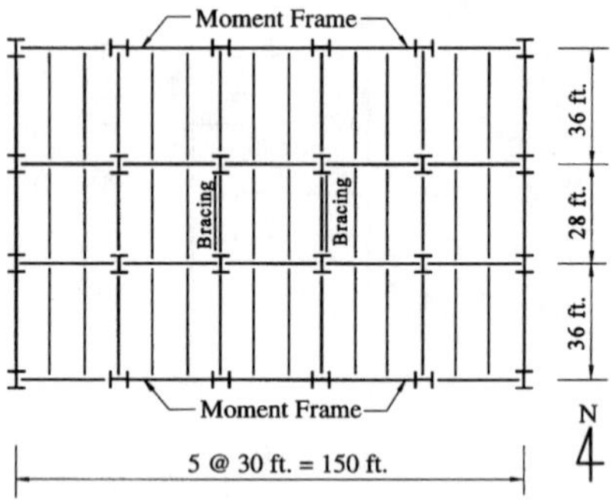

Figure 6.43 Plan of the AISC frame.

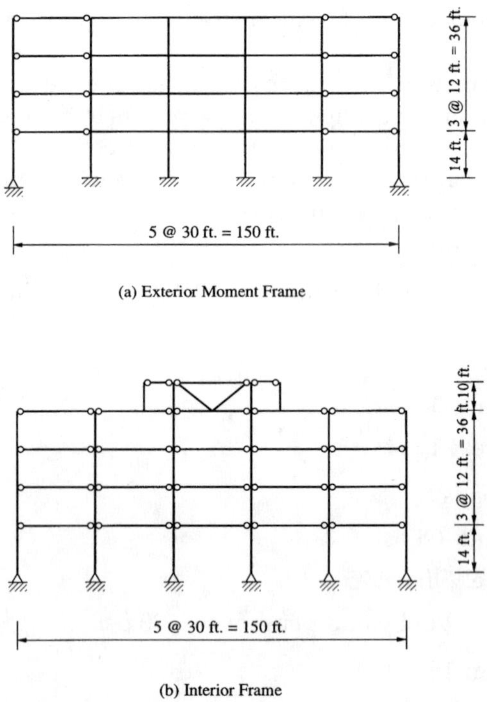

Figure 6.44 Typical sections of the AISC frame.

(2) $1.2D + 1.6S + 0.5L$

(3) $1.2D + 1.3W + 0.5(L + S)$

where

D = dead load
L = live load
S = snow load
W = wind load

The live load reduction is considered in the calculation of loads.

Analyses

The moment frame shown in Figure 6.44(a) with leaning columns in the east-west direction was analyzed using advanced analysis. The yield stress is 36 ksi. The column sizes were changed in every two stories. The analysis and design processes were carried out by the three proposed advanced methods. The geometric imperfection is $\Psi = 1/500$ and the equivalent notional load is $0.002 \sum P_u$. The further reduced tangent modulus of $0.85E_t$ was used.

Member Sizes

Herein, the load-carrying capacity at the ultimate state rather than the formation of the first plastic hinge was taken as the strength of the structural system. Most member sizes of the frame are governed by the wind load combination. It is the upper story girders that are governed by gravity loads. All three methods result in the same selection of members. These are compared to those given by conventional LRFD methods (Figure 6.45) where the sizes are given by Maleck et al. [7]. The proposed procedures generally result in members one or two sizes smaller than the LRFD method could provide. This is because the AISC frame is highly redundant and thus benefits greatly by allowing inelastic moment redistribution to be considered.

Serviceability

The overall drift of the first-order analysis is calculated as H/366 under wind loads of 1.0W. Therefore, this frame does not satisfy the drift limit H/400. The maximum inter-story drift is H/357, which does satisfy the drift limit of H/300. The column sizes must be increased to meet the drift requirements. If the interior column sizes in the first and second stories are increased from W14x68 to W16x67 sections, the drift is H/404 and thus does not exceed the limit.

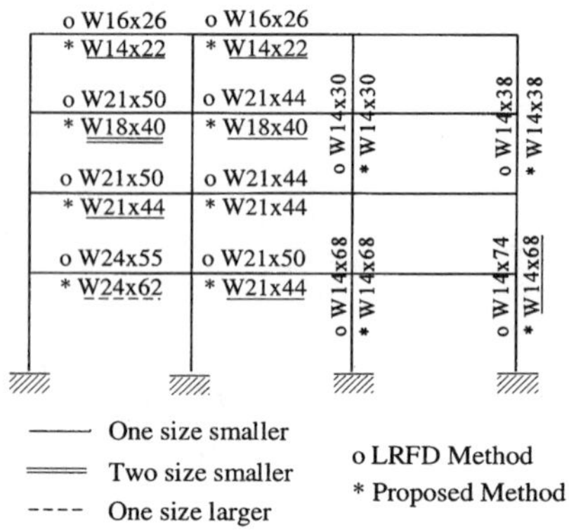

Figure 6.45 Comparison of member sizes of the AISC frame.

6.6 *Semi-Rigid Frames*

6.6.1 *Two-Story One-Bay Semi-Rigid Frame*

Figure 6.46 shows a two-story one-bay semi-rigid frame. Each story is 12 ft high and 25 ft wide. The spacing of the frames is 25 ft. The frame is subjected to a distributed gravity and concentrated lateral loads. The roof beam connections are top- and seat- L6x4.0x3/8x7 angle with double web angles of L4x3.5x1/4x5.5 made of A36 steel. The floor beam connections are top- and seat-angles L6x4x9/16x7 with double web angles of L4x3.5x5/16x8.5. All fasteners are A325 3/4" Diameter bolts. All members are assumed to be continuously braced laterally.

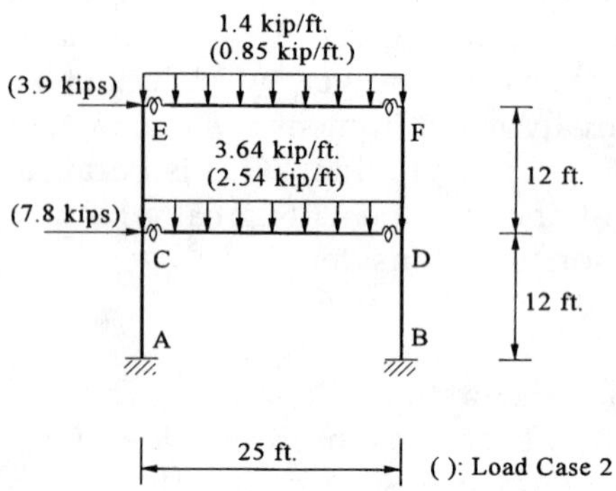

Figure 6.46 Configuration and load condition of two-story one-bay semi-rigid frame.

Design by Advanced Analysis

Step 1: Load Condition and Preliminary Member Size
The load conditions are given as follows.

Roof loads:

Dead load = 20 psf

Live load = 20 psf

Floor loads

Dead load = 68 psf

Live load = 40 psf

Wind load

Wind pressure = 20 psf

The load calculations are

Roof loads

$D = (0.02)(25) = 0.5\ \text{k/ft}$

$L = (0.02)(25) = 0.5\ \text{k/ft}$

Floor loads

$D = (0.068)(25) = 1.7\ \text{k/ft}$

$L = (0.040)(25) = 1.0\ \text{k/ft}$

Wind load

$W = (0.02)(25)(6) = 3$ kips at roof level

$W = (0.02)(25)(12) = 6$ kips at roof level

For simplicity, the live load reduction is neglected.
The load combinations are:

$1.2D + 1.6L$ (Load Case 1)

$1.2(0.5) + 1.6(0.5) = 1.4\ \text{k/ft}$ at roof level

$1.2(1.7) + 1.6(1.0) = 3.64\ \text{k/ft}$ at floor level

$1.2D + 0.5L + 1.3W$ (Load Case 2)

$1.2(0.5) + 0.5(0.5) = 0.85\ \text{k/ft}$ at roof

$1.2(1.7) + 0.5(1.0) = 2.54\ \text{k/ft}$ at floor

$1.3(3) = 3.9$ kips horizontal at roof level

$1.3(6) = 7.8$ kips horizontal at floor level

The resulting load conditions are shown in Figure 6.46.

The initial member sizes are selected as W8x28, W16x40, and W12x22 for the columns, the floor beam, and the roof beam.

Step 2: Structural Modeling

Each column is modeled by one element and the beams are modeled by two elements. The distributed gravity loads are converted to concentrated loads on the beam as shown in Figure 6.47. In explicit imperfection modeling, the geometric imperfection is obtained by multiplying the column height by 0.002. In the equivalent notional load method, the notional load is 0.002 times the total gravity load plus the lateral load. In the further reduced tangent modulus method, the program automatically accounts for geometric imperfection effects. Although users can choose any of these three models, the further reduced tangent modulus model is the only one presented herein. The incremental loads are computed by dividing the concentrated load by the scaling factor of 20 Figure 6.48).

Step 3: Analysis

The three parameters of the connections can be computed by the use of the computer program 3PARA. Corresponding to the input format in Table 2.3, the input data CONN.DAT may be generated in Tables 6.18a and 6.18b. The output CONN.OUT file is shown in Tables 6.19a and 6.19b.

Table 6.18a Input Data CONN.DAT of Connection for Roof Beam

2	36.0	29,000			
7	0.375	0.875	2.5	1.25	12
5.5	0.25	0.6875	2.5		

Table 6.18b Input Data CONN.DAT of Connection for Floor Beam

2	36.0	29,000			
7	0.5625	1.0625	2.5	1.25	16
8.5	.3125	.75	2.5		

Table 6.19a CONN.OUT of the Connection at Roof Level

```
INPUT DATA
==========

     CONNECTION TYPE:
     ----------------

        TOP- AND SEAT- ANGLE WITH WEB-ANGLE CONNECTION

     MATERIAL PROPERTY:
     ------------------

        Fy = 36.0000    E  = 29000.0

     DIMENSIONS RELATED TO TOP ANGLE:
     --------------------------------

        lt =  7.0000    tt =  0.3750    kt =  0.8750

        gt =  2.5000    W  =  1.2500    d  = 12.0000


     DIMENSIONS RELATED TO BOTTOM ANGLE:
     -----------------------------------

        ls =  7.0000    ts =  0.3750    ks =  0.8750

        gs =  2.5000    W  =  1.2500


     DIMENSIONS RELATED TO WEB ANGLE:
     --------------------------------

        da =  5.5000    ta =  0.2500    ka =  0.6875

        ga =  2.5000


 OUTPUT OF THREE PARAMETERS
 ==========================

        INITIAL STIFFNESS, Rki =   90886.97

        ULTIMATE MOMENT,    Mu  =     446.24

        SHAPE PARAMETER,    n   =      1.403
```

Table 6.19b CONN.OUT of the Connection at Floor Level

```
INPUT DATA
==========

    CONNECTION TYPE:
    ----------------

       TOP- AND SEAT- ANGLE WITH WEB-ANGLE CONNECTION

    MATERIAL PROPERTY:
    ------------------

       Fy = 36.0000    E  = 29000.0

    DIMENSIONS RELATED TO TOP ANGLE:
    --------------------------------

       lt =  7.0000    tt =  0.5625    kt =  1.0625

       gt =  2.5000    W  =  1.2500    d  = 16.0000


    DIMENSIONS RELATED TO BOTTOM ANGLE:
    -----------------------------------

       ls =  7.0000    ts =  0.5625    ks =  1.0625

       gs =  2.5000    W  =  1.2500


    DIMENSIONS RELATED TO WEB ANGLE:
    --------------------------------

       da =  8.5000    ta =  0.3125    ka =  0.7500

       ga =  2.5000


 OUTPUT OF THREE PARAMETERS
 ==========================

       INITIAL STIFFNESS, Rki =  607384.31

       ULTIMATE MOMENT,   Mu  =    1361.03

       SHAPE PARAMETER,   n   =      0.927
```

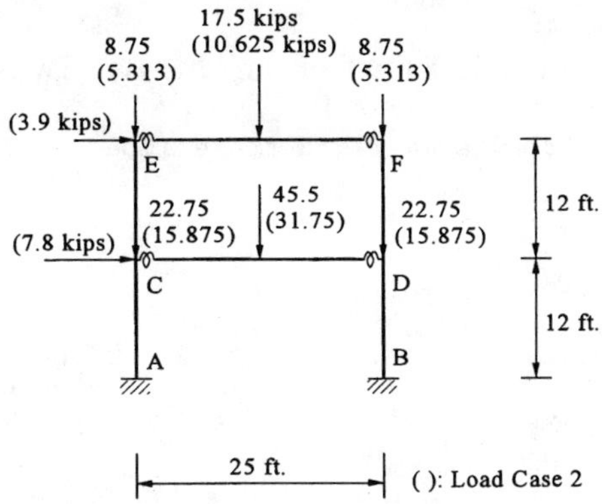

Figure 6.47 Concentrated load condition converted from the distributed load for the two-story one-bay semi-rigid frame.

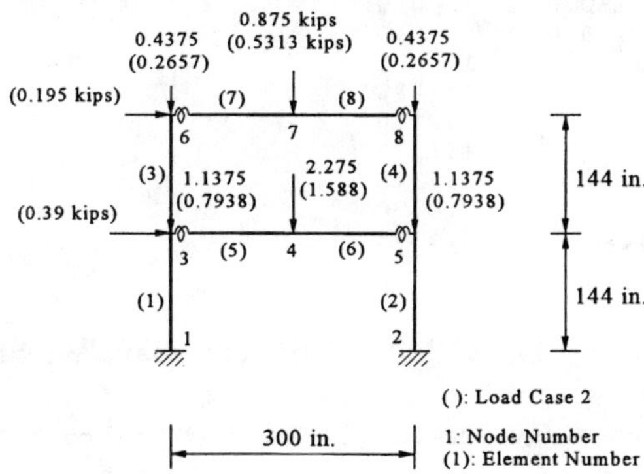

Figure 6.48 Structural modeling of the two-story one-bay semi-rigid frame.

Referring to the input instructions (Section 5.3.2), the input data is written in the form shown in Table 6.20. Programs DATAGEN, INPUT, and PAAP are executed sequentially by typing "RUN".

Step 4: Check of Load-Carrying Capacity

From the output P.OUT2, the ultimate load-carrying capacities of the load combination 1 and 2 are determined as 44.8 kips and 39.2 kips on Node 4. Compared to the applied loads of 45.5 kips and 31.75 kips, the initial member sizes are adequate.

Step 5: Check of Serviceability

When the lateral loads (Figure 6.46) are regarded as 1.3W, the lateral displacement at roof level corresponding to 1.0W is computed as 0.972 in. from the computer output P.OUT2. The drift ratio is 1/296, which

Table 6.20a Input Data P.DAT of the One-Bay Two-Story Semi-Rigid Frame (Load Case 1)

```
One-bay two-story semi-rigid frame (Load case 1)
     3    1
     8    2  100
     2    3    0
     4    8    0
     1    1361.0  607384.0      0.927
     2     446.0   90887.0      1.403
     1      8.25      98.0       27.2   29000.0       36.0    1
     2     11.80     518.0       72.9   29000.0       36.0
     3      6.48     156.0       29.3   29000.0       36.0
     1    5    1    1
     2    6    2    1
     3    7    1    2
     4    8    2    2
     1       0.0     144.0    1    1    3
     2       0.0     144.0    1    2    5
     3       0.0     144.0    1    3    6    2    2
     5     150.0       0.0    2    3    4    2
     7     150.0       0.0    3    6    7    2
     1    1    1    1    2
     3              -1.1375
     4              -2.2750
     5              -1.1375
     6              -0.4375
     7              -0.8750
     8              -0.4375
```

Table 6.20b Input Data P.DAT of the One-Bay Two-Story Semi-Rigid Frame (Load Case 2)

```
One-bay two-story semi-rigid frame (Load case 2)
     3    1
     8    2  100
     2    3    0
     4    8    0
     1    1361.0  607384.0      0.927
     2     446.0   90887.0      1.403
     1      8.25      98.0       27.2   29000.0       36.0    1
     2     11.80     518.0       72.9   29000.0       36.0
     3      6.48     156.0       29.3   29000.0       36.0
     1    5    1    1
     2    6    2    1
     3    7    1    2
     4    8    2    2
     1       0.0     144.0    1    1    3
     2       0.0     144.0    1    2    5
     3       0.0     144.0    1    3    6    2    2
     5     150.0       0.0    2    3    4    2
     7     150.0       0.0    3    6    7    2
     1    1    1    1    2
     3      0.39    -0.7938
     4              -1.5880
     5              -0.7938
     6     0.195    -0.2657
     7              -0.5313
     8              -0.2657
```

does not meet the limitation 1/400. The initial member sizes must be increased.

Design by the AISC-LRFD Method

In fact, the AISC-LRFD Specification does not provide the design procedure for semi-rigid frames. Therefore, the procedure presented in Section 1.3.8 is used herein.

The preliminary member sizes of the column, the roof beam, and the floor beam are selected as W8x28, W12x22, and W16x45. The design procedures presented in Section 1.3.8 are used for semi-rigid frame design [5]. Three parameters, the modified initial stiffness, and the secant stiffness of connections are given in Table 6.21. K-factors for sway and non-sway frames are calculated in Table 6.22. The amplification factors B_1 and B_2, the design moment, and the unit values of the interaction equations are summarized in Tables 6.23a and 6.23b. The column members are governed by load combination 2 with a maximum unit value of the interaction equation equal to 0.988. The beam sizes are controlled by load combination 1 with a maximum unit value of 0.887.

Table 6.21 Estimated Parameters for Connections

Parameters	Connections at roof level	Connections at floor level
Initial stiffness R_{ki}	90,887 k-in/rad	607,384 k-in/rad
Ultimate moment M_u	446 k-in	1,361 k-in
Shape parameter n	1.403	0.927
Modified initial stiffness R_{ko}	57,071 k-in/rad	288,375 k-in/rad
Secant stiffness R_{kb}	43,147 k-in/rad	152,411 k-in/rad

Table 6.22 Estimated K-factors

	Non-sway			Sway		
Column	G_{BOT}	G_{TOP}	K-factor	G_{BOT}	G_{TOP}	K-factor
AC	0	0.971	0.63	0	2.25	1.32
BD	0	0.971	0.63	0	2.25	1.32
CE	0.971	2.0	0.83	2.25	4.05	1.84
DF	0.971	2.0	0.83	2.25	4.05	1.84

Comparison of Results

The member sizes given by both advanced analysis and LRFD methods are compared in Figure 6.49. While the column and roof beam sizes are the same for both methods, the floor beam is one size smaller in advanced analysis because it allows for inelastic moment redistribution.

Table 6.23a Member Capacity Checks for Load Case 1

Column		Non-sway		Sway					
	P_u	M_i	M_j	M_i	M_j	B_1	B_2	M_u	Unit value
AC	63	295	589	0	0	1	1.09	589	0.887
DB	63	−295	−589	0	0	1	1.09	−589	0.887
CE	17.5	681	487	0	0	1	1.05	681	0.821
DF	17.5	−681	−487	0	0	1	1.05	−681	0.821

Beam	Non-sway			Sway				
	M_i	M_c	M_j	M_i	M_j	B_2	M_u	Unit value
CD	−1,270	2,140	−1,270	0	0	1.09	2,140	0.802
EF	−487	826	−487	0	0	1.05	826	0.87

Table 6.23b Member Capacity Checks for Load Case 2

Column		Non-sway		Sway					
	P_u	M_i	M_j	M_i	M_j	B_1	B_2	M_u	Unit value
AC	42.3	210	419	−497	−342	1	1.06	−316	0.457
DB	42.4	−210	−419	−500	−345	1	1.06	−784	0.988
CE	10.5	463	302	−142	−139	1	1.03	317	0.388
DF	10.7	−463	−302	−141	−139	1	1.03	−608	0.719

Beam	Non-sway			Sway				
	M_i	M_c	M_j	M_i	M_j	B_2	M_u	Unit value
CD	−882	1,500	−882	485	−486	1.06	1,500	0.562
EF	−302	495	−302	139	−139	1.03	495	0.521

6.6.2 Two-Story Four-Bay Semi-Rigid Frame

Figure 6.50 shows a two-story four-bay semi-rigid frame. Each story is 12 ft high and 25 ft wide. The spacing of the frames is 25 ft. The frame is subjected to a distributed gravity and concentrated lateral loads. All connections are the same as those of the two-story one-bay semi-rigid frame in Section 6.6.1.

Design by the Advanced Analysis

Step 1: Load Condition and Preliminary Member Size

The load conditions and calculations are the same as those of Section 6.6.1. The resulting load conditions are shown in Figure 6.50. The initial member sizes are selected as W8x21, W12x22, and W16x40 for the columns, the roof beams, and the floor beams.

Step 2: Structural Modeling

Each column is modeled with one element and beam is modeled with two elements. The distributed gravity loads are converted to equivalent concentrated loads on the beam as shown in Figure 6.51. Imperfection models are the same as those in Section 6.6.1. For simplicity, the further reduced tangent modulus model is the only one presented herein. The incremental loads are computed by dividing the concentrated load by the scaling factor of 20 (Figure 6.52).

Step 3: Analysis

Three parameters of connections are the same as those in Tables 6.19a and 6.19b of Section 6.6.1 because the connection configuration and the initial beam depth are identical. The input data may be listed in Tables 6.24a and 6.24b. Programs are executed by entering the batch file command RUN. The program will continue to analyze with increasing load steps up to the ultimate state.

Step 4: Check of Load-Carrying Capacity

As shown in output file P.OUT2, the ultimate load-carrying capacities of the load combination 1 and 2 are 45.0 kips and 42.5 kips at Node $7 \sim 13$ (Figure 6.52). Compared to the applied loads 45.5 kips and 31.75 kips, the initial member sizes are adequate.

Step 5: Check of Serviceability

The lateral displacement at roof level corresponding to 1.0W is computed as 0.392 in. from the computer output P.OUT2. The drift ratio is 1/735, which satisfies the limitation 1/400. The preliminary member sizes are satisfactory.

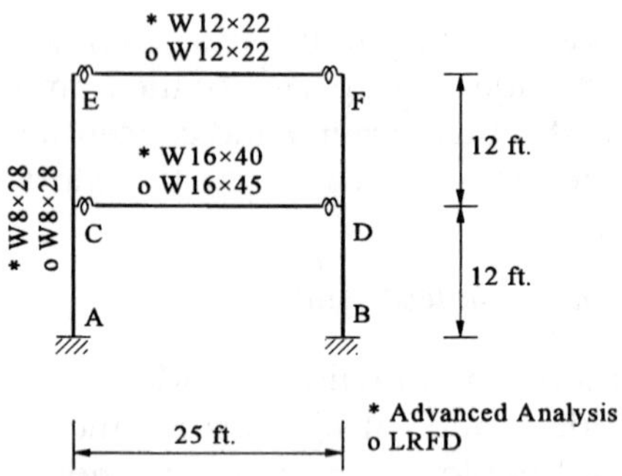

Figure 6.49 Comparison of member sizes of the two-story one-bay semi-rigid frame.

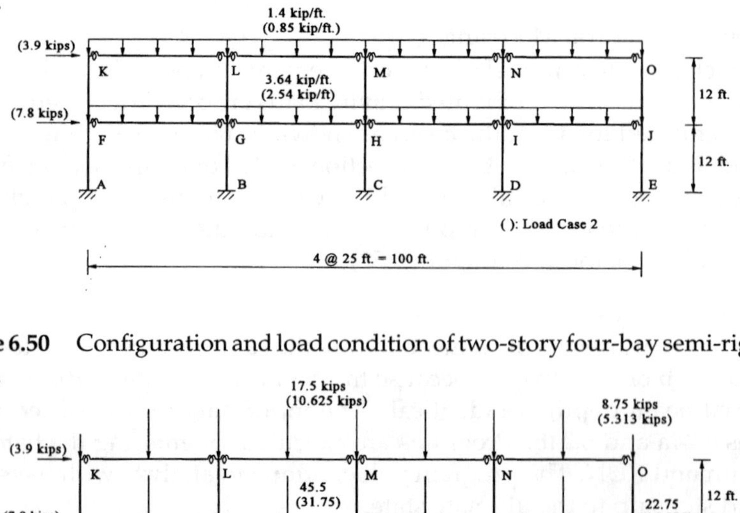

Figure 6.50 Configuration and load condition of two-story four-bay semi-rigid frame.

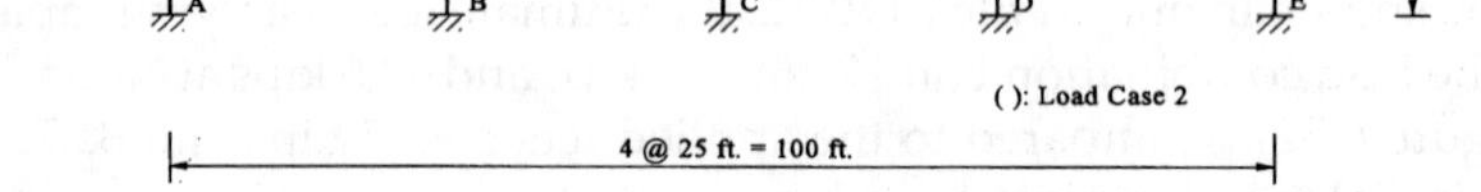

Figure 6.51 Concentrated load condition converted from the distributed load for the two-story four-bay semi-rigid frame.

Table 6.24a Input Data P.DAT of the Four-Bay Two-Story Semi-Rigid Frame (Load Case 1)

```
Four-bay two-story semi-rigid frame (Load case 1)
    3    1
   23    5  100
    2    3    0
   16   26    0
    1    1361.0  607384.0       0.927
    2     446.0   90887.0       1.403
    1      6.16      75.3        20.4   29000.0       36.0    1
    2      6.48     156.0        29.3   29000.0       36.0
    3      11.8     518.0        72.9   29000.0       36.0
    1   11    1    1
    2   12    2    1
    3   13    1    1
    4   14    2    1
    5   15    1    1
    6   16    2    1
    7   17    1    1
    8   18    2    1
    9   19    1    2
   10   20    2    2
   11   21    1    2
   12   22    2    2
   13   23    1    2
   14   24    2    2
   15   25    1    2
   16   26    2    2
    1       0.0     144.0     1    1    6
    2       0.0     144.0     1    2    8
    3       0.0     144.0     1    3   10
    4       0.0     144.0     1    4   12
    5       0.0     144.0     1    5   14
    6       0.0     144.0     1    6   15
    7       0.0     144.0     1    8   17    3    2
   10       0.0     144.0     1   14   23
   11     150.0       0.0     3    6    7    8
   19     150.0       0.0     2   15   16    8
    1    1    1    1    5
    6             -1.1375
    7             -2.2750              7
   14             -1.1375
   15             -0.4375
   16             -0.8750              7
   23             -0.4375
```

Table 6.24b Input Data P.DAT of the Four-Bay Two-Story Semi-Rigid Frame (Load Case 2)

```
Four-bay two-story semi-rigid frame (Load case 2)
      3     1
     23     5  100
      2     3    0
     16    26    0
      1    1361.0  607384.0      0.927
      2     446.0   90887.0      1.403
      1      6.16      75.3       20.4   29000.0        36.0     1
      2      6.48     156.0       29.3   29000.0        36.0
      3      11.8     518.0       72.9   29000.0        36.0
      1    11     1     1
      2    12     2     1
      3    13     1     1
      4    14     2     1
      5    15     1     1
      6    16     2     1
      7    15     1     1
      8    16     2     1
      9    19     1     2
     10    20     2     2
     11    21     1     2
     12    22     2     2
     13    23     1     2
     14    24     2     2
     15    25     1     2
     16    26     2     2
      1       0.0     144.0     1     1     6
      2       0.0     144.0     1     2     8
      3       0.0     144.0     1     3    10
      4       0.0     144.0     1     4    12
      5       0.0     144.0     1     5    14
      6       0.0     144.0     1     6    15
      7       0.0     144.0     1     8    17     3     2
     10       0.0     144.0     1    14    23
     11     150.0       0.0     3     6     7     8
     19     150.0       0.0     2    15    16     8
      1     1     1     1     5
      6      0.39   -0.7938
      7             -1.5880                 7
     14             -0.7938
     15     0.195   -0.2657
     16             -0.5313                 7
     23             -0.2657
```

Table 6.25 Estimated K-factors

Column	Non-sway			Sway		
	G_{BOT}	G_{TOP}	K-Factor	G_{BOT}	G_{TOP}	K-Factor
AF	0	0.911	0.62	0	1.89	1.28
BG	0	0.501	0.59	0	1.04	1.17
CH	0	0.501	0.59	0	1.04	1.17
DI	0	0.501	0.59	0	1.04	1.17
EJ	0	0.911	0.62	0	1.89	1.28
FK	0.911	1.74	0.81	1.89	3.32	1.73
GL	0.501	0.956	0.75	1.04	1.82	1.44
HM	0.501	0.956	0.75	1.04	1.82	1.44
IN	0.501	0.956	0.75	1.04	1.82	1.44
JO	0.911	1.74	0.81	1.89	3.32	1.73

Table 6.26a Member Capacity Checks for Load Case 1

Column		Non-sway		Sway					
	P_u	M_i	M_j	M_i	M_j	B_1	B_2	M_u	Unit value
AF	59.7	215	428	0	0	1	1.15	428	0.942
BG	129	−20.2	−42	0	0	1	1.15	−42	0.719
CH	125	0	0	0	0	1	1.15	0	0.655
DI	129	20.2	42	0	0	1	1.15	42	0.719
EJ	59.7	−215	−428	0	0	1	1.15	−428	0.942
FK	17	541	463	0	0	1	1.06	541	0.877
GL	35.3	−52.4	−29.5	0	0	1	1.06	−52.4	0.158
HM	34.9	0	0	0	0	1	1.06	0	0.0974
IN	35.3	52.4	29.5	0	0	1	1.06	52.4	0.158
JO	17	−541	−463	0	0	1	1.06	−541	0.877

Beam	Non-sway			Sway				
	M_i	M_c	M_j	M_i	M_j	B_2	M_u	Unit value
FG	−968	2,020	−1,800	0	0	1.15	2,020	0.977
GH	−1,700	1,750	−1,620	0	0	1.15	1,750	0.843
HI	−1,620	1,750	−1,700	0	0	1.15	1,750	0.843
IJ	−1,800	2,020	−968	0	0	1.15	2,020	0.977
KL	−463	779	−596	0	0	1.06	779	0.974
LM	−567	741	−568	0	0	1.06	741	0.926
MN	−568	741	−567	0	0	1.06	741	0.926
NO	−596	779	−463	0	0	1.06	779	0.974

Table 6.26b Member Capacity Checks for Load Case 2

Column		Non-sway		Sway					Unit
	P_u	M_i	M_j	M_i	M_j	B_1	B_2	M_u	value
AF	39	153	304	−163	−124	1	1.1	168	0.465
BG	87.4	−14.9	−30.6	−195	−168	1	1.1	−229	0.689
CH	84.4	0	0	−195	−166	1	1.1	−215	0.658
DI	86.9	14.9	30.6	−200	−172	1	1.1	−205	0.662
EJ	41.4	−153	−304	−171	−131	1	1.1	−447	0.856
FK	10	365	288	−34.8	−37.4	1	1.04	328	0.532
GL	21.5	−35.7	−17.9	−74.4	−67.6	1	1.04	−113	0.188
HM	21.3	0	0	−70.7	−66.1	1	1.04	−73.3	0.143
IN	25.5	35.7	17.9	−73.7	−67.4	1	1.04	−52	0.119
JO	10.6	−365	−288	−32.6	−37	1	1.04	−398	0.64

Beam	Non-sway			Sway				Unit
	M_i	M_c	M_j	M_i	M_j	B_2	M_u	value
FG	−668	1,420	−1,260	159	−128	1.1	1,420	0.684
GH	−1,190	1,220	−1,130	114	−118	1.1	−1,260	0.607
HI	−1,130	1,220	−1,190	119	−116	1.1	−1,320	0.636
IJ	−1,260	1,420	−668	130	−163	1.1	1,420	0.684
KL	−288	472	−362	37.4	−34.4	1.04	472	0.589
LM	−344	451	−346	33.2	−33.1	1.04	451	0.564
MN	−346	451	−344	33.1	−33.1	1.04	451	0.564
NO	−362	472	−288	34.2	−37	1.04	472	0.589

Design by the AISC-LRFD Method

The initial member sizes are W8x21 exterior columns, W8x24 interior columns, W12x19 roof beams, and W16x36 floor beams. The design may be performed by the procedure described in Section 1.3.8 [5] The three parameters, the modified initial stiffness, and the secant stiffness of connections are listed in Table 6.21. K-factors of sway and non-sway frames are calculated in Table 6.25. The amplification factors B_1 and B_2, the design moment, and the unit values of the interaction equations are summarized in Table 6.26.

The exterior columns are governed by the load combination 1 with a maximum unit value of the interaction equation equal to 0.942. The interior columns are governed by the load combination 1 with a maximum unit value of 0.719. The roof beams and floor beams are governed by the load combination 1 and the unit values are equal to 0.974 and 0.977.

Comparison of Results

The member sizes by the advanced analysis and the LRFD method are compared in Figure 6.53. The beam members are one size larger in the advanced analysis method, and the interior columns are one size smaller.

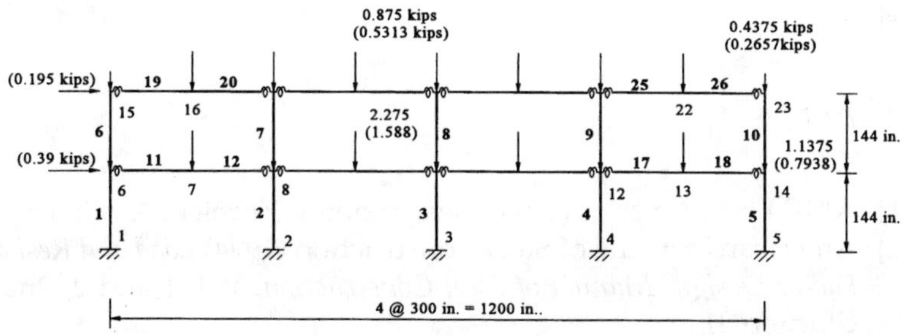

Figure 6.52 Structural modeling of the two-story four-bay semi-rigid frame.

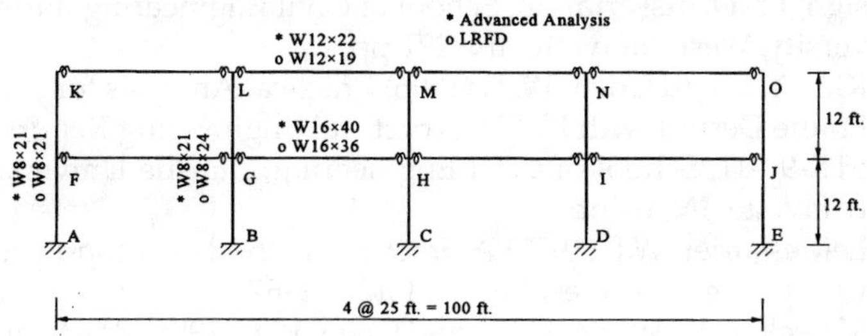

Figure 6.53 Comparison of member sizes of the two-story four-bay semi-rigid frame.

6.7 Summary

In this chapter, design examples are presented to compare practical advanced analysis and conventional LRFD design procedures. The design examples cover simple structures, trusses, braced, unbraced, and semi-rigid frames. The three models used are: (1) explicit imperfection, (2) equivalent notional load, and (3) further reduced tangent modulus.

Users may select the load-carrying capacity corresponding to either the formation of first plastic hinge or the ultimate-state. Using the former, plastic hinge results in generally the same member sizes as the LRFD method. Use of the collapsed ultimate state, however, leads to smaller member sizes due to the benefit of the inelastic moment redistribution.

Input data for each example are provided to be easily reproducible. Detailed instructions are provided to guide users through this new method of design. Use of proposed advanced analysis can eliminate the tedious and sometimes confused member capacity checks in the conventional LRFD method, and often leads to significant savings in design.

References

[1] AISC Lecture Series (1991) Design notes, Zureick, A., Chicago, IL.

[2] American Institute of Steel Construction (1994) *Load and Resistance Factor Design, Manual of Steel Construction*, Vol. 1 and 2, 2nd ed., Chicago, IL.

[3] BOCA (1990) The BOCA (Building Officials & Code Administration) National Building Code, 11th ed., Illinois.

[4] Kim, S.E. (1996) Practical Advanced Analysis for Steel Frame Design, Ph.D. dissertation, School of Civil Engineering, Purdue University, West Lafayette, IN, 271 pp.

[5] Kim, Y. S. and Chen, W. F. (1996) Practical Analysis for Semi-Rigid Frame Design with LRFD, Structural Engineering Report No. CE-STR-96-11, School of Civil Engineering, Purdue University, West Lafayette, IN, 61 pp.

[6] LeMessurier, W.J. (1977) A practical method of second-order analysis. II. Rigid Frames, *Eng. J.*, 14(2), 49-67.

[7] Maleck, A.E., White, D.W., and Chen, W.F. (1995) Practical application of advanced analysis in steel design, Proceedings, 4th Pacific Structural Steel Conference, *Steel Struct.*, 119-126.

[8] Ziemian, R.D. and McGuire, W. (1992) A method for incorporating live load reduction provisions in frame analysis, *Eng. J.*, AISC, 29, 1-3.

Index